Thermodynamics and Statistical Mechanics of Macromolecular Systems

The structural mechanics of proteins that fold into functional shapes, polymers that aggregate and form clusters, and organic macromolecules that bind to inorganic matter can be understood only through statistical physics and thermodynamics.

This book reviews the statistical mechanics concepts and tools necessary for the study of structure formation processes in macromolecular systems that are essentially influenced by finite-size and surface effects. Readers are introduced to molecular modeling approaches, advanced Monte Carlo simulation techniques, and systematic statistical analyses of numerical data. Applications to folding, aggregation, and substrate adsorption processes of polymers and proteins are discussed in great detail. Particular emphasis is placed on the reduction of complexity by coarse-grained modeling, which allows for the efficient, systematic investigation of structural phases and transitions.

Providing insight into modern research at this interface between physics, chemistry, biology, and nanotechnology, this book is an excellent reference for graduate students and researchers.

Michael Bachmann is Associate Professor in the Department of Physics and Astronomy at the University of Georgia. His major fields of interest include theoretical physics, computational physics, statistical physics, biophysics, and chemical physics.

Thermodynamics and Statistical Mechanics of Macromolecular Systems

MICHAEL BACHMANN

The University of Georgia

CAMBRIDGE
UNIVERSITY PRESS

University Printing House, Cambridge CB2 8BS, United Kingdom

One Liberty Plaza, 20th Floor, New York, NY 10006, USA

477 Williamstown Road, Port Melbourne, VIC 3207, Australia

314-321, 3rd Floor, Plot 3, Splendor Forum, Jasola District Centre, New Delhi - 110025, India

79 Anson Road, #06-04/06, Singapore 079906

Cambridge University Press is part of the University of Cambridge.

It furthers the University's mission by disseminating knowledge in the pursuit of
education, learning and research at the highest international levels of excellence.

www.cambridge.org
Information on this title: www.cambridge.org/9781107014473

First published 2014

A catalogue record for this publication is available from the British Library

Library of Congress Cataloging in Publication data
Bachmann, Michael
Thermodynamics and statistical mechanics of macromolecular systems / Michael Bachmann.
pages cm
ISBN 978-1-107-01447-3 (hardback)
1. Macromolecules–Thermodynamics. 2. Biomolecules–Structure.
3. DNA–Structure. 4. Statistical mechanics.
I. Title.
QP801.P64B33 2014
547′7–dc23 2013032171

ISBN 978-1-107-01447-3 Hardback

Dedicated to my family

Contents

Preface and outline

The idea to write this book unfolded when I more and more realized how equally frustrating and fascinating it can be to design research projects in molecular biophysics and chemical physics – frustrating for the sheer amount of inconclusive and contradicting literature, but fascinating for the mechanical precision of the complex interplay of competing interactions on various length scales and constraints in conformational transition processes of biomolecules that lead to functional geometric structures. Proteins as the "workhorses" in any biological system are the most prominent examples of such biomolecules.

The ability of a "large" molecule consisting of hundreds to tens of thousands of atoms to form stable structures spontaneously is typically called "cooperativity." This term is not well defined and could easily be replaced by "emergence" or "synergetics" – notions that have been coined in other research fields for the same mysterious feature of macroscopic ordering effects. There is no doubt that the origin of these net effects is of "microscopic" (or better nanoscopic) quantum nature. By noting this, however, we already encounter the first major problem and the reason why heterogeneous polymers such as proteins have been almost ignored by theoretical scientists for a long time. From a theoretical physicist's point of view, proteins are virtually "no-no's." Composed of tens to thousands of amino acids (already inherently complex chemical groups) linearly lined up, proteins reside in a complex, aqueous environment under thermal conditions. They are too large for a quantum-chemical treatment, but too small and too specific for a classical, macroscopic description. They do not at all fulfill the prerequisites of the thermodynamic limit and do not scale. In consequence, the standard statistical theory of phase transitions is not directly applicable, although many aspects of molecular structure formation processes resemble those known from phase transitions. Since 20 types of amino acids occur frequently in bioproteins, the number of possible compositions is astronomically large, but only of the order of 100 000 highly specific types of bioproteins are functional in the human cell system. Beside this obviously elementary evolutionary aspect, the heterogeneous composition (which causes glass-like behavior) and their high specialization level raise the question, to what extent folding properties can be generic at all. This is actually one of the key questions. A negative answer is not very likely; nature has always proven that even the most complex structures possess symmetries (in a more general context), which explain their stability. Stability is necessary, because these molecular systems exist and function in a thermal environment. It is even appropriate to formulate the whole problem in the following way: it is the interplay and balance between system and environment that stabilizes the structure of the system. Having said that, there is no reasonable way to try to understand any structure formation process without including thermodynamics and, therefore, statistical mechanics.

Another apparent problem is that analytical approaches virtually fail to explain processes of heterogeneous systems, leaving computer simulations the only available tool for theoretical studies. Since protein folding is a relatively slow process (microseconds to seconds), it is almost impossible to use molecular dynamics simulations, operating on nanosecond timescales, for folding studies. Alternatively, Monte Carlo methods are inefficient, if the surrounding water molecules are explicitly simulated. The models are generally not well defined and computer simulations on atomic scales often require large-scale supercomputing resources. The abovementioned key question of generics affects the possibility and limitation of using much more efficient coarse-grained models. For these reasons, studies of biomolecular systems remain a true challenge to theoretical and computational biologists, chemists, and physicists. However, the fact that, among others, neurodegenerative diseases such as Alzheimer's and all virus infections are associated with structural properties of biomolecules makes it worth the efforts to research macromolecular systems of such scale.

This book is a "research book" for the interdisciplinary community. This means it offers many approaches to deal with molecular systems by means of statistical mechanics and computer simulation, yet it will give no precise answers to the above questions. It shall provide young scientists from all affected disciplines of natural and technological sciences with the background to get started, but it also addresses senior scientists by promoting alternative views. The book could also be of value as a compendium as it includes widely accepted research results, in particular for homopolymer systems.

More specifically, we are going to discuss thermodynamic properties of conformational transitions for single- and multiple-chain polymer and protein systems, with particular focus dedicated to molecular folding, aggregation, and adsorption processes to solid substrates. In most of the presented examples, we will investigate the structural transitions by statistical analyses of simplified models. This is based on the idea that in cooperative processes like structural transitions, the collective action of the mechanical degrees of freedom allows for a reduction of the phase space. In other words, the essential features of these transitions are expected to be described qualitatively correctly by models in which a strongly reduced number of effective degrees of freedom is considered only. This reduction of mechanical complexity is called coarse-graining and has proven to be extremely successful in the understanding of complex phenomena and phase transitions of macroscopic systems. If the analogy of structural transitions in rather small molecular systems and phase transitions of macroscopic systems holds true, then coarse-grained approaches can also be valuable tools for the description of molecular behavior. Coarse-grained modeling and simulation will, therefore, play a vital role in this book.

The finiteness of molecular systems, the geometric nature of the structural transitions, and the constraints (e.g., stiff bonds) that affect the mechanical motion render a theoretical treatment typically very difficult. For this reason, the design of efficient algorithms is inevitable for unraveling the properties of structural transitions of molecular systems. We will discuss various examples throughout this book, where sophisticated computer simulation methodologies were employed to obtain the statistical information needed for a thermodynamic analysis of such transitions. Therefore, a short review of modern simulational methods is also included, as it is considered to be beneficial for readers who wish to get started or who would simply like to know where the results discussed in this book originated from.

In the first chapter of this book, we begin with an introduction of the molecular structure and the modeling of linear macromolecules. Fundamental aspects of thermodynamics and statistical mechanics, with emphasis on finite-size effects and their statistical analysis, are reviewed in Chapter 2. In Chapter 3, properties of the complete sequence and conformation space are systematically analyzed for short lattice proteins by exact enumeration of a minimalistic hydrophobic–polar heteropolymer model. Computer simulations of larger systems require efficient algorithms. Such algorithms are reviewed in Chapter 4 and important aspects of analyses of finitely long time series of data generated by these algorithms are discussed. As a first application, the study of homopolymer freezing and collapse transitions on regular lattices is the subject of Chapter 5. In this regard, the influence of surface and finite-size effects upon crystallization of elastic flexible and semiflexible polymers is addressed in detail in Chapters 6 and 7, respectively. Returning to proteins, characteristic folding properties of proteins and the classification of folding channels are investigated in Chapters 8 and 9. Generic local geometries like secondary structures induced by constraints that effectively reflect many-body effects are discussed in Chapter 10 by introducing tube-like polymers. The extension of coarse-grained modeling to multiple-chain systems is described in Chapter 11, where also analyses of aggregation transitions of short heteropolymers in different statistical ensembles are presented. In Chapter 12, we unravel the hierarchical nature of phase transitions by discussing the exemplified aggregation transition of homopolymers. Pseudophase diagrams of adsorption processes of lattice and off-lattice homopolymers to solid substrates are investigated in detail in Chapter 13. An introductory, simple example for substrate-specific binding of peptides to solid substrates is studied in detail in Chapter 14.

I am indebted to Wolfhard Janke for many years of successful cooperation, advice, and support. For constant support and extremely helpful discussions, I am also thankful to David P. Landau, Kurt Binder, Hagen Kleinert, and Gerhard Gompper. The collaboration with Thomas Neuhaus, Anders Irbäck, Joan Adler, Axel Pelster, and Qianqian Cao is also very much appreciated. It is a particular pleasure to thank my long-term collaborators Thomas Vogel, Stefan Schnabel, Karsten Goede, Monika Möddel, Christoph Junghans, Jonathan Groß, Daniel T. Seaton, and Tristan Bereau for the joint successful work, without which it would not have been possible to write this book. Many other people have also actively and passively helped streamline my thoughts in the exciting field of structural biophysics, which I am also quite thankful for. I would also like to thank Érica de Mello Silva and Paulo H. L. Martins at the Universidade Federal de Mato Grosso, Cuiabá, for their kind hospitality during my current visit to Brazil.

Michael Bachmann
Athens, GA (USA)
June 2013

Introduction

1.1 Relevance of biomolecular research

More than 30 million people, infected by the *human immunodeficiency virus* (HIV), suffer from the *acquired immune deficiency syndrome* (AIDS);[1] 2 billion humans carry the *hepatitis B* virus (HBV) within themselves, and in more than 350 million cases the liver disease caused by the HBV is chronic and, therefore, currently incurable.[2] These are only two examples of worldwide epidemics due to virus infections. Viruses typically consist of a compactly folded nucleic acid (single- or double-stranded RNA or DNA) encapsulated by a protein hull. Proteins in the hull are responsible for the fusion of the virus with a host cell. Virus replication, by DNA and RNA polymerase in the cell nucleus and protein synthesis in the ribosome, is only possible in a host cell. Since regular cell processes are disturbed by the virus infection, serious damage or even the destruction of the fine-tuned functional network within a biological organism can be the consequence.

Another class of diseases is due to structural changes of proteins mediated by other molecules, so-called prions. As there is a strong causal connection between the three-dimensional structure of a protein and its biological function, refolding can cause the loss of functionality. A possible consequence is the death of cells. Examples for prion diseases in the brain are *bovine spongiform encephalopathy* (BSE)[3] and its human form *Creutzfeldt–Jakob disease* (CJD).[4]

A further source for damaging cellular networks is protein misfolding followed by amyloid aggregation. In the case of *Alzheimer's disease* (AD), which is a neurodegenerative disease and the most common type of *dementia*, amyloid beta (Aβ) peptides in sufficiently high concentration experience structural changes and tend to form aggregates. Following the amyloid hypothesis, it is believed that these aggregates (which can also take fibrillar forms) are neurotoxic, i.e., they are able to fuse into cell membranes of neurons and open calcium ion channels. It is known that extracellular Ca^{2+} ions intruding into a neuron can promote its degeneration. About 24 million, mainly elderly, people are currently affected.[5]

---

[1] UNAIDS/WHO *AIDS Epidemic Update: December 2007.*

[2] WHO Fact Sheet No. 204 (2000).

[3] The economic impact of BSE is disastrous. Following *USA Today* from August 4, 2006, US beef exports declined from $3.8 billion in 2003, before the first mad cow was detected in the USA, to $1.4 billion in 2005.

[4] The WHO Fact Sheet No. 180 (2002) reports a rate of one per million people.

[5] Alzheimer's Disease International, Global Perspective **16**(3), 1 (2006).

This exemplified collection of diseases manifests the extraordinary importance of molecular, biologically motivated, research. It comprises the understanding of the synthesis of biomolecules such as proteins based on the genetic code (gene expression), the characteristics and specificity of folding and aggregation processes, as well as the unraveling of the dynamic and kinetic aspects of biological processes. In almost all such processes functional proteins are involved. Of course, treating patients suffering from these diseases is a *medical* problem, but revealing the nature of mechanisms behind the functioning of living organisms is an interdisciplinary task for the whole ensemble of natural sciences.

This is also the case when it comes to nanotechnological applications. Science and technology meet and partly fuse at smallest length scales. In particular, learning from biological systems – from complex networks of biological functionality or from structural properties of single molecules – has become more and more important and relevant for special-purpose applications on micrometer or even nanometer scales. This also includes the interaction of soft materials with solid matter, where systematic research is just in its early stages. Examples for such hybrid systems with enormous potential for future applications are, among many others, biosensors in the form of adhesion-specific nanoarrays for the identification of proteins in solution and also nanoelectronic circuits on polymer basis. On the experimental side, the progress in the development of high-resolution equipment and new experimental techniques not only allows for detection of what is happening on atomic scales, but also enables local manipulation of molecules which is essential for the design of specific applications. On the other hand, computational capacities have reached a level that now makes it possible to study macromolecular systems more systematically in simulations of suitable models by means of sophisticated numerical methods. Simulations are particularly relevant for investigations of processes that are currently still inaccessible to experimental research.

At this point, the challenge for theoretical physics, in particular, is twofold: first, the modeling and analysis of specific molecular structures at atomic scales and second, the generalization of the conformational (pseudophase) transitions accompanying structure formations processes within a mesoscopic frame. Both approaches facilitate the systematic understanding of molecular processes that is typically difficult to achieve in experiments.

Protein folding, peptide aggregation, polymer collapse, crystallization, and adsorption of polymers and proteins to nanoparticles and solid substrates have an essential feature in common: in all these processes, structure formation is guided by a collective, cooperative behavior of the molecular subunits lining up to build chain-like macromolecules. In this process, polymers and proteins experience conformational transitions related to thermodynamic phase transitions. For chains of finite length, an important difference of crossovers between conformational (pseudo)phases is, however, that these transitions are typically rather smooth processes, i.e., thermodynamic activity is not necessarily signalized by strong entropic or energetic fluctuations. The interest in properties of finite-length polymers and proteins has grown rapidly within the past few years, not only because of the technological advances on the nanometer scale, but in particular due to the fact that the thermodynamics of small-scale systems is the key to the understanding of the biomechanical principles of signal exchange and transport processes being relevant for life, as, for example, receptor–ligand binding between proteins or molecular flow through nanopores.

Since proteins play a key role in all biological processes, we will discuss their properties more specifically in the following.

1.2 Proteins

In order to get an impression of the complexity of the task of describing the relationship between chemical composition, geometric structure, and the function of individual macromolecules, we will now look more closely at one of the most prominent examples that represents an entire class of biomolecules: proteins.

1.2.1 The trinity of amino acid sequence, structure, and function

Proteins are highly specialized macromolecules performing essential functions in a biological system, such as control of transport processes, stabilization of the cell structure, and enzymatic catalyzation of chemical reactions; others act as molecular motors in the complex machinery of molecular synthetization processes. Chemically, proteins are built up of sequences of amino acid residues linked by peptide bonds. The polypeptide chain consists of a linear backbone with the amino-acid-specific side chains attached to it. The atomic composition of the protein backbone is shown in Fig. 1.1. Typical proteins consist of up to $N = 50, \ldots, 3000$ residues. The 20 different types of amino acids occurring in bioproteins are shown in Fig. 1.2. The side chains of these amino acids govern the specificity of each amino acid in protein folding processes and differ in chemical and physical properties under the influence of the surrounding solvent. Solubility in an aqueous environment is dependent on the occurrence of polar groups in the side chain, such as, e.g., the hydroxylic groups in serine and threonine. Hydrophobic side chains are insoluble and not very reactive in a polar environment. Typical large and strongly hydrophobic side chains such as, for example, phenylalanine or tryptophan possess aromatic rings. Others, like alanine or leucine with methine (-CH), methylene (-CH$_2$), or methyl (-CH$_3$) groups in the side chain, are aliphatic, i.e., these nonaromatic side chains only contain hydrogen and carbon atoms.

Primarily only arginine and lysine (positively charged) and aspartic and glutamic acid (negatively charged) contribute explicitly to the total charge of a protein in a neutral environment. In addition, histidine is typically positively charged in a slightly acidic environment, but neutral in neutral solution.

The frequency and the sequential arrangement of hydrophobic, polar, and charged amino acid residues in the amino acid sequence (also called the *primary structure*) of a protein are mainly responsible for the formation of a stable and unique native conformation. Protein conformations or segments of it are typically classified on different length scales. Parts of protein conformations that form local symmetric substructures are called *secondary structures*. These include helices, planar sheets (or strands), and turns. Substructures of this

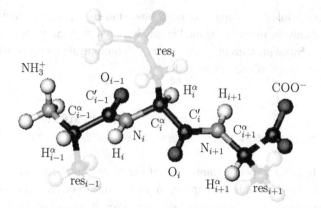

Atomic composition of the protein backbone. Amino acids are connected by the peptide bond between C′ and N. Side chains or amino acid residues ("res") are usually connected to the backbone by a bond with the C^α atom (except proline, which has a second covalent bond to its backbone nitrogen).

type are common to all line-like objects and generally can be considered as the underlying geometry of linear polymers. The formation of secondary structures is not necessarily connected with the formation of hydrogen bonds, but these structures are essentially stabilized by hydrogen bonds. The whole conformation of a single protein, including aligned secondary structures, defines its *tertiary structure*. The tertiary fold typically consists of a very compact core of hydrophobic residues that is screened from the aqueous environment by a shell of polar amino acids being in direct contact with the solvent. This assembly of separate polar and hydrophobic parts is characteristic for proteins – it reduces entropy and thus ensures stability. Eventually, in large proteins or protein compounds, different hydrophobic domains can form widely independently. The global shape of the macromolecule that can include composites of individual tertiary domains is generally classified as *quaternary structure* [1–3].

The understanding of general aspects of the folding of proteins into native conformations is particularly essential, as the shape of the stable three-dimensional geometrical structure of a protein often determines the biological function of a protein – or its malfunction, if the protein has misfolded, refolded, or denatured.

The spectrum of protein functions in a biological organism is manifold. Proteins are involved in almost all cell processes. Ion channels and nanopores formed by membrane proteins control ion and water flow into and out of the cell [4]. Figure 1.3 shows the atomic structure of the membrane protein aquaporine, a pore embedded in the cell membrane that is permeable for water molecules. The efficiency is extreme. Up to 3 billion water molecules can rush through this pore per second [5, 6]. Other cell proteins like actin, for example, are responsible for the mechanical stability of the cell backbone. Actin polymerizes to filaments and these filaments can form stable networks. Besides cell stability, these networks also enable transport processes along these "tracks," e.g., vesicle transport mediated by myosin proteins. The interplay between actin and myosin is also important for the ability of muscle tissue to contract. Biochemical reactions are

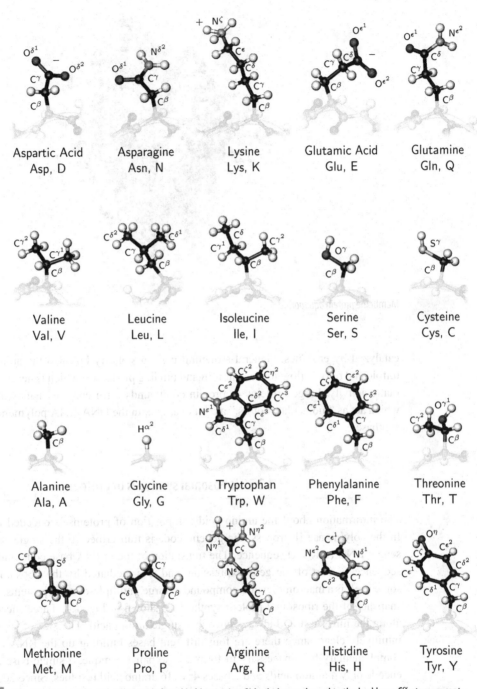

Fig. 1.2   The side chains of the 20 amino acids found in bioproteins. Side chains are bound to the backbone C^α atom, except proline, which has an extra bond to the backbone nitrogen N. Heavy atoms have been assigned standard labels [1]. There is no side chain for glycine, the free C^α bond is saturated by the hydrogen $H^{\alpha 2}$.

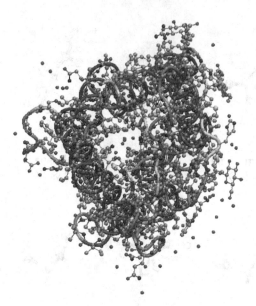

Fig. 1.3   Membrane protein *aquaporine*.

catalyzed by enzymes. General structural protein stability but also the ability to locally unfold and refold allow for receptor–ligand binding processes, which is necessary for enzymatic activity. Large proteins or protein compounds form complex nanoscale machines, which act as "molecular motors," as, for example, in the DNA/RNA polymerases and ATP synthase.

1.2.2 Ribosomal synthesis of proteins

The information about the amino acid composition of proteins is encoded in the DNA. In the polymerase II process, the genetic code is transcribed to the single-stranded messenger RNA (mRNA) sequence. The translation of the mRNA base code into amino acid sequences is part of the gene expression, and it is mediated by the ribosome. The ribosome itself is a macromolecular compound of large, multiple-domain proteins. A schematic snapshot of the ribosomal protein synthesis is shown in Fig. 1.4. Three successive bases along the mRNA strand always encode a single amino acid. The size of such a codon is intuitively clear: since there are four different bases building up the RNA code and the complementary base necessary to form a base pair is unique, a single-base codon could encode only 4 amino acids and 2 bases $4^2 = 16$ amino acid residues. Since 20 amino acids were identified in typical bioproteins, 3 bases are required to form a codon. The $4^3 = 64$ possibilities not only allow multiple, redundant codes for amino acids (which is a first kind of genetic error correction), but also enable the definition of start and stop codons, which are necessary to separate the codes for the different proteins within the linear RNA sequence [7, 8].

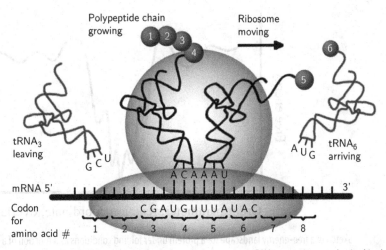

Fig. 1.4   The sequence of amino acids building up proteins is, based on the genetic DNA expression, synthesized in the ribosome.

After the ribosome has read a codon off the mRNA strand, a transfer RNA (tRNA) molecule connects to the mRNA codon. A tRNA molecule mainly contains a three-base section, the anticodon, which is complementary to a specific codon at the mRNA, and the associated amino acid residue. Thus the tRNA molecules serve as translators between the base codons and the amino acids. The ribosome separates the amino acid from the tRNA and attaches it to the already synthesized part of the protein sequence. This process continues until a stop codon is reached. Eventually, the protein is released into the aqueous solvent within the cell. It is widely believed that in this moment the protein is still unstructured and the formation of the functional structure is a spontaneous folding process. Larger proteins that would exhibit an increased tendency to misfold in the complex and crowded environment are often encapsulated in chaperons that assist in the folding process.

1.2.3 From sequence to function: The protein folding process

Anfinsen's refolding experiments [9] showed that the native conformation is *not* a result of the synthetization process in the ribosome. Rather, it is a dynamical process that strongly depends on intrinsic properties such as the amino acid sequence, but it is also influenced by the solvent properties and temperature of the surrounding solvent. Typical folding times are of the order of milliseconds to seconds. One of the most substantial problems in the understanding of protein folding is the "strategy" the protein follows in finding the unique conformation (geometrical structure), the so-called native fold. Thermodynamically, this native conformation represents the state of minimal free energy. Therefore, protein folding is not simply an energetic minimization process – it is also affected by entropic forces. This means that the folding trajectory is a stochastic process from a mostly random initial structure toward the global free-energy minimum, thereby circumventing free-energy barriers. The free-energy landscape of a protein, considered as a function of the protein's

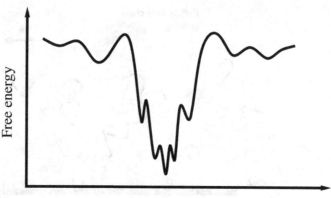

Order parameter, reaction coordinate, overlap, ...

Fig. 1.5 Sketch of a free-energy landscape for a protein under folding conditions as a function of a single cooperativity parameter, which is often referred to as a reaction coordinate in analogy to chemical reaction kinetics, as an order parameter by considering the folding process as an analog of a thermodynamic phase transition, or as a kind of overlap parameter similar to what is often used in descriptions of metastable systems such as spin glasses.

degrees of freedom, is commonly assumed to be extremely rugged and complex and it seems paradoxical that the protein is able to find the "needle-in-a-haystack" structure by a stochastic search process within a relatively short time period.

Protein folding is, however, also a process of high cooperativity, i.e., structure formation requires a collective arrangement of at least a subset of the degrees of freedom. For this reason, it is expected that a single or a few cooperativity parameter(s) – comparable to order parameters in thermodynamic phase transitions – allow(s) for the discrimination between dominating macrostates, i.e., the structural "phases." A strongly simplified sketch of such a free-energy landscape as a function of a single cooperativity parameter is shown in Fig. 1.5. For stability reasons, the single funnel-like global free-energy valley is sufficiently deep to prevent thermal unfolding. Local folding processes can cause weakly stable or metastable conformations that slow down the folding process. Thus, the folding channel is not necessarily smooth and local free-energy minima can be present in the funnel.

The assumption of the existence of a reduced set of relevant collective degrees of freedom thus enables a generalized view of folding processes as structural or conformational transitions and their classification. Indeed, there has been enormous progress in this direction within the past years, and candidates with comparatively simple folding trajectories were identified. This regards, e.g., single-exponential or downhill folding, where no barriers hinder and slow down the folding process. Another prominent example is two-state folding, a "first-order-like," "discontinuous" transition with a single barrier. More complex are "folding-through-intermediates" events with more than one free-energy barrier or even metastability, where the native fold is degenerate and, therefore, the formation of different structures is (almost) likely probable. The latter case is obviously important for proteins involved in mechanical or motoric processes, where local refolding can be necessary for fulfilling a specific biological function.

1.3 Molecular modeling

Structure formation at the atomic level is, in principle, a traditional quantum-chemical many-body problem. Amino acids occurring in bioproteins contain between 7 (glycine) and 24 (tryptophan) atoms.[6] Thus, typical bioproteins consist of hundreds to tens of thousands of atoms. In general, the structural properties of macromolecules depend on two classes of chemical bonds: *covalent* and *noncovalent* bonds. Covalent bonds are based on common electron pairs shared between atoms and stabilize the chemical composition of the molecule. On the other hand, noncovalent bonds[7] are based on much weaker effective interactions due to screening, polarization effects, or dipole moments, partly induced by the surrounding polar solvent. These interactions are responsible for the three-dimensional structure of macromolecules in solvent.

1.3.1 Covalent bonds

The formation of a covalent bond between atoms is a pure quantum-mechanical effect and due to an effective pairwise attraction between electrons in outer, unsaturated shells of two atoms. This spin-dependent exchange interaction overcompensates the electrostatic repulsion between the electrons and results in an electron pair, which is shared by the atoms involved. Covalent bonds are very stable and a thermal decomposition, e.g., at room temperature, is extremely unlikely. The dissolution energy of biochemically relevant covalent bonds lies between 50 kcal/mole (disulfide bridges S–S) and 170 kcal/mole (C $=$ O double bonds). For comparison, the thermal energy at room temperature $T_r = 300$ K is $RT_r \approx 0.6$ kcal/mole.[8] This energy is also not sufficient to excite vibrations of covalent bonds at room temperature. Therefore, the effective bond lengths, i.e., the distances between the atom cores, are rigid. Furthermore, the bond angles between two successive covalent bonds are relatively rigid. In any event, weak vibrational fluctuations are thermally excitable, but the typical fluctuation widths are comparatively small and usually do not exceed 5°. In proteins, covalent bonds obviously stabilize the sequence of the amino acids linked by covalent peptide bonds, i.e., its primary structure. Although torsional degrees of freedom also are affected by covalent bonds, a subset of torsional angles is widely flexible: the so-called *dihedral torsion angles*. Figure 1.6 shows a conformation of a small peptide, with the end-groups NH_3^+ and COO^- and the amino acid phenylalanine (Phe) highlighted. The Ramachandran dihedral angles in the backbone are typically denoted ϕ (torsional angle between atoms C'_{i-1}, N_i, C_i^α, and C'_i of the $(i-1)$th and ith amino acid) and ψ (between N_i, C_i^α, C'_i, N_{i+1}). The angles ϕ and ψ are comparatively flexible in the interval $-180° < \phi, \psi \leq 180°$, because the torsional barriers imposed by the electronic properties

---

[6] Numbers of atoms refer to uncharged amino acids within a polypeptide chain.

[7] Typically associated with *nonbonded interactions*.

[8] The gas constant in molar units is $R = N_A k_B \approx 1.99 \times 10^{-3}$ kcal/K mole, where $N_A \approx 6.02 \times 10^{23}$ mole^{-1} is the Avogadro constant and $k_B \approx 3.30 \times 10^{-27}$ kcal/K is the Boltzmann constant.

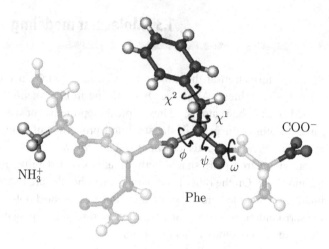

Fig. 1.6 Definition of the backbone dihedral angles ϕ, ψ, and ω. Exemplified for phenylalanine, also the only two side-chain degrees of freedom χ^1 and χ^2 are denoted. The convention is that the torsional angles can have values between $-180°$ and $+180°$, counted from the N-terminus (NH_3^+) to the C-terminus (COO^-) according to the right-hand rule and in the side chains starting from the C^α atom.

of the covalent bonds are rather weak. An exception is proline, where the particular geometry of the side chain restricts ϕ to a value close to $-75°$. The angle ω is associated with the torsion of the peptide bond C_i^α, C_i', N_{i+1}, and C_{i+1}^α. However, the sp2 hybridizations of the C' and N valence electrons and a p-electron uninvolved in the hybridizations that form an electron cloud surrounding the C'-N peptide bond entail a large torsional barrier. Thus, $\omega \approx 180°$, and C_i^α, C_i', N_{i+1}, and C_{i+1}^α form an almost planar *trans* conformation. Proline is special as it is bound to three massive radicals instead of the usual two. For this reason, there is also a non-negligible amount (about 10%) of proline involving peptide bonds in *cis* conformation ($\omega \approx 0°$) [2]. Phenylalanine possesses two torsional side-chain angles, χ^1 and χ^2, that can be thermally activated under physiological conditions. Depending on the type of amino acid and its atomic composition, the number of torsional side-chain angles varies.

Thus, the three-dimensional geometric structure of proteins is little dependent on covalent bonds, and it is rather due to the much weaker effects between nonbonded atoms. Nonetheless, the rigidity of covalent bond lengths and bond angles affects the process of structure formation (e.g., the folding of a protein into the native state with lowest free energy). Generally, the steric constraints promote frustration and metastability and thus the existence of stable native conformations is a very particular property of the comparatively small number of functional bioproteins selected by evolution. However, the majority of all possible amino acid sequences suffers from degeneration effects, e.g., nonfunctional metastability, and also from only weakly stable conformations under physiological conditions. For this reason, such sequences play only a very minor role in biological systems (molecular motors, for example, require only small activation barriers for motion, but refolding often affects only small parts of the structure).

1.3.2 Effective noncovalent interactions and nanoscopic modeling: Toward a semiclassical all-atom representation

Noncovalent inter-atomic interactions are induced by van der Waals forces, electrostatic potentials between partial charges of atoms, torsional potentials, dipole–dipole arrangements (hydrogen bond formation), and effective inter-monomeric forces, as, for example, caused by the hydrophobic effect, which is due to the interaction of the protein with the surrounding solvent. The standard approaches to calculate ground-state energies, transition probabilities, and quantum-mechanical expectation values for small molecules are quantum chemistry (QC) and density-functional theory (DFT), often in combination with analytical and numerical methods. The bottleneck of these methods is that the electron distributions of heterogeneous macromolecules such as proteins are so complex that these approaches typically fail to predict a precise energy spectrum of the whole system.

The knowledge of the energy spectrum is quite important for the most relevant questions in protein-folding studies as it affects the topology of the free-energy landscape as a function of a small set of relevant "folding coordinates" (similar to reaction coordinates in physical chemistry or order parameters in the terminology of statistical physics), its barriers, and thus the possible folding trajectories under "normal" conditions.[9] Thus, protein folding is a thermodynamic process in the presence of kinetic barriers and is often considered as a "conformational transition," i.e., it exhibits features of thermodynamic phase transitions, although prerequisites of real phase transitions – for example the existence of a thermodynamic limit – are not satisfied. Conformational transitions are thus rather crossovers between different classes of structures (e.g., random coils, helices, and sheets). It should be noted, however, that for small molecules and short segments of large molecules, as, for example, isolated amino acids, QC and DFT are quite useful and results from these calculations have entered into force fields used in semiclassical models.

Since a macromolecular quantum-chemical analysis is virtually impossible, the most promising approach employed in the past is investigating dynamics and statistics of semiclassical models. These models are usually based on an energy function consisting of effective potentials for noncovalent interactions. "Effective" here means that quantum-mechanical effects as well as the influence of the surrounding solvent enter into parameters describing individual properties of amino acids, their side chains, but also properties of individual atoms regarding their actual position in the molecule. Due to the large number of constraints (atomic van der Waals radii, fixed covalent bond lengths, energetic torsional barriers), the complexity of this parameter field is enormous and the modeling of the noncovalent interactions is a substantial problem. The parametrization of the associated potentials is frequently based on structural data from NMR or X-ray experiments, physico-chemical estimations of enthalpies, bioinformatical database analyses, but also on quantum-chemical and other theoretical approaches. The main difficulty and major source of error is that in all cases the results for a specific system or a subset of proteins stored in

---

[9] We refer to "normal" conditions as the region of the space of the external parameters temperature and pH value, where the aqueous solvent surrounding the protein is fluid and neutral under normal pressure.

a database are assumed to be reliable for *all* proteins. This finally sets up what is called the "force field" which in certain models comprises of the order of $\mathcal{O}(10^3)$ parameters.

1.4 All-atom peptide modeling

Although the nature of nonbonded atomic and molecular interactions is known in principle, semiclassical models of proteins differ not only in the parametrization of the potentials, but also in the form of the effective energetic contributions. The problem that is common to most all-atom approaches is that these models are often gauged against a subset of the protein data bank (pdb) [10]. Within this subset, these models can be applied with some success to structure predictions. In structural predictions for sequences not in the gauge set, these models often fail. A frequent feature of such models is, for example, the overweighting of a certain type of secondary structures, either helices or strands. Another general problem is the enormous complexity of these models and the huge parameter sets, resulting in slow dynamics in computer simulations. This is particularly apparent in molecular dynamics simulations of protein folding ranging from extremely (CPU) time-consuming to simply impossible. Reasonable results currently can be obtained only in computer simulations of small peptides – but the understanding of their folding and aggregation behaviors could actually be the key to deeper insights into generic aspects of structure formation of large bioproteins. In standard peptide models, all atoms and covalent bonds, bond and torsion angles of the molecule are assigned individual parameters mimicking quantum-mechanical effects. Examples of such parameters are van der Waals radii and partial charges of the atoms, lengths of covalent bonds between atoms, angles between successive covalent bonds, torsional angles and torsional barriers. Atomic parameters typically depend on the position of the respective atom in the amino acid and the chemical composition of the amino acid residue; even different parametrizations of the same type of atom are distinguished. One such example is hydrogen, whose energetic properties depend on the chemical group it belongs to (e.g., if it is part of an aliphatic, aromatic, hydroxylic or carboxylic, amide or amine group, bound to sulfur or to C^δ of proline).

In such models, each atom i, located at the position $\mathbf{r}_i$, carries a partial charge q_i. Covalent bonds between atoms, according to the chemical structure of the amino acids, are considered rigid, i.e., bond lengths are kept constant, as well as bond angles between covalent bonds, and certain rigid torsion angles. In some models, this constraint is weakened by allowing the bonds and bond angles to fluctuate slightly. Distances between nonbonded atoms i and j are defined as $r_{ij} = |\mathbf{r}_i - \mathbf{r}_j|$ and measured in Å in the following. The set of degrees of freedom covers all dihedral torsion angles $\boldsymbol{\xi} = \{\xi_\alpha\}$ of the αth residue's backbone ($\phi_\alpha, \psi_\alpha, \omega_\alpha$) and side chain ($\chi = \chi_\alpha^{(1)}, \chi_\alpha^{(2)}, \ldots$). The model incorporates electrostatic Coulomb interactions between the partial atomic charges (all energies in kcal/mole),[10]

$$E_C(\boldsymbol{\xi}) = 332 \sum_{i,j} \frac{q_i q_j}{\varepsilon r_{ij}(\boldsymbol{\xi})}, \tag{1.1}$$

---

[10] The dielectric constant of water at room temperature is $\varepsilon \approx 80$.

effective atomic dipole–dipole interaction modeled via Lennard-Jones potentials

$$E_{\mathrm{LJ}}(\boldsymbol{\xi}) = \sum_{i,j} \left(\frac{A_{ij}}{r_{ij}^{12}(\boldsymbol{\xi})} - \frac{B_{ij}}{r_{ij}^{6}(\boldsymbol{\xi})} \right), \tag{1.2}$$

O-H and N-H hydrogen-bond formation,

$$E_{\mathrm{HB}}(\boldsymbol{\xi}) = \sum_{i,j} \left(\frac{C_{ij}}{r_{ij}^{12}(\boldsymbol{\xi})} - \frac{D_{ij}}{r_{ij}^{10}(\boldsymbol{\xi})} \right), \tag{1.3}$$

and considers dihedral torsional barriers (if any):

$$E_{\mathrm{tor}}(\boldsymbol{\xi}) = \sum_{l} U_l \left(1 \pm \cos(n_l \xi_l) \right). \tag{1.4}$$

The total energy of a conformation, whose structure is completely defined by the set of dihedral angles $\boldsymbol{\xi}$, is

$$E_0(\boldsymbol{\xi}) = E_{\mathrm{C}}(\boldsymbol{\xi}) + E_{\mathrm{LJ}}(\boldsymbol{\xi}) + E_{\mathrm{HB}}(\boldsymbol{\xi}) + E_{\mathrm{tor}}(\boldsymbol{\xi}). \tag{1.5}$$

The hundreds of parameters $q_i, A_{ij}, B_{ij}, C_{ij}, D_{ij}, U_l$, and n_l form a force field. Additional sophisticated model parameter sets must be imposed, if the interaction of the protein with solvent molecules, ions, and counterions shall explicitly be considered. For simulations in implicit solvent, peptide models can simply be extended by a solvation-energy contribution, which is, e.g., given by [11]

$$E_{\mathrm{solv}}(\boldsymbol{\xi}) = \sum_i \sigma_i A_i(\boldsymbol{\xi}), \tag{1.6}$$

where A_i is the solvent-accessible surface area of the ith atom for a given conformation and σ_i is the solvation parameter for the ith atom. The values for σ_i depend on the type of the ith atom and are parametrized in a separate force field. The total potential energy of the molecule then reads $E_{\mathrm{tot}}(\boldsymbol{\xi}) = E_0(\boldsymbol{\xi}) + E_{\mathrm{solv}}(\boldsymbol{\xi})$.

All force-field-based models are inherently problematic, because the attempt to accurately mimic both local and multibody interactions, which are essentially based on quantum effects, by semiclassical effective interactions must fail, at least in general. In certain cases, an appropriate fine-tuning of the parameter sets is possible and can lead to reasonable results, but applying the same parameter set to a different problem might cause false predictions. The main reason is that the energetic margins for certain substructures of a protein are so small that "fine-tuning" of a single interaction and its parameter subset easily unbalances another subset. The force-field parameters are not independent, because the interactions cannot be decoupled, if their energy scales are of similar magnitude (which for many systems affects almost all energetic contributions in these models). The interaction with the environment adds to the problem and the computational challenges to simulate these models under the essential influence of thermal fluctuations seems to rule out these modeling attempts completely. Nonetheless, simulations of these models are currently the only way to gain some insight into structural processes and function of proteins and protein complexes at nanoscopic scales.[11]

1.5 The mesoscopic perspective

Another, completely different vista is opened by employing minimalistic, coarse-grained protein models. Coarse-graining of models, where relevant length scales are increased by reducing the number of microscopic degrees of freedom, has proven to be very successful in polymer science. Although specificity is much more sensitive for proteins, since details (charges, polarity, etc.) and differences of the amino acid side chains can have strong influences on the fold, mesoscopic approaches are also of essential importance for the basic understanding of conformational transitions affecting the folding process. It is also the only possible approach for systematic analyses such as the evolutionarily significant question why only a few sequences in nature are "designing" and thus relevant for selective functions. On the other hand, what is the reason why proteins prefer a comparatively small set of target structures, i.e., what explains the preference of designing sequences to fold into the *same* three-dimensional structure? Many of these questions are still widely unanswered yet. Actually, the complexity of these questions requires a huge number of comparative studies of complete classes of peptide sequences and structures that cannot be achieved by means of computer simulations of microscopic models. Currently only two approaches are promising. One is the bioinformatics approach of designing and scoring sequences and structures (and also possible combinations of receptors and ligands in aggregates), often based on database scanning according to certain criteria. Another, more physically motivated approach makes use of coarse-grained models, where only a few specific properties of the monomers enter into the models. Frequently, only two types of amino acids are distinguished: hydrophobic (H) and polar (P) residues, giving the class of corresponding models the name "hydrophobic–polar" (HP) models (see Fig. 1.7). In the simplest case, the HP peptide chain is a linear, self-avoiding chain of H and P residues on a regular lattice [12, 13]. Such models allow a comprising analysis of both the conformation *and* sequence space, e.g., by exactly enumerating all combinatorial possibilities. Other important aspects in lattice model studies are the identification of lowest-energy conformations of comparatively long sequences and the characterization of the folding thermodynamics.

Since lattice models suffer from undesired effects of the underlying lattice symmetries, simple hydrophobic–polar *off-lattice* models were introduced. One such model is the AB model, where, for historical reasons, A symbolizes hydrophobic and B polar regions of the protein, whose conformations are modeled by polymer chains in continuum space governed by effective bending energy and van der Waals interactions [14]. These models allow for the analysis of different mutated sequences with respect to their folding

---

[11] Optimally, to increase confidence, results obtained with different force fields should be compared with each other, or, if possible, verified by experiments. The value of this kind of empirical research relies almost completely on redundancy and consistency.

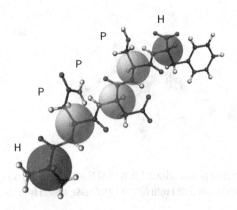

Fig. 1.7 Coarse-graining peptides in a "united atom" approach. Each amino acid is contracted to a single "C^{α}" interaction point. The effective distance between adjacent, bonded interaction sites is about 3.8 Å. In the coarse-grained hydrophobic–polar models considered here, the interaction sites have no steric extension. The excluded volume is modeled via type-specific Lennard-Jones pair potentials. In hydrophobic–polar (HP) peptide models, only hydrophobic (H) and polar (P) amino acid residues are distinguished.

characteristics. Here, the idea is that the folding transition is a kind of pseudophase transition, which can in principle be described by one or a few order-like parameters. Depending on the sequence, the folding process can be highly cooperative (single-exponential), less cooperative depending on the height of a free-energy barrier (two-state folding), or even frustrating due to the existence of different barriers in a metastable regime (crystal or glassy phases). These characteristics known from functional proteins can be recovered in the AB model, which is computationally much less demanding than all-atom formulations and thus enables more systematic theoretical analyses.

It is a common feature of such coarse-grained models that these enable a broader view on the general problem of protein folding, but for precise, specific predictions, their applicability is limited. In analogy to magnetic systems, they are rather comparable with the Ising model for ferromagnets or the Edwards–Anderson–Ising model for spin glasses. It should also be remarked that, due to their nontrivial simplicity, coarse-grained models are also a perfect testing ground for newly developed algorithms.

1.5.1 Why coarse-graining...? The origin of the hydrophobic force

Functional proteins in a biological organism are typically characterized by a unique three-dimensional molecular structure, which makes the protein selective for individual functions. In most cases, the free-energy landscape is believed to exhibit a rough shape with a large number of local minima and, for functional proteins, a deep, funnel-like global minimum. This assumed complexity is the reason why it is difficult to understand how the random-coil conformation of covalently bonded amino acids spontaneously folds into a well-defined stable "native" conformation. Furthermore, it is expected that there are only

(a)   (b)

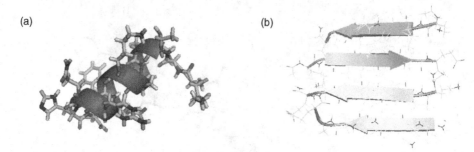

Fig. 1.8   (a) The C-peptide of ribonuclease A consists of 13 amino acids and is a typical α-helix former. (b) 7-residue segments of the Aβ peptide, associated with Alzheimer's disease, tend to form planar shapes, so-called β-strands.

a small number of folding paths from any unfolded conformation to this final fold. Protein folding follows a strict hierarchy at different length scales. The primary structure is provided by the ribosome. Since subsequent amino acids are uniformly linked by a covalent peptide bond independent of the geometric structure of the protein, the typical length scale of the primary structure is a single amino acid. The next level are secondary structures like α-helices, β-sheets, and turns. These substructures, which are probably simply intrinsic geometries of any line-like object, are stabilized by backbone hydrogen bonding between dipoles of different amino acids. Therefore, the scale of secondary structures is determined by their typical segment size, which is of the order of a few amino acids. Consequently, secondary-structure formation is often the first step in protein folding. This is followed by the formation of single-domain tertiary structures. The length scale of tertiary structures corresponds to the effective diameter of the entire, globular fold. While secondary structures are local, tertiary structures are global.

In fact, the process of tertiary-structure formation is what renders protein folding special. The main driving force for the folding of a complex domain, i.e., of up to hundreds of amino acids, is an effective cooperative interaction between many amino acid side chains and it is strongly influenced by the solubility properties (in particular its polarization) of the aqueous solvent the protein resides in. Roughly, amino acid side chains can be classified as polar, hydrophobic, and neutral. While polar residues favor contact with polar water molecules, hydrophobic acids avoid contact with water, which results in an effective attraction between hydrophobic side chains. In consequence, this attractive force leads to a formation of a highly compact hydrophobic core, which is screened from the solvent by a shell of polar amino acids. For very large proteins, the final stage in the folding process is the quaternary structure, where the size of a protein domain is the typical length scale. In proteins, the size of individual secondary-structure segments like α-helices and β-strands (for examples see Fig. 1.8) is typically rather small. The reason is that proteins are "interacting polymers," i.e., the amino acids interact with each other and form a globular or tertiary shape. This is due to the fact that amino acids noticeably differ only in their side chains. Adjacent amino acid backbones are connected via the peptide bond, and electric dipoles formed by backbone atoms are typically involved in hydrogen-bond formation. Backbone–backbone interaction provides the symmetry of secondary structures. However,

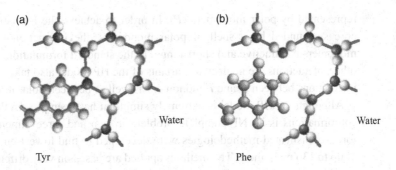

(a)   (b)

Water   Water

Tyr   Phe

Fig. 1.9   (a) Tyrosine (Tyr) is an amino acid with an OH group in the side chain. Thus, it is hydrophilic as the OH dipole can form a hydrogen bond with a polar solvent molecule (water). (b) Phenylalanine (Phe), on the other hand, is a typical example of a hydrophobic amino acid. The CH_2-C_6H_5 side chain does not contain a polar group. Phe in the surface-accessible protein shell would disturb the hydrogen-bond network of the solvent, rendering it an energetically disfavored configuration.

the interaction between the nonbonded side chains is non-uniform and strongly dependent on the side chain type. Roughly, two significantly different classes of side chains occur: hydrophilic ones that favor contact with a surrounding polar solvent like water, and hydrophobic side chains, which are nonpolar, thus disfavoring contact with water molecules (for representatives of the two classes see Fig. 1.9). Therefore, the effective force that leads to the formation of a compact hydrophobic core surrounded by a screening shell of polar amino acids is called hydrophobic force. For spontaneously folding single-domain proteins it is the essential driving force in the tertiary folding process.

1.5.2 Coarse-grained hydrophobic–polar modeling

The formation of tertiary hydrophobic-core structures is a complex process. Although atomic details, e.g., van der Waals volume exclusion separating side chains in linear and ring structures, polarizability, and partial charges, noticeably influence the folding process and the native fold, it should be possible to understand certain aspects of the folding characteristics, at least qualitatively, by means of coarse-grained models which are based on a few effective parameters. Minimalistic hydrophobic–polar lattice and off-lattice heteropolymer models, suitable for addressing these questions, are introduced in the following.

Hydrophobic–polar lattice proteins

The simplest model for a qualitative description of protein folding is the lattice hydrophobic–polar (HP) model [12]. In this model, the continuous conformational space is reduced to discrete regular lattices and conformations of proteins are modeled as self-avoiding walks restricted to the lattice. Assuming that the hydrophobic interaction is the most essential force toward the native fold, sequences of HP proteins consist of only two types of monomers (or classes of amino acids): amino acids with high hydrophobicity are treated as hydrophobic monomers (H), while the class of polar (or hydrophilic) residues is represented by polar monomers (P). In order to achieve the formation of a hydrophobic core surrounded by a shell of polar monomers, the interaction between hydrophobic monomers is attractive and short-range. In the standard formulation of the model [12], all other interactions are neglected. Variants of the HP model also take into account (weaker) interactions between H and P monomers as well as between polar monomers [13].

Although the HP model is extremely simple, it has been proven that identifying native conformations is an NP-complete problem in two and three dimensions [15].[12] Therefore, sophisticated methodologies were developed to find lowest-energy states for chains of up to 136 monomers. The methods applied are based on very different algorithms, ranging from exact enumeration in two dimensions [16, 17] and three dimensions on cuboid (compact) lattices [13, 18–20], and hydrophobic-core construction methods [21, 22] over genetic algorithms [23–27], Monte Carlo simulations with different types of move sets [28–31], and generalized ensemble approaches [32] to Rosenbluth chain-growth methods [33] of the 'Go with the Winners' type [34–40]. With some of these algorithms, thermodynamic quantities of lattice heteropolymers can be studied as well [19, 32, 36, 39–41].

In the HP model, a monomer of an HP sequence $\boldsymbol{\sigma} = (\sigma_1, \sigma_2, \ldots, \sigma_N)$ is characterized by its residual type ($\sigma_i = P$ for polar and $\sigma_i = H$ for hydrophobic residues), the position $1 \leq i \leq N$ within the chain of length N, and the spatial position $\mathbf{x}_i$ to be measured in units of the lattice spacing. A conformation is then symbolized by the vector of the coordinates of successive monomers, $\mathbf{X} = (\mathbf{x}_1, \mathbf{x}_2, \ldots, \mathbf{x}_i, \ldots, \mathbf{x}_N)$. The distance between the ith and the jth monomer is denoted by $x_{ij} = |\mathbf{x}_i - \mathbf{x}_j|$. The bond length between adjacent monomers in the chain is identical to the spacing of the used regular lattice with coordination number q.[13] These covalent bonds are thus not stretchable. A monomer and its nonbonded nearest neighbors may form so-called contacts. Therefore, the maximum number of contacts of a monomer within the chain is $(q - 2)$ and $(q - 1)$ for the monomers at the ends of the chain. To account for the excluded volume, lattice proteins are self-avoiding, i.e., two monomers cannot occupy the same lattice site. The total energy for an HP protein reads in energy units ε_0 (we set $\varepsilon_0 = 1$ in the following)

$$E_{\mathrm{HP}} = \frac{1}{2}\varepsilon_0 \sum_{i \neq j} C_{ij} U_{\sigma_i \sigma_j}, \qquad (1.7)$$

where $C_{ij} = (1 - \delta_{i+1\,j})\Delta(x_{ij} - 1)$ with

$$\Delta(z) = \begin{cases} 1, & z = 0, \\ 0, & z \neq 0 \end{cases} \qquad (1.8)$$

is a symmetric $N \times N$ matrix called a *contact map* and

$$U_{\sigma_i \sigma_j} = \begin{pmatrix} u_{HH} & u_{HP} \\ u_{HP} & u_{PP} \end{pmatrix} \qquad (1.9)$$

---

[12] Roughly, a computational problem is called NP-complete (where NP refers to "nondeterministic polynomial"), if no algorithm is known that solves the problem in polynomial time $t \sim \mathcal{O}(N^\alpha)$, where N is the system size and α a finite constant.

[13] The coordination number of a lattice is the number of nearest neighbors of a given lattice site, e.g., for a d-dimensional hypercubic lattice $q = 2d$ and for an fcc lattice $q = 12$.

is the 2×2 interaction matrix. Its elements $u_{\sigma_i \sigma_j}$ correspond to the energy scales of HH, HP, and PP contacts. For labeling purposes we shall adopt the convention that $\sigma_i = 0 \,\hat{=}\, P$ and $\sigma_i = 1 \,\hat{=}\, H$.

In the simplest formulation [12], only the attractive hydrophobic interaction is nonzero,

$$u_{HH}^{HP} = -1, \quad u_{HP}^{HP} = u_{PP}^{HP} = 0 \quad \text{(HP model)}. \tag{1.10}$$

Therefore,

$$U_{\sigma_i \sigma_j}^{HP} = -\delta_{\sigma_i H} \delta_{\sigma_j H}. \tag{1.11}$$

This parametrization, which we will traditionally call the *HP model* in the following, has been extensively used to identify ground states of HP sequences, some of which are believed to show up qualitative properties comparable with realistic proteins whose 20-letter sequence was transcribed into the 2-letter code of the HP model [21, 23, 42–44].

In this simple form of the standard HP model, the lowest-energy states are usually highly degenerate and therefore the number of designing sequences (i.e., sequences with unique ground state – up to the usual translational, rotational, and reflection symmetries) is very small, at least on the three-dimensional simple cubic (sc) lattice. Incorporating additional inter-residue interactions, symmetries are broken, degeneracies are smaller, and the number of designing sequences increases [19, 20]. Based on the Miyazawa–Jernigan matrix [45] of inter-residue contact energies between real amino acids, an additional attractive nonzero energy contribution for contacts between H and P monomers is more realistic [13] and the elements of the interaction matrix (1.9) are set to

$$u_{HH}^{MHP} = -1, \quad u_{HP}^{MHP} = -1/2.3 \approx -0.435, \quad u_{PP}^{MHP} = 0 \quad \text{(MHP model)}, \tag{1.12}$$

corresponding to Ref. [13]. The factor 2.3 is a result of an analysis for the inter-residue energies of contacts between hydrophobic amino acids and contacts between hydrophobic and polar residues [45], which motivated the relation $2u_{HP} > u_{PP} + u_{HH}$ [13]. We refer to this variant as the *MHP model* (mixed HP model).

Going off-lattice: Heteropolymer modeling in continuum

The lattice models discussed in the previous section suffer from the fact that the results for the finite-length heteropolymers typically depend on the underlying lattice type. It is difficult to separate realistic effects from artifacts induced by the use of a certain lattice structure. This problem can be avoided, in principle, by studying off-lattice heteropolymers, where the degrees of freedom are continuous. On the other hand, this advantage is partly counterbalanced by the increasing computational efforts for sampling the relevant regions of the conformational state space. In consequence, a precise analysis of statistical properties of off-lattice heteropolymers by means of sophisticated Monte Carlo methods can reliably be performed only for chains much shorter than those considered in the lattice studies. In the following, we focus on hydrophobic–polar heteropolymers described by the so-called AB model [14], where A monomers are hydrophobic and residues of type B are polar (or hydrophilic).

We denote the spatial position of the ith monomer in a heteropolymer consisting of N residues by $\mathbf{x}_i$, $i = 1, \ldots, N$, and the vector connecting nonadjacent monomers i and j by $\mathbf{r}_{ij} = \mathbf{x}_i - \mathbf{x}_j$. For covalent bond vectors, we set $|\mathbf{b}_i| \equiv |\mathbf{r}_{i\,i+1}| = 1$. The bending angle between monomers k, $k + 1$, and $k + 2$ is ϑ_k ($0 \leq \vartheta_k \leq \pi$) and $\sigma_i = A, B$ symbolizes the type of the monomer. In the AB model [14], the energy of a conformation is given by

$$E_{AB} = \frac{1}{4} \sum_{k=1}^{N-2} (1 - \cos \vartheta_k) + 4 \sum_{i=1}^{N-2} \sum_{j=i+2}^{N} \left(\frac{1}{r_{ij}^{12}} - \frac{C(\sigma_i, \sigma_j)}{r_{ij}^{6}} \right), \qquad (1.13)$$

where the first term is the bending energy and the sum runs over the $(N - 2)$ bending angles of successive bond vectors. The second term partially competes with the bending barrier by a potential of Lennard-Jones type. It depends on the distance between monomers being nonadjacent along the chain and accounts for the influence of the AB sequence on the energy. The long-range behavior is attractive for pairs of like monomers and repulsive for AB pairs of monomers:

$$C(\sigma_i, \sigma_j) = \begin{cases} +1, & \sigma_i, \sigma_j = A, \\ +1/2, & \sigma_i, \sigma_j = B, \\ -1/2, & \sigma_i \neq \sigma_j. \end{cases} \qquad (1.14)$$

The AB model is a C^α type model in that each residue is represented by only a single interaction site (the "C^α atom"). Thus, the natural dihedral torsional degrees of freedom of realistic protein backbones are replaced by virtual bond and torsion angles. The large torsional barrier of the peptide bond between neighboring amino acids is in the AB model effectively taken into account by introducing the bending energy.

Although this coarse-grained picture will obviously not be sufficient to reproduce microscopic properties of specific realistic proteins, it qualitatively exhibits sequence-dependent features known from nature, as, for example, tertiary folding pathways known from two-state folding, folding through intermediates, and metastability [46, 47]. The discussion of the capability of mesoscopic models in polymeric structure formation processes will be a central aspect throughout the following chapters.

1.6 Polymers

Bioproteins form a large class of molecules, but are only a subset of the much larger class of polymers. Other biologically relevant groups of molecules include the single- and double-stranded deoxyribonucleic acids (DNA) and ribonucleic acids (RNA), the carriers and messengers of genetic information, respectively. Other classes accommodate synthetic peptides and polymers which have substantial relevance in nanotechnology.

1.6.1 DNA and RNA

DNA and RNA chains are essentially linear sequences of sugar molecules (deoxyribose in DNA and ribose in RNA), linked by a negatively charged phosphate group (phosphodiester

Fig. 1.10 Typical helix conformation of double-stranded DNA. The sugar molecules line up on the backbone rails (depicted as strands), whereas the bases lie in the interior of the helix. Any pair of bases is joined by hydrogen bonds.

bridge). Attached to each of the sugars is one of the four specific bases adenine, guanine, cytosine, or thymine (DNA only; in RNA thymine is replaced by uracil). The monomer unit consisting of the sugar, base, and phosphate group is called a nucleotide and the linear chain is a polynucleotide.

The bases contain -NH and -NH$_2$ groups that are capable of forming hydrogen bonds with oxygen or nitrogen atoms of another base. In consequence, polynucleotides can arrange in helical double-strand conformations (Fig. 1.10) or a single strand can form a loop and aggregate with itself to form a double-stranded conformation. The hydrogen bonds stabilize these secondary structures locally, but they are sufficiently weak so that they can easily be broken up. This is essential for gene regulation and expression processes such as replication (DNA polymerase), transcription (RNA polymerase), and finally, translation (ribosomal synthesis of proteins).

The shortest length scale of double-stranded DNA is fixed by the pitch of the double helix that it typically forms. The pitch is defined as the length of a complete helix turn. For B-DNA helices (which the famous Watson–Crick helix belongs to),[14] it is about 34 Å, which corresponds to 10 base pairs. On larger scales, DNA behaves like a semiflexible polymer, i.e., the bending flexibility of the chain is limited. This scale is associated with the persistence length, which is defined as the length of the contour along which bond-bond correlations have decayed. For double-helical DNA, a typical value is 500 Å. In cells, DNA segments of the same size wrap around histone cores, forming solenoid supercoils. The DNA/histone assembly is called a nucleosome. On larger scales, nucleosomes can be considered as monomeric units joined by unstructured linker DNA. A chromatin fiber forms. Eventually, a long chromatin fiber with a length of about 2 cm tightly packs and creates a chromosome.

---

[14] There are also other helix forms of DNA that differ in their helical parameters, such as handedness (right, left), number of base pairs per turn, pitch, helix diameter, etc.

1.6.2 Modeling free DNA

The simplest model for free (or linker) DNA is the wormlike-chain or Kratky–Porod model [48]. It is based on the assumption that changing the contour of a linear chain by bending costs energy. If we describe the contour of length l by introducing the contour parameter $s \in [0, l]$, an infinitesimal segment of the contour (arc length) can be expressed in local coordinates by

$$ds^2 = dx^2(s) + dy^2(s) + dz^2(s), \tag{1.15}$$

or short $ds = \sqrt{d\mathbf{x}^2(s)}$. The contour integral can then be written as

$$l = \int_0^l ds = \int_0^l ds \sqrt{\frac{d\mathbf{x}^2(s)}{ds^2}} = \int_0^l ds \, |\mathbf{u}(s)|. \tag{1.16}$$

Here we have introduced the tangent vector

$$\mathbf{u}(s) = \frac{d\mathbf{x}(s)}{ds}, \tag{1.17}$$

which, according to Eq. (1.16), has unit length, $|\mathbf{u}| = 1$. The contour is bent at s, if

$$\frac{d\mathbf{u}}{ds} = \lim_{\Delta s \to 0} \frac{\mathbf{u}(s + \Delta s) - \mathbf{u}(s)}{\Delta s} \neq \mathbf{0}, \tag{1.18}$$

or

$$\mathbf{u}(s + \Delta s) \cdot \mathbf{u}(s) = \cos \theta \neq 1, \tag{1.19}$$

i.e., the bending angle is nonzero: $\theta \neq 0$. The left-hand side of Eq. (1.19) reminds us of the spin–spin interaction, as described by the Heisenberg model of a topologically one-dimensional, discrete chain of n spin vectors $\mathbf{S}_i$, embedded into three-dimensional space, with $|\mathbf{S}_i| = 1$: $E = -J \sum_{i=1}^{n-1} \mathbf{S}_i \cdot \mathbf{S}_{i+1}$, where J is the coupling strength between neighboring spins. Spin alignment is obviously energetically favorable; bending along the spin chain costs energy. We can transfer this idea to the polymer model, discretize the contour by splicing n finite segments $\mathbf{b}_i$ of length $|\mathbf{b}_i| = b$ under the constraint $nb = l$, and re-express the left-hand side of Eq. (1.19) by

$$-\mathbf{b}_i \cdot \mathbf{b}_{i+1} = \frac{1}{2}[(\mathbf{b}_{i+1} - \mathbf{b}_i)^2 - 2b^2] = -b^2 \cos \theta_i. \tag{1.20}$$

If we fix the energy scale in a way that the entirely straight chain has zero bending energy, the discrete semiflexible, wormlike-chain polymer model can be written as

$$E_{\text{bend}} = \kappa \sum_{i=1}^{n-1} (1 - \cos \theta_i) = \frac{1}{2} \kappa b \sum_{i=1}^{n-1} b \left(\frac{\mathbf{b}_{i+1} - \mathbf{b}_i}{b} \right)^2, \tag{1.21}$$

where κ is a material parameter associated with the bending stiffness of the discrete polymer. For the continuous formulation, we substitute $b \to \Delta s$, $\mathbf{b}_i \to \mathbf{u}(s)$, $\mathbf{b}_{i+1} \to \mathbf{u}(s + \Delta s)$, such that in the limit $\Delta s \to 0$ the sum in Eq. (1.21) can be replaced by an integral. The continuous form of the wormlike-chain model then reads:

$$E_{\text{bend}} = \frac{1}{2}\overline{\kappa} \int_0^l ds \left[\frac{d\mathbf{u}(s)}{ds}\right]^2. \tag{1.22}$$

In this representation, the bending stiffness is given by $\overline{\kappa} = \lim_{b\to 0, n\to\infty} \kappa$ (but still $nb = l = \text{const}$).

Given the simplicity, it is rather surprising that various dynamic and structural properties of DNA and RNA can successfully be described by this model. One essential feature is that the exponential decay of the bond–bond correlation function defines a characteristic length scale, the persistence length ξ, as we have already mentioned,

$$\langle \mathbf{u}(s)\mathbf{u}(s') \rangle \sim e^{-|s-s'|/\xi}. \tag{1.23}$$

In three-dimensional embedding space, it is $\xi = \overline{\kappa}/k_B T$, where k_B is the Boltzmann constant and T is the temperature. Thus, since $\overline{\kappa}$ is a material parameter and if T is fixed (e.g., room temperature), this result implies that $\xi = \text{const}$, independent of l. Therefore, free DNA, RNA chains, and also some protein complexes such as myosin fibers and actin filaments, contained in the cytoskeleton of cells, can typically be well characterized and identified by their persistence length.

1.6.3 Flexible, attractively self-interacting polymers

Double-stranded DNA can be described well by the wormlike-chain model, because the double helix imposes a stiffness large enough that a segment of the chain does not interact with any other part of the same chain on large length scales dominated by the persistence length.[15] However, there are also polymers consisting of units that enable effective molecular attraction between nonbonded parts of the chain. We have already discussed the proteins as a class of polymers, in which nonbonded interaction between different monomers significantly contributes to the folding properties. Generally, structural changes that lead to qualitatively different conformations, so-called conformational transitions, require an effective attractive interaction between nonbonded monomers. These effective attractive forces typically compete with repulsive, short-range volume exclusion effects. Therefore, in contrast to rather stiff polymers such as DNA, only flexible polymers can undergo such major structural changes on comparatively short length scales.

The most unspecific among the nonbonded interactions is caused by van der Waals forces, which correspond to effective, electric dipole–dipole interactions between atomic or molecular units. In the absence of stronger interactions, such as the Coulomb interaction between charged units, van der Waals forces are most essential for structure formation processes. In fact, for neutral, flexible polymers with no hydrogen-bond supporting units (an example is polyethylene $-[C_2H_4]_n-$), van der Waals forces are the only driving forces toward compact structures. Because of its importance we will discuss the electrostatic origin of these forces in more detail in the following.

---

[15] In addition, DNA nucleotides are negatively charged, such that nonbonded units repel each other by Coulomb interaction.

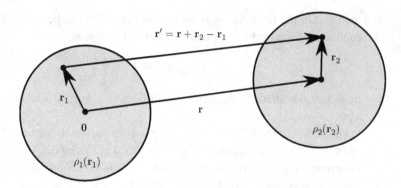

Fig. 1.11 Two neutral charge distributions $\rho_1(\mathbf{r}_1)$ and $\rho_2(\mathbf{r}_2)$ interacting with each other by van der Waals forces.

We start off by writing down the Coulomb potential between two charge distributions $\rho_1(\mathbf{r}_1)$ and $\rho_2(\mathbf{r}_2)$, which are considered to be neutral, i.e., the net charge is zero:

$$Q_{1,2} = \int d^3 r_{1,2} \rho_{1,2}(\mathbf{r}_{1,2}) = 0. \tag{1.24}$$

In this context, we are only interested in the pairwise interaction of charge elements located in different charge clouds, whose centers are separated by the distance vector $\mathbf{r}$ (see Fig. 1.11). Then, the Coulomb potential reads:

$$V(\mathbf{r}) = \frac{1}{4\pi\varepsilon_0} \int d^3 r_1 \int d^3 r_2 \frac{\rho_1(\mathbf{r}_1)\rho_2(\mathbf{r}+\mathbf{r}_2)}{|\mathbf{r}-(\mathbf{r}_1-\mathbf{r}_2)|}. \tag{1.25}$$

For brevity, we introduce the direction vector $\mathbf{r}' = \mathbf{r} - (\mathbf{r}_1 - \mathbf{r}_2)$ between the charge elements $\rho(\mathbf{r}_1)d^3 r_1$ and $\rho(\mathbf{r}+\mathbf{r}_2)d^3 r_2$ and define also $\mathbf{l} = \mathbf{r}_1 - \mathbf{r}_2$. Since we are interested in the long-range effects, we further assume that $|\mathbf{r}| \gg |\mathbf{l}|$, which enables us to perform a Taylor expansion of $1/|\mathbf{r}-\mathbf{l}|$. Remembering that a vectorial Taylor expansion of a scalar function is given by

$$f(\mathbf{r}-\mathbf{l}) = f(\mathbf{r}) - \nabla f(\mathbf{r}) \cdot \mathbf{l} + \frac{1}{2}\sum_{ij} \frac{\partial^2}{\partial x_i \partial x_j} f(\mathbf{r}) l_i l_j + \cdots, \tag{1.26}$$

with Cartesian components $\mathbf{r} = (x_1, x_2, x_3)$ and $\mathbf{l} = (l_1, l_2, l_3)$, we obtain the multi-pole expansion

$$\frac{1}{\mathbf{r}'} = \frac{1}{r} + \frac{1}{r^3}\mathbf{r}\cdot\mathbf{l} + \frac{1}{2}\left[\frac{3}{r^5}(\mathbf{r}\cdot\mathbf{l})^2 - \frac{1}{r^3}\mathbf{l}^2\right] + \mathcal{O}(1/r^5). \tag{1.27}$$

Replacing $\mathbf{l}$ by $\mathbf{r}_1 - \mathbf{r}_2$ again and inserting the expansion into the integral (1.25), we notice that several terms vanish because of the charge neutrality condition (1.24), e.g.,

$$\frac{1}{r^3}\mathbf{r}\cdot\int d^3 r_1 \mathbf{r}_1 \rho_1(\mathbf{r}_1) \int d^3 r_2\, \rho_2(\mathbf{r}_2) = 0. \tag{1.28}$$

Therefore, the only remaining contribution up to order $1/r^3$ is the famous dipole–dipole interaction energy

$$V(r) = \frac{1}{r^3} \int d^3 r_1 \int d^3 r_2\, \rho_1(\mathbf{r}_1)\rho_2(\mathbf{r}_2) \left[\mathbf{r}_1\cdot\mathbf{r}_2 - 3(\mathbf{e}_r\cdot\mathbf{r}_1)(\mathbf{e}_r\cdot\mathbf{r}_2)\right], \tag{1.29}$$

where we have introduced the unit vector in the direction of the distance vector: $\mathbf{e}_r = \mathbf{r}/r$. The integrals do not depend on r and thus we conclude from classical electrostatics that the dipole–dipole interaction strength decreases with the third power of the distance. Yet, this is not consistent with observations and the reason is that, although we consider ranges well beyond atomic dimensions, not all quantum fluctuation effects cannot be ignored. Even at large distances, there is always a residual overlap of the electron wave functions. However, it is not sufficient to form electron pairs (no covalent bond is formed), so that electron exchange interaction may be excluded.

For this purpose, let us assume that the charge distributions belong to two bodies, which can be atoms or molecular units. By employing the quantum-mechanical correspondence principle, we replace the two-body potential (1.29) by an operator that acts in the combined state space of both bodies: $V(r) \rightarrow \hat{H}_1(\mathbf{r}_1, \mathbf{r}_2)$. We assume that the dipole–dipole interaction is weak compared with the individual energetic states of the bodies, i.e., we consider the dipole–dipole operator $\hat{H}_1$ as a perturbation, such that the Hamiltonian of the entire system can be written as

$$\hat{H} = \hat{H}_0 + \hat{H}_1, \tag{1.30}$$

where $\hat{H}_0$ is the Hamilton operator of the unperturbed system (no explicit interaction between the two bodies). The eigenstates of the two-body system are given as product states of the individual eigenstates. We simply symbolize these by $|\psi_{12}^{(\alpha)}\rangle_0 = |\psi_1^{(\alpha_1)}\rangle|\psi_2^{(\alpha_2)}\rangle$, where α formally refers to the complete set of quantum numbers that uniquely characterizes the state of the combined space. If we denote the energy eigenvalues of $\hat{H}_0$ in this state by $E_0^{(\alpha)}$, the Schrödinger equation of the unperturbed two-body system reads

$$\hat{H}_0|\psi_{12}^{(\alpha)}\rangle_0 = E_0^{(\alpha)}|\psi_{12}^{(\alpha)}\rangle_0. \tag{1.31}$$

The energy is simply the sum of the energies in both systems. We assume that the unperturbed system resides in its ground state; the ground-state energy is denoted by $E_0^{(0)}$. The Schrödinger equation for the interacting system cannot be solved exactly, but since the interaction is supposed to be weak, we expect that perturbation theory yields a reasonable result for the energetic change of the ground state caused by the interaction.

The perturbation expansion for the ground-state energy $E^{(0)}$ of the weakly interacting system is given by

$$E^{(0)} = E_0^{(0)} + {}_0\langle\psi_{12}^{(0)}|\hat{H}_1|\psi_{12}^{(0)}\rangle_0 - \sum_\alpha \frac{\left|{}_0\langle\psi_{12}^{(\alpha)}|\hat{H}_1|\psi_{12}^{(0)}\rangle_0\right|^2}{E_0^{(\alpha)} - E_0^{(0)}} + \cdots \tag{1.32}$$

We further assume that higher-order terms are negligible and the series expansion can be truncated.[16] The zeroth-order term is obviously the total ground-state energy of the non-interacting two-body system. Inserting the product states and the perturbation (1.29) into the first-order correction, it separates into terms of the form $\prod_i \langle\psi_i^{(0)}|\hat{\mathbf{r}}_i|\psi_i^{(0)}\rangle$. For atomic

---

[16] This assumption implies that the sum in this form converges to the "true" (but unknown) value of the ground-state energy. In most cases, this assumption is actually not satisfied. Starting at a certain order the correction can become larger than the first-order contribution – the series diverges. Truncating such a series after it has "swapped phases" can only yield reasonable results, if resummation methods (Padé, Borel, variational approaches, etc.) are applied, which are based on the fact that the "true" result must be finite.

bodies these terms are precisely zero, because we have assumed that each system is in its ground state (S state), in which case the wave function is spherically symmetric. However, we can even more generally assume that residual angular momenta in asymmetric molecular groups statistically average out under the influence of thermal fluctuations.

Thus, only the second-order term in the expansion (1.32) yields a nonvanishing result. The numerical value of the sum is material-specific, i.e., it will depend on the individual properties of the bodies. However, what is more interesting is that the energy correction decays with $1/r^6$. Since the energies of excited states are higher than the ground-state energy, we even can conclude that this correction must be negative, i.e., both bodies attract each other with a distance-dependent interaction

$$\Delta E(r) \sim -1/r^6. \tag{1.33}$$

This is a generic result and in this form valid for all bodies. If the bodies approach each other because of the so-called attractive van der Waals force originating from this effective potential and come too close, the electronic clouds of the bodies will repel each other. For this reason, the attractive potential (1.33) cannot be valid below a certain finite distance of the bodies (see Fig. 1.12). Therefore, it is reasonable to add a repulsive term that accommodates volume exclusion at short distance. It is convenient but rather arbitrary to choose $1/r^{12}$. This yields the famous Lennard-Jones potential [49]

$$V_{LJ}(r) = 4\varepsilon \left[\left(\frac{\sigma}{r} \right)^{12} - \left(\frac{\sigma}{r} \right)^6 \right], \tag{1.34}$$

where the van der Waals distance σ as the basic length scale for the interaction range and ε, fixing the energy scale of the dipole–dipole interaction, have been introduced. The parameters are chosen to fix the zero potential $V_{LJ}(r = \sigma) = 0$ and the potential minimum $V_{LJ}(r_{min} = 2^{1/6}\sigma) = -\varepsilon$. Since the distance σ is associated with the minimum distance between nonbonded atomic or molecular units, the effective size of an atom/molecule is typically parametrized by the van der Waals radius $r_{vdW} = \sigma/2$ (Fig. 1.12).

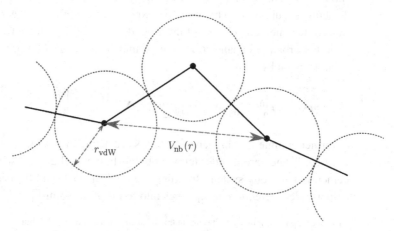

Fig. 1.12 Illustration of the nonbonded interaction $V_{nb}(r)$ between monomers in a polymer chain. The nonbonded interaction is typically modeled by the Lennard-Jones potential (1.34).

The potential (1.34) is the most commonly used model for nonbonded atomic and molecular interaction. It enables us to define the class of flexible self-interacting polymers by the property that pairs of nonbonded monomers only interact via van der Waals forces described by the Lennard-Jones potential. Unfortunately, the form of the potential is not particularly handy for analytic calculations. However, because of its effective short interaction range, it enables efficient algorithmic implementations of molecular models that are employed in computer simulations.

Nonetheless, the structural analysis of Lennard-Jones-based flexible polymer models is not at all simple, and the simulation can be computationally extremely demanding for long chains, but also for small systems, which are dominated by finite-size effects.

The investigation of particularly long chains of interacting flexible polymers with thousands to millions of monomers is only possible if the idea of an effectively attractive interaction between nonbonded monomers at short range is drastically simplified. In such cases, one typically models the flexible polymer as a self-avoiding walk (SAW) on an underlying basic lattice (e.g., the cubic lattice in three dimensions). We have already discussed this in a similar way for lattice proteins, when we introduced the HP model (1.7). The interacting self-avoiding walk (ISAW), mimicking a flexible lattice polymer, can therefore be considered as a lattice protein with hydrophobic monomers only. Only nearest-neighbor contacts between nonbonded monomers energetically contribute. Hence, the ISAW energy function of any conformation with coordinates $\mathbf{X} = (\mathbf{x}_1, \mathbf{x}_2, \ldots, \mathbf{x}_N)$ is simply given by

$$E(\mathbf{X}) = -\varepsilon \sum_{i,j>i+1} \Delta(x_{ij} - 1), \tag{1.35}$$

where the double sum runs over all pairs of nonbonded monomers. We adopted $\Delta(z)$ from Eq. (1.8) to make sure that only nearest-neighbor monomers on the lattice interact with each other. By introducing the number of nonbonded nearest-neighbor contacts $n_{NN}(\mathbf{X})$, the energy function can also be written as

$$E(\mathbf{X}) = -\varepsilon n_{NN}(\mathbf{X}), \tag{1.36}$$

which is a particularly convenient form if the model is extended by other interactions, because the statistical average of $n_{NN}(\mathbf{X})$ can then be used as an order parameter for phases of compact polymers. Since in this model all nearest-neighbor contacts are attractive, a homopolymer can be considered as a "homogeneous heteropolymer" with hydrophobic monomers only. Not surprisingly, Eq. (1.36) is thus a special case of the HP model (1.7).

1.6.4 Elastic polymers

For some classes of polymers, under certain conditions, it is also necessary to model longitudinal bond vibrations. The elastic bond potential that describes the confinement of bonded monomers must be repulsive in both limits of too short and too large distances between the monomers. Fluctuations are only possible in a finite region about the equilibrium bond length r_0.

Often, the bond is simply considered as a harmonic spring and, therefore, a harmonic approximation is used to model bond fluctuations (Rouse model),

$$V_{\text{bond}}^{\text{harm}}(r_{ii+1}) = \frac{1}{2}K(r_{ii+1} - r_0)^2, \qquad (1.37)$$

where r_{ii+1} is the distance between bonded monomers i and $i + 1$ and K is the spring constant. For this reason, coarse-grained representations of elastic polymers are typically referred to as bead-spring models in contrast to bead-stick models for polymers, for which bond vibrations can be neglected (an example is the peptide bond in proteins).

To explicitly take into account that the bond is confined and thus the fluctuation width is finite, the potential of bonded monomers can be introduced as

$$V_{\text{bond}}^{\text{sym}}(r_{ii+1}) = \begin{cases} V_{\text{FENE}}(r_{ii+1}) & \text{if } |r_{ii+1} - r_0| \leq R, \\ \infty & \text{if } |r_{ii+1} - r_0| > R. \end{cases} \qquad (1.38)$$

The bond potential within the confinement region $|r_{ii+1} - r_0| \leq R$ (the total symmetric fluctuation width is $2R$ and centered about r_0) is typically modeled by the finitely extensible nonlinear elastic (FENE) potential [50], which we introduce here in the form [51]

$$V_{\text{FENE}}(r_{ii+1}) = -\frac{1}{2}KR^2 \ln\{1 - [(r_{ii+1} - r_0)/R]^2\}. \qquad (1.39)$$

For small bond vibrations ($|r_{ii+1} - r_0| \ll R$), the leading term of the Taylor expansion yields the harmonic potential (1.37).

If one is interested in studies of the limit $K \to 0$, all interactions between the monomers of the dissolved chain must be nonbonded interactions, which are typically modeled by the Lennard-Jones potential (1.34). Thus, a *bonded* Lennard-Jones potential is sometimes added to the FENE potential [52, 53]:

$$V_{\text{bond}}(r_{ii+1}) = V_{\text{bond}}^{\text{sym}}(r_{ii+1}) + 4\varepsilon\left[\left(\frac{\sigma}{r_{ii+1}}\right)^{12} - \left(\frac{\sigma}{r_{ii+1}}\right)^6\right]. \qquad (1.40)$$

This bond potential is obviously asymmetric. Another frequently used variant is an asymmetric form of the bond potential that consists of the combination of the repulsive part of the Lennard-Jones potential and the attractive part of the FENE potential [50, 54]:

$$V_{\text{bond}}^{\text{comb}}(r_{ii+1}) = \begin{cases} 4\varepsilon\left[\left(\frac{\sigma}{r_{ii+1}}\right)^{12} - \left(\frac{\sigma}{r_{ii+1}}\right)^6 + \frac{1}{4}\right] & \text{if } r_{ii+1} \leq 2^{1/6}\sigma, \\ 0 & \text{if } r_{ii+1} \geq 2^{1/6}\sigma, \end{cases}$$
$$+ \begin{cases} -\frac{1}{2}KR_0^2 \ln[1 - (r_{ii+1}/R_0)^2] & \text{if } r_{ii+1} \leq R_0, \\ \infty & \text{if } r_{ii+1} \geq R_0. \end{cases} \qquad (1.41)$$

All these variants of the bond potential are generic, which means that elasticity is captured as a property of chemical, covalent bonds, but this is done in a rather unspecific way. For most investigations of structural phases of polymers on coarse-grained level, details will not be relevant, although it should also be pointed out that surface properties of crystalline or amorphous structures of finite polymers are influenced by differences in inherent length scales of competing bonded and nonbonded interactions. Such length scales are given, for example, by respective locations of potential minima.

Lattice simulations of polymers with variable bond length are done best by using the bond-fluctuation model [55, 56]. In three dimensions on a simple-cubic lattice, a monomer is placed in the center of a unit cell of this underlying lattice. Because of self-avoidance no other monomer can occupy the same site, i.e., the chain is not allowed to cross itself. The bond-fluctuation model is designed in a way that such crossings are inherently excluded. It can be shown that, if we introduce the basis set of 6 vectors

$$\mathbf{b}^{(1)} = \begin{pmatrix} 2 \\ 0 \\ 0 \end{pmatrix}, \ \mathbf{b}^{(2)} = \begin{pmatrix} 2 \\ 1 \\ 0 \end{pmatrix}, \ \mathbf{b}^{(3)} = \begin{pmatrix} 2 \\ 1 \\ 1 \end{pmatrix}, \ \mathbf{b}^{(4)} = \begin{pmatrix} 2 \\ 2 \\ 1 \end{pmatrix}, \ \mathbf{b}^{(5)} = \begin{pmatrix} 3 \\ 0 \\ 0 \end{pmatrix}, \ \mathbf{b}^{(6)} = \begin{pmatrix} 3 \\ 1 \\ 0 \end{pmatrix},$$

(1.42)

each allowed bond vector between bonded monomers i and $i + 1$ can only be one that belongs to this set:

$$\{\mathbf{r}_{i\,i+1}\} = \bigcup_{k=1}^{6} \mathcal{P}\,\mathbf{b}^{(k)}.$$

(1.43)

In this representation, $\mathcal{P}$ is an operator that allows all exchanges of spatial components and permutations of signs for each component of the vector it is applied to. Thus, $\mathcal{P}\,\mathbf{b}^{(k)} = \{(\pm b_l^{(k)}, \pm b_m^{(k)}, \pm b_n^{(k)})\}$, where $(l, m, n) = P(1, 2, 3)$ is any permutation of $\{1, 2, 3\}$. The set (1.43) contains 108 different bond vectors ($k = 1, 5$: 6, $k = 2, 3, 4, 6$: 24). The possible bond lengths are $r_{i\,i+1} = 2$, $\sqrt{5}$, $\sqrt{6}$, 3, and $\sqrt{10}$.

2 Statistical mechanics: A modern review

2.1 The theory of everything

In the past century theoretical physicists developed two "fundamental" theories: the theory of gravity based on the concept of general relativity for stellar systems on large scales and quantum mechanics, developed to explain physical effects on small, i.e., atomic, scales. Quantum mechanics is inevitably connected to Heisenberg's uncertainty principle and it is thus, *by definition*, a statistical theory. A quantum system is considered to be in a state $|\psi(t)\rangle$, and its time evolution is described by the Schrödinger equation

$$i\hbar \frac{\partial}{\partial t}|\psi(t)\rangle = \hat{H}|\psi(t)\rangle, \tag{2.1}$$

where $\hat{H} = E\hat{1}$ is the system-specific Hamilton operator the energy E of the system is associated to; $\hbar$ is Planck's constant. For $\hat{H}$ and thus E being time-independent, the formal solution of Eq. (2.1) reads $|\psi(t)\rangle = \exp[-iE(t - t_0)/\hbar]|\psi_0\rangle$ with $|\psi_0\rangle \equiv |\psi(t_0)\rangle$. Re-insertion into Eq. (2.1) yields

$$\hat{H}|\psi_0\rangle = E|\psi_0\rangle, \tag{2.2}$$

known as the stationary Schrödinger equation. This theory would be completely deterministic, if the stationary state $|\psi_0\rangle$ would be known or could uniquely be identified or prepared by any measurement. In quantum mechanics, the measurement of a quantity A is represented by a Hermitean operator that is applied to the quantum state: $\hat{A}|\psi_0\rangle$. If $\{|a_n\rangle \,|\, n \text{ discrete}\}$ represents the orthonormalized, discrete set of eigenstates of $\hat{A}$, then $\hat{A}|a_n\rangle = a_n|a_n\rangle$ with the real eigenvalue a_n. Introducing the dual state vector $\langle a_n|$, orthonormalization requires $\langle a_n|a_m\rangle = \delta_{nm}$. The probability interpretation can be generalized to any point in time of the dynamics.

The unique preparation of a certain quantum state is, however, impossible. Expanding $|\psi_0\rangle$ in eigenstates of $\hat{A}$, $|\psi_0\rangle = \sum_n c_n|a_n\rangle$, and thus $c_n = \langle a_n|\psi_0\rangle$ which is a complex number. If $|\psi_0\rangle$ is normalized, the real number $|c_n|^2 = \langle a_n|\psi_0\rangle\langle a_n|\psi_0\rangle^*$ is interpreted as the probability that the system is in the nth eigenstate $|a_n\rangle$ after the measurement of $\hat{A}$. Since the system must be in any eigenstate of $\hat{A}$ after the measurement, $\sum_n |c_n|^2 = 1$. An average or expectation value of $\hat{A}$ is obtained by calculating $\langle \hat{A}\rangle = \sum_n a_n|c_n|^2 = \langle \psi_0|\hat{A}|\psi_0\rangle$.

The uncertainty principle requires that our information, even after "preparation" of a system state, must remain incomplete. This is because two Hermitean operators do not necessarily commute, $\hat{A}\hat{B} \neq \hat{B}\hat{A}$, and thus the result of the simultaneous measurement depends on the order of operations. The most prominent example is the position-momentum uncertainty of a free particle of mass m, $\hat{x}\hat{p} \neq \hat{p}\hat{x}$: knowing the initial coordinates exactly, no information about the momentum can be gained, i.e., according to Eq. (2.1) the information about the particle position is completely lost in the next moment, $|x(t + \Delta t)\rangle = \exp(-iE\Delta t/\hbar)|x(t)\rangle$, provided $E = p^2/2m \neq 0$.[1]

The consequence of these general considerations is that only statistical information is naturally provided by quantum mechanics. Quantum states of small systems fluctuate; only the statistical information, represented by expectation values and probability densities, is relevant. This also means that a single experiment is useless: even possessing exact knowledge about the full spectrum of quantum levels the electron in a hydrogen atom can occupy does not help at all to "guess" the electron state in a single H atom in a single experiment. Only repeating an experiment many times or considering many atoms simultaneously in an experiment will provide the necessary information to obtain the spectrum that resembles the theoretically predicted one. And even this will not be perfect, as spectral lines have a broadness and the spectrum is not discrete; it rather resembles a distribution with peaks at the right positions. This is because the measurement is performed with macroscopic equipment that interacts with the system. Also, the system is never isolated. Even in a vacuum, particle and field fluctuations occur that influence "the system" – a lesson learned from quantum field theory.

Thus what is really measured in experiments of systems on atomic scales is not the pure quantum mechanics of the system itself. Rather, the results reflect the complex and cooperative interaction of many agents. Cooperativity of many particles not only stabilizes our (necessarily restricted) information about the quantum system, it also stabilizes the matter. Macroscopic systems like solids are stable because of the collective behavior of the huge number of small quantum systems interacting with each other. On the other hand, the destabilization of a macroscopic system, such as the melting of a solid, again requires a cooperative effect that leads to a macroscopic phase transition. A laser works only because of the spontaneous, coherent, and timely emission of radiation.

In conclusion, although the quantum nature of small systems is apparent, it only promotes the formation of stable structures of sizes beyond nanoscopic scales, if many particles *cooperatively* interact with each other. However, the amount of data that would be required to characterize this macrostate on a quantum level can be neither calculated because of its giant extent nor measured because of fundamental and experimental uncertainties. Thus, is it really necessary to go down to the quantum level to identify the system parameters which somehow contain the condensed information of the collective behavior of the individual quantum states? Thus far, we have only talked about quantum fluctuations but, at nonzero temperatures, thermal fluctuations are also relevant. Thus, obviously, a theory that allows for the explanation of macroscopic phases and the transitions between these, i.e., a physical "theory of everything" [57], must be of statistical nature.

---

[1] If $E = 0$, $p = 0$, i.e., the momentum is sharply known. In this case, the particle position x is completely unknown. It is impossible to construct a state that violates the uncertainty relation. The wave function that corresponds to the quantum state with the smallest uncertainty $\Delta x \Delta p = \hbar/2$ is the Gaussian wave packet.

2.2 Thermodynamics and statistical mechanics

2.2.1 The thermodynamic limit

One of the fundamental and universal conclusions from thermodynamics and its under-lying basis, statistical mechanics, is the existence of competing principles of order and disorder. While on the one hand a physical system tends to reduce its energy E, which typically leads to the formation of stable, symmetric structures with large-scale order (e.g., solids), the thermal environment of the system provokes distortions or fluctuations that can result in a less ordered state (as in liquids) or, in the extreme case, force complete disorder (gas phase). This tendency to increase the accessible phase space volume, i.e., allowing the system to change its configuration by thermal fluctuations, is closely connected to the entropy S of the system. Seen from this point of view, it is not surprising that the empirical laws of thermodynamics rule these two quantities: the First Law states the conservation of (internal) system energy and the Second Law associates the direction of the evolution of processes with the system entropy, which includes the important case of an isolated system that maximizes its entropy in equilibrium. Eventually, the Third Law declares the entropy to be constant at absolute zero temperature, independently of the control parameters of the system. At absolute zero, the system is in its energetic ground state $E_{\min}$ and since ground states are nondegenerate in natural systems, the entropy is not only minimal at $T = 0$, it is exactly zero.

There is an additional law, the Zeroth Law, that propagates the definition of a measure for the thermal state of the system in equilibrium: the temperature. It is only consequent to define it in a way that relates to each other the competing effects of order and disorder:

$$T(E) = \left(\frac{\partial S(E, V, N)}{\partial E} \right)_{N,V}^{-1}, \tag{2.3}$$

where the subscript at the parentheses denotes the conserved quantities which here are the volume V and the particle number N. Undoubtedly one of the most powerful theories – it is generally applicable to all natural systems – thermodynamics suffers from several conceptual problems.

The physical basis of the axioms may be intuitively apparent, but its quantitative descrip-tion is purely empirical, and at this level it is not based on a fundamental theory. This particularly regards the precise definition of the entropy. Thermodynamics cannot explain the diversity of structures belonging to the same thermodynamic phase; fluctuations are not considered. A macrostate is parametrized by only a small number of macroscopic parame-ters such as energy, entropy, temperature, pressure, volume, and chemical potential. Due to its long history and its empirical nature, the interpretation of thermodynamics was always closely connected to the available experimental methods and the systems that could be investigated. Thermodynamic systems were typically considered as "large," at least large enough to neglect the individual microstates of each of the particles and the interactions among them. However, all of these contributions enable the necessary cooperativity to form an ordered or disordered macrostate. In other words, thermodynamics makes heavy use of cooperativity without allowing for the introduction of this concept.

The microscopic basis of thermodynamics is provided by statistical mechanics. Interestingly, for a long time, it has not been considered as a stand-alone theory with all of its consequences, but rather as the microscopic theory behind thermodynamics. Thus, systems remained "large," only the discussion of the thermodynamic limit ($N \to \infty$, $V \to \infty$, but $N/V = $ const.) seemed to be of interest, which is justified for very large macroscopic systems, of course. However, finite-size effects, which by definition vanish in the thermodynamic limit, were not taken seriously for a long time. The identification of smallness effects in experiments has been and is still difficult.

On the theoretical side, the situation was comparable. The idea of universality initiated the introduction of statistical field theory and renormalization group theory, which provide information only in the somewhat diffuse "scale-free" regime near macroscopic phase transitions. To satisfy this huge interest in macroscopic effects, computer simulations were also performed. Since simulations often are only possible for finite systems, the hierarchical finite-size scaling theory was developed *to get rid of the finite-size effects*. This works fine for phase transitions where the entropy is still dominant in both phases, which is the case in second-order or continuous transitions. If the entropy changes rapidly during the transition, as in first-order or discontinuous phase transitions, and becomes small in the low-temperature phase, *energetic effects* dominate this type of transition. An example is crystallization, which does not allow for a standard finite-scaling approach. Nucleation processes are governed by the competition of particles at the surface to the environment and in the bulk, and this renders each finite system unique.

Biomolecules are finite systems that gained much attention recently. These are rather small polymers and because of their functionalized chemical composition many of them sustain their size for all time. A finite-size scaling approach does not make sense at all. The folding of proteins, for example, is only due to finiteness effects under thermodynamic conditions. It is, therefore, necessary to develop or adapt concepts for understanding cooperative effects also for systems on rather small and mesoscopic length scales (nm to μm) [58].

2.2.2 Thermodynamics of closed systems: The canonical ensemble

For the molecular systems considered throughout this book, it is sufficient to concentrate on the situation where the (small) system of interest is coupled to a "large" heat bath which provides the constant canonical temperature T. We further assume that the system-accessible volume V is constant and also is the size of the system, parametrized by the particle number N. This is the standard canonical (or NVT) ensemble. Thermodynamically, one speaks of a closed system (since $dV = 0$ and $dN = 0$), which together with the heat bath forms a thermally isolated system. By definition, the total energy E_{tot} cannot change and thus $dE_{tot} = dE + dE_{bath} + dE_{int} = 0$. The total energy is not necessarily an extensive variable as long as the interaction between the system and the heat bath, E_{int}, cannot be neglected. Extensivity would mean that the energy of the combined system is simply the sum of the energy of the system (E) and of the heat bath (E_{bath}). Typically, the coupling between the system and the heat bath is considered to be "sufficiently weak" to justify the assumption of extensitivity of the energy. Later on, in the discussion of aggregation processes of small systems, we will see that the energy is non-extensive if surface effects at the boundary of coupled small systems are relevant. In equilibrium, heat bath and system decouple entropically, such that $dS_{tot} = dS + dS_{bath} = 0$. The entropy is introduced as an extensive quantity for reasons that become more obvious below.

The canonical temperature T is for the finite system not necessarily identical to the caloric or microcanonical temperature $T_{micro}(E)$ defined in Eq. (2.3). One must keep this in mind to better understand why it is difficult to clearly identify transition temperatures in the subsequent discussions of structural transitions experienced by small molecular systems. The reason is that for constant heat bath temperature, the system energy E cannot be constant, since energy from the heat bath is pumped into the system or energy is dissipated from the system and absorbed by the heat bath. These energy transfers are mediated by the system–heat bath coupling degrees of freedom. It is therefore more useful to introduce an average energy, called the internal energy U.

In the beginning of this section, the entropy $S = S(E, V, N)$ was introduced as a function of the natural variables E, V, and N (all other thermodynamic quantities depend on the natural variables). Replacing E by U and inverting the relation for S, we obtain for the internal energy $U = U(S, V, N)$ or, in differential form,

$$dU(S, V, N) = \left(\frac{\partial U}{\partial S}\right)_{N,V} dS + \left(\frac{\partial U}{\partial V}\right)_{S,N} dV + \left(\frac{\partial U}{\partial N}\right)_{S,V} dN. \qquad (2.4)$$

The macroscopic variables U, S, V, and N are all extensive variables. Ratios of extensive variables are intensive variables and thus do not linearly depend on the system size. In analogy to Eq. (2.3), it is therefore suitable to express the intensive variable "canonical temperature" by the derivative of U with respect to S,

$$T = \left(\frac{\partial U}{\partial S}\right)_{N,V}; \qquad (2.5)$$

TdS is associated with the heat exchange in the system. Since $-pdV$ is the mechanical work necessary to change the volume, $p = -(\partial U/\partial V)_{S,N}$ corresponds to the pressure. The chemical potential is defined as $\mu = (\partial U/\partial N)_{S,V}$. Substituting these expressions, one recovers the differential form of the First Law:

$$dU(S, V, N) = TdS - pdV + \mu dN. \qquad (2.6)$$

Thus, in the canonical ensemble $dU = TdS$, i.e., the internal energy can only be changed by heat transfer. Since U is intuitively an extensive variable and T is desired to be an intensive variable, the entropy had to be introduced as an extensive variable.

Since the entropy is difficult to use as a control parameter in an experiment and it is also not a variable being representative for the ensemble (as T, V, and N are), it would be useful to replace S by T. Since in Eq. (2.6) S is an independent variable and T is not, we have to perform a Legendre transformation from the internal energy $U(S, V, N)$ to a thermodynamic potential called "free energy" by

$$F(T, V, N) = U(S(T, V, N), V, N) - TS(T, V, N), \qquad (2.7)$$

or differentially $dF = -SdT - pdV - \mu dN$. In the canonical ensemble, where $dF = 0$ in equilibrium, the free energy is the central thermodynamic potential, i.e., at the phase transition between two phases A and B, $F_A = F_B$, which defines the transition point.

2.2.3 Thermodynamic equilibrium and the statistical nature of entropy

The introduction of the entropy in the context of thermodynamics was necessary to account for the tendency of a thermal environment to "disorder" a system. However, everything is based on the empirical rule given by the Second Law stating that $TdS \geq dU + pdV - \mu dN$. In a thermally isolated system, where $dU = 0$, $dV = 0$, and $dN = 0$, the entropy increases until it has reached a maximum value that defines the thermal equilibrium state, where $dS = 0$. Hence, in the closed (or canonical) system, $dS \geq dU/T$ and thus $(\partial S/\partial U)_{N,V} \geq T^{-1}$. In thermal equilibrium, the equality corresponds to the expression used to introduce the temperature in Eq. (2.5), and T is a unique measure associated with the equilibrium state.

A prominent consequence of this relation is the heat transfer if two isolated systems I and II, which may be in equilibrium at temperatures T_I and T_{II}, respectively, are thermally coupled. After the coupling, the systems shall be closed systems, i.e., heat exchange is possible, but there is no particle transfer and no volume change. The combined system remains isolated, which entails $dU_{tot} = 0$. This corresponds to the canonical situation of coupling a system to a heat bath: we expect the hotter system to transfer energy to the colder system until the equilibrium state of the total system is reached, where $dS_{tot} = 0$. The differential total entropy is given by

$$dS_{tot} = dS_I + dS_{II} = \frac{1}{T_I}dU_I + \frac{1}{T_{II}}dU_{II}. \qquad (2.8)$$

Since the systems are coupled in a way that does not allow for the interaction of the systems, $dU_{int} = 0$, and thus $dU_{tot} = dU_I + dU_{II} = 0$, which implies $dU_{II} = -dU_I$. Inserting this into Eq. (2.8) yields

$$dS_{tot} = \left(\frac{1}{T_I} - \frac{1}{T_{II}}\right)dU_I, \qquad (2.9)$$

which gives insight into the direction of heat transfer. As long as, for example, $T_I > T_{II}$, the internal energy of system I decreases, $dU_I < 0$ (and $dU_{II} > 0$), since the Second Law requires $dS_{tot} \geq 0$ for the combined, isolated system. In the irreversible non-equilibrium process of heat exchange, the temperatures in systems I and II change with time until the equilibrium is reached, where the temperatures are in both systems identical and the entropy indeed becomes maximal, $dS_{tot} = 0$.

The somehow curious conclusion from these considerations is that with the entropy a fundamental quantity is introduced into thermodynamics which, however, is difficult to imagine and hardly accessible in direct experimental measurements. Therefore, the origin of this quantity cannot be of macroscopic nature and must be provided by a microscopic theory. The main assumptions postulated for this quantity are that it represents the thermal disorder of a system, it is extensive, and it takes its maximum value in the equilibrium state of the system. In the example discussed earlier of the two coupled systems, we have learned that energy is transferred from the hotter to the colder system. In consequence, the originally colder system takes up heat and the internal energy increases; but also the entropy of this system must increase.[2] The increases of internal energy, entropy, and temperature combined with the empirical observation that symmetries or structures in the system become smaller or "melt" lead to the conclusion that the system state is getting more disordered under these conditions.

It is therefore plausible to associate the entropy to the number of microstates that are accessible to the system. A microstate is a "snapshot" of the individual classical or quantum-mechanical states of the individual particles in the whole system. Such an individual state is in atomic systems the quantum state of the nth particle $|a_n\rangle$; on larger scales, it can be sufficient to characterize it classically by the mechanical particle state given by position and momentum, $s_n = (x_n, p_n)$. The microstates of an N-particle system are then given by the state vectors $|a_1 a_2 \ldots a_n \ldots a_N\rangle$ or $(s_1, s_2, \ldots, s_n, \ldots, s_N)$, respectively. Typically, an equilibrium ensemble of microstates dominates at a given temperature and represents the macrostate.

Let us denote the number of all possible microstates in a system by $\mathcal{W}$. Returning to the example of the two coupled systems, the respective numbers of microstates are $\mathcal{W}_\mathrm{I}$ and $\mathcal{W}_\mathrm{II}$ before the systems are coupled. Since each microstate in system I can be coupled with $\mathcal{W}_\mathrm{II}$ microstates in system II (or vice versa), the total number of possible microstates in the coupled system is multiplicative: $\mathcal{W}_\mathrm{tot} = \mathcal{W}_\mathrm{I}\mathcal{W}_\mathrm{II}$. Since the entropy shall be an extensive quantity, Boltzmann defined it as the logarithm of $\mathcal{W}$:

$$S = k_\mathrm{B} \ln \mathcal{W}, \qquad (2.10)$$

where k_B is the Boltzmann constant. It is introduced into this relation to give S a physical dimension. This famous formula connects macroscopic thermodynamics with microscopic statistics. Since in the canonical ensemble a macrostate is expressed by T, V, and N, also $\mathcal{W}$ shall depend on these quantities. However, since the actual choice of a microstate does not only have an a priori probability $\sim \mathcal{W}^{-1}$, it will also depend on the energy of the microstate and its energetic degeneracy. It is therefore desirable to generalize Eq. (2.10). For simplicity and without loss of generality, we concentrate on a discrete system with microstates i that have energies E_i. A microstate has a normalized probability of realization p_i with $\sum_i p_i = 1$. For a given temperature, we have to take into account that the microstates that form the ensemble that represents the macrostate of the system at this temperature may possess different energies. A microstate-based entropy definition would not be sufficiently

---

[2] The entropy of the originally hotter system actually decreases, but the total entropy increases. The total entropy would remain unchanged only in the trivial case that the temperatures of the separate systems were already identical before the coupling. A decrease of the total entropy in an isolated system is impossible; otherwise the Second Law would be violated.

general. Therefore, we average over all possible microstates and weight their contributions to the entropy with the probability of their realization, p_i. Then, the expression (2.10) is replaced by

$$S = -k_B \langle \ln p \rangle, \tag{2.11}$$

with the definition of the statistical average $\langle O \rangle = \sum_i O_i p_i$ for any quantity O. This statistical definition of the entropy and the principle of maximum entropy in equilibrium allows to determine the probability p_i that a certain microstate is realized.

Let us now introduce the functional

$$\Phi[p] = \frac{1}{k_B} S[p] + \sum_m \lambda_m C_m[p], \tag{2.12}$$

where $S[p]/k_B = -\sum_i p_i \ln p_i$ is the entropy functional and $C_m[p]$ is a constraint as, e.g., the normalization of the probability is. The constraints are coupled into the functional by constant Lagrange multipliers λ_m, such that $\lambda_m = \partial \Phi[p]/\partial C_m[p]$.

The maximization of the entropy under constraints is then performed by setting the variation of the functional (2.12) to zero:

$$\delta \Phi[p] = 0. \tag{2.13}$$

Since the microstate probabilities will depend on the constraints introduced, different thermodynamic situations correspond to different statistical "ensembles" of microstates. The most prominent ensembles of microstates are discussed in the following.

Microcanonical ensemble

The most obvious constraint is the normalization of the probability p_i, which can simply be written as $C_1[p] = \sum_i p_i - 1$. Inserting this into Eq. (2.12) and performing the variation (2.13) leads to

$$\sum_i \delta p_i (-\ln p_i - 1 + \lambda_1) = 0. \tag{2.14}$$

Since arbitrary independent variations δp_i can be performed for different microstates i, the left-hand side only vanishes if the expression in the parentheses is zero. Hence, $p_i = \exp(\lambda_1 - 1) = \text{const.}$, i.e., all microstates have the same probability of realization. Since the normalization constraint requires $\sum_i p_i = \exp(\lambda_1 - 1) \sum_i 1 = 1$, it is useful to define the total number of available microstates, which is called the partition sum and reads in this ensemble trivially $Z_{micro} = \sum_i 1 = \exp(1 - \lambda_1)$. Thus, the probability for a microstate is a uniform distribution:

$$p_i = \frac{1}{Z_{micro}} = \text{const.} \tag{2.15}$$

For a physical system it is always convenient to group microstates with respect to their energies E_i, as the energy is a natural quantity that represents the state of a system. However, it is necessary to assume that energetic states can be and are typically degenerate, i.e., different microstates i and j can possess the same energy $E_i = E_j$. However, the degeneracy or the *number of microstates* g_E for a given energy itself will vary with the energy such that $g_E = \sum_i \delta_{EE_i}$ is an energetic distribution.

In this light, the general microstate probability (2.15) is not very helpful for the ensemble of all microstates. However, since we can assume that in a sufficiently small energy interval $E - \Delta E/2 < E < E + \Delta E/2$ (where $0 < \Delta E/|E| \ll 1$), the number of states will not noticeably change with E. Then, the assumption that all microstates in this *microcanonical ensemble* possess almost the same energy $E \approx$ const is justified and the probability of realization for all microstates in the given energy shell is identical, as Eq. (2.15) states. Assuming for simplicity $E =$ const, the partition sum is in the microcanonical ensemble identical with the number of states $Z_{\mathrm{micro}} = \sum_k g_{E_k} \delta_{EE_k} = g_E = \sum_i \delta_{EE_i}$, where k numbers the energy levels and i labels the microstates as before.

Thermodynamically, the microcanonical ensemble corresponds to a large isolated system, i.e., to a system with almost no energetic fluctuations in which case indeed $U \approx E \approx$ const. Since in equilibrium S has reached its maximum value, this system possesses a caloric temperature $T_{\mathrm{micro}} = (\partial S_E / \partial E)^{-1}$. Since $p_i \equiv p_E = g_E^{-1}$ (under the idealized assumption $E =$ const), we obtain

$$S_E = k_B \ln g_E, \tag{2.16}$$

i.e., the temperature T_{micro} is statistically related to the change of the number of states with energy E. Because of these properties, a microcanonical ensemble of microstates ideally represents a heat bath with fixed temperature T_{micro}.

In case of a continuous system, the number of states g_E is replaced by the density of states $g(E)$, i.e., the number of states per energy. Since a microcanonical energetic macrostate is defined by the ensemble of system configurations (or conformations) with energies $E - \Delta E/2 < E < E + \Delta E/2$, the number of states in this ensemble is given by $g_E = g(E)\Delta E$ in the limit $\Delta E \to 0$. The correct Boltzmann entropy formula would therefore read:

$$S(E) = k_B \ln \int_{E-\Delta E/2}^{E+\Delta E/2} dE\, g(E) = k_B \ln g(E)\Delta E = k_B \ln g(E) + k_B \ln \Delta E. \tag{2.17}$$

The latter term is a constant and cancels, if only entropy differences are considered. This is what is typically of most interest (e.g., in the microcanonical statistical analysis of structural transitions) and, thus, one often simply writes for the microcanonical entropy

$$S(E) = k_B \ln g(E), \tag{2.18}$$

although this is not correct in the strict sense (note that $g(E)$ has the physical dimension 1/energy). In computer simulations, the problem regularizes itself, because the continuous energy space needs to be discretized and ΔE corresponds to the width of an "energy bin." Precisely, in computer simulations numbers of states are calculated rather than densities of states, but nevertheless it is common to consider g and S as functions of E.

Canonical ensemble

Embedding a (smaller) closed system into a heat bath, it will by heat transfer finally assume the heat bath temperature in equilibrium. This does not mean that in equilibrium the energy exchange between heat bath and system stops. Via the heat bath coupling, the system can gain energy from the heat bath by fluctuations and lose energy to it by dissipation. Thus, the system energy E_i of a current system microstate i will change if by the interaction with the heat bath a new microstate j is formed which possesses a different energy $E_j \neq E_i$. Whereas the energetic fluctuations and thus E can in principle take any value, the internal energy of the system, U, is constant in thermal equilibrium at a given constant temperature T, because $dU = TdS$ and S takes its maximum value in equilibrium.[3] Hence, it is useful to introduce the statistical average $\langle E \rangle$ and to relate it to the internal energy: $U = \langle E \rangle = $ const. For the derivation of the microstate probability p_i from the principle of maximum entropy, it is therefore necessary to introduce into the functional (2.12) a second constraint $C_2[p] = \langle E \rangle - U = \sum_i p_i E_i - U$ in addition to the normalization constraint C_1.

Variation of the functional (2.12) on equal footing as before and setting the variation to zero leads to $p_i = \exp(1 - \lambda_1 - \lambda_2 E_i)$. The Lagrange multiplier λ_1 is determined from the normalization condition for p_i and we define from it the partition sum of the canonical ensemble as $Z_{\text{can}} = \exp(\lambda_1 - 1) = \sum_i \exp(-\lambda_2 E_i)$. Thus, for the microstate probability,

$$p_i = \frac{1}{Z_{\text{can}}} e^{-\lambda_2 E_i}. \tag{2.19}$$

Since λ_2 and Z_{can} are constants with respect to the energy, the canonical microstate probability p_i is an exponentially decaying function of the microstate energy, in contrast to the uniformly distributed microstate probability in the microcanonical ensemble. The internal or mean energy is given by $U = \langle E \rangle = \sum_i E_i \exp(-\lambda_2 E_i)/Z_{\text{can}}$.

In order to determine λ_2, we insert expression (2.19) into the entropy functional, which yields

$$S = k_{\text{B}} \ln Z_{\text{can}} + \lambda_2 k_{\text{B}} U. \tag{2.20}$$

Multiplying by T and comparing this result with Eq. (2.7), we can easily identify the terms from our statistical consideration with the macroscopic quantities from thermodynamics. Thus, the free energy of the system is related to the partition sum via

$$F(T, V, N) = -k_{\text{B}} T \ln Z_{\text{can}}(T, V, N) \tag{2.21}$$

and the Lagrange multiplier λ_2 corresponds to the inverse thermal energy, which is typically called $\beta = 1/k_{\text{B}} T$. Finally, the canonical microstate probability can be written as

$$p_i = \frac{1}{Z_{\text{can}}} e^{-E_i/k_{\text{B}} T}, \tag{2.22}$$

and the partition sum is $Z_{\text{can}} = \sum_i \exp(-\beta E_i)$. Although p_i is the most general probability in the canonical ensemble, it is often useful to introduce the canonical energy distribution

---

[3] The canonical temperature T of the closed system is identical to the microcanonical heat bath temperature, i.e., $T = T_{\text{micro}}^{\text{heat bath}}$. However, this does not mean that also the canonical and microcanonical system temperatures coincide; actually $T_{\text{micro}}^{\text{system}} = T$ is only valid in the thermodynamic limit, where the relative energetic fluctuations ($\Delta E/E$) of infinitely large systems vanish.

by grouping all microstates with the same energy. Since for a given temperature the partition sum is a constant independently of the grouping of microstates, we can rewrite it as

$$Z_{\text{can}} = \sum_i e^{-\beta E_i} = \sum_E \sum_i \delta_{E E_i} e^{-\beta E_i} = \sum_E e^{-\beta E} \sum_i \delta_{E E_i} = \sum_E g_E e^{-\beta E}, \quad (2.23)$$

where g_E is again the number (or in continuum the density) of states. It is therefore convenient to introduce the canonical energy distribution by

$$p_{\text{can},E} = \frac{1}{Z_{\text{can}}} g_E e^{-\beta E}, \quad (2.24)$$

or for a system with continuous (as in a classical system) or quasi-continuous (as in a quantum system with many close energy levels like in a solid) energy space, $p_{\text{can}}(E)dE = Z_{\text{can}}^{-1} g(E) e^{-\beta E} dE$, in which case the partition sum is expressed by the integral

$$Z_{\text{can}} = \int dE g(E) e^{-\beta E} \quad (2.25)$$

and is called the partition function.

The internal or mean energy can now be written in the form

$$U(T,V,N) = \langle E \rangle (T,V,N) = \frac{1}{Z_{\text{can}}(T,V,N)} \sum_E g_E(V,N) e^{-\beta(T)E} \quad (2.26)$$

$$= k_B T^2 \frac{\partial}{\partial T} \ln Z_{\text{can}}(T,V,N) \quad (2.27)$$

and obviously depends on the temperature T. In order to verify the expression (2.5) for the canonical temperature as introduced in the thermodynamical formalism, we invert the relation (2.26) to get $T = T(U,V,N)$. Then, the entropy (2.20) can also be written as a function in these variables and reads

$$S(U,V,N) = k_B \ln Z_{\text{can}}(T(U,V,N),V,N) + \frac{U}{T(U,V,N)} \quad (2.28)$$

and the derivative with respect to U indeed reproduces the expected result (2.5). However, since U is constant in the canonical ensemble whereas E is not, T and T_{micro} need not be identical. The energy distribution (2.24) is exponentially suppressed for high energy values which contribute only significantly at high temperatures. At low temperatures, $p_{\text{can},E}$ is dominated by microstates with small energies. Thus, if the system is not in a macrostate close to a phase transition with phase coexistence, $p_{\text{can},E}$ will typically possess a single peak with a maximum at a certain energy. This peak becomes sharper with increasing system size and in the thermodynamic limit microstates with the peak energy E_{max} dominate the whole ensemble. Then, $\langle E \rangle \approx E_{\text{max}}$, such that $\partial S/\partial U \approx \partial S/\partial E_{\text{max}}$, i.e., microcanonical and canonical temperatures coincide. Thus, in the thermodynamic limit, the canonical and the microcanonical ensemble yield identical quantitative information about the thermodynamic behavior of the system.

The grand canonical ensemble

Another frequently required demand is the fluctuation of the number of particles N in the system and to keep the chemical potential μ constant. This is the physical situation that is statistically provided by the grand canonical ensemble. We have already introduced the constraints with respect to the normalization of the microstate probability and the constant mean value of energy. A system that exchanges particles with its environment can only be in equilibrium with the surrounding "particle bath" if the average number of system particles is constant. This requires the introduction of a third constraint $C_3[p] = \langle \mathcal{N} \rangle - N = \sum_i \mathcal{N}_i p_i - N$, where $\mathcal{N}_i$ is the number of particles in the ith microstate. This constraint is coupled into the functional (2.12) with the Lagrange multiplier λ_3. Variation of the functional $\Phi[p]$ under all these constraints yields $p_i = \exp(1 - \lambda_1 - \lambda_2 E_i - \lambda_3 \mathcal{N}_i) = \exp(-\lambda_2 E_i - \lambda_3 \mathcal{N}_i)/Z_{\text{grand}}$ and $TS = k_B T \ln Z_{\text{grand}} + \lambda_2 k_B TU + \lambda_3 k_B TN$. In order to connect this relation to thermodynamics and for the identification of the Lagrange multipliers, we introduce in analogy to the free energy of closed systems the grand canonical potential $\Omega(T, V, \mu)$ as the Legendre transformation of the free energy with respect to N and μ: $\Omega(T, V, \mu) = F(T, V, N(T, V, \mu)) - \mu N(T, V, \mu) = U(T, V, \mu) - TS(T, V, \mu) - \mu N(T, V, \mu)$. Comparison with the result from the variation gives $\Omega(T, V, \mu) = -k_B T \ln Z_{\text{grand}}(T, V, \mu)$, $\lambda_2 = 1/k_B T = \beta$, and $\lambda_3 = -\mu/k_B T = -\mu\beta$. Hence, the microstate probability in the grand canonical ensemble reads

$$p_i = \frac{1}{Z_{\text{grand}}} e^{-\beta(E_i - \mu \mathcal{N}_i)} \tag{2.29}$$

with the partition sum $Z_{\text{grand}} = \sum_i \exp[-\beta(E_i - \mu \mathcal{N}_i)]$. The entropy is finally given by

$$S(T, V, \mu) = k_B \ln Z_{\text{grand}}(T, V, \mu) + \frac{1}{T}[U(T, V, \mu) - \mu N(T, V, \mu)]. \tag{2.30}$$

Substituting T by $T(U, V, \mu)$ entails

$$S(U, V, \mu) = k_B \ln Z_{\text{grand}}(T(U, V, \mu), V, \mu)$$
$$+ \frac{1}{T(U, V, \mu)}[U - \mu N(T(U, V, \mu), V, \mu)]. \tag{2.31}$$

The grand canonical temperature can thus be expressed via

$$T \equiv T_{\text{grand}} = (\partial S/\partial U)_{V,\mu}^{-1}. \tag{2.32}$$

The derivation of the ensembles presented in this section is completely general and also valid for quantum systems. However, it is not complete as quantum statistics of quantum particles such as photons, electrons, protons, etc., requires the additional consideration of fermionic and bosonic particle symmetries, respectively, in dependence of their particle spin. In all applications presented in this book, we only consider classical or semiclassical systems, where the quantum effects are "hidden" in the parametrization of the effective potentials used in the models. The assumption is that mesoscopic systems such as macromolecules and molecular aggregates behave sufficiently cooperatively to allow for the investigation of the net effect only, but not the individual quantum-mechanical contributions. This is an idealization, but it is based on the fact that for a correct physical description of macroscopic systems, a precise quantum-mechanical treatment is not necessarily required.

This brings us back to the fundamental "theory of everything." The demand for it is a profound philosophical problem and it is under serious debate. For the understanding of the dynamics of a macroscopic system, Newton's equations of motion are completely sufficient. On the other hand, the question of why this macroscopic system is stable requires the consideration of the fundamental principles valid at very short length scales, i.e., in the quantum regime. However, from our current understanding of physics, quantum mechanics in its current form is not the desired fundamental theory, as it does not allow for the quantization of gravity, which is the dominant force in interactions of astronomically large systems, where Einstein's field equations apply. Interestingly, the classical physics at intermediate length and energy scales is a limiting case of both quantum physics and gravitation. It is the cooperative behavior of many particles or sections of a macroscopic system that governs its physical properties and, therefore, since cooperativity and statistics are inevitably connected, statistical mechanics is probably the most fundamental theoretical concept in the natural sciences.

2.3 Thermal fluctuations and the statistical path integral

In the thermodynamic description, the capability of a system to react to excitations or changes of environmental parameters such as temperature or pressure is expressed by so-called response quantities like the heat capacity or the compressibility, respectively. The heat capacity, for example, quantifies the capacity of heat storage by the system. For a system with constant volume such as in the canonical ensemble, it is defined as the amount of heat exchange TdS while the temperature is changed, $C_V(T) = T(\partial S/\partial T)_{N,V} = (\partial U/\partial T)_{N,V}$. From the latter, the statistical expression for the heat capacity

$$C_V(T) = \frac{d\langle E\rangle(T)}{dT} = \frac{1}{k_B T^2}\left[\langle E^2\rangle(T) - \langle E\rangle^2(T)\right] = \frac{1}{k_B T^2}\left\langle[\langle E\rangle(T) - E]^2\right\rangle \qquad (2.33)$$

can easily be derived.[4] Thus, the heat capacity corresponds to the fluctuations of energy and as such to the width of the energetic distribution $p_{can,E}(T)$. The larger the energetic fluctuation width, the larger is also the number of energetic states that can be thermally excited and, therefore, the larger is the heat capacity. The fluctuation formula (2.33) can be generalized and the fluctuation of any quantity O can be defined via the temperature derivative of its mean value

$$\frac{d\langle O\rangle}{dT} = \frac{1}{k_B T^2}\left(\langle OE\rangle - \langle O\rangle\langle E\rangle\right), \qquad (2.34)$$

---

[4] The heat capacity is an extensive quantity, but it is often common to convert it into an intensive quantity by dividing out an extensive system property such as mass or volume. In our context, we will sometimes refer to $c_V = C_V/N$, where N is the system size (e.g., the number of monomers in a polymer chain), as the specific heat.

which is particularly interesting if O can be considered as a suitable order parameter that allows for the quantitative separation of phases. The temperature, where the fluctuations become maximal, is an estimate for the transition temperature. However, for finite systems, different fluctuation quantities typically signal the same transition at noticeably different fluctuation peak temperatures. Only in the thermodynamic limit, these peak temperatures converge toward a single phase transition temperature.

In the discussion of the canonical ensemble, we have already introduced the partition function $Z_{\text{can}} = \int dE\, g(E) \exp(-\beta E)$ for a system with continuous energy space. This is the typical situation for a system with continuous mechanical phase–space degrees of freedom, position and momentum. Let's denote by $\mathbf{X} = (\mathbf{x}_1, \mathbf{x}_2, \ldots, \mathbf{x}_n, \ldots, \mathbf{x}_N)$ the $3N$-dimensional vector of the three-dimensional coordinates $\mathbf{x}_n$ and by $\mathbf{P} = (\mathbf{p}_1, \mathbf{p}_2, \ldots, \mathbf{p}_n, \ldots, \mathbf{p}_N)$ the $3N$-dimensional vector of the momenta $\mathbf{p}_n$. Then, the canonical partition function, which accounts for the thermally weighted fluctuations of the phase–space variables, is written as

$$Z_{\text{can}} = \int \mathcal{D}P\mathcal{D}X \exp\left[-\beta H(\mathbf{P}, \mathbf{X})\right], \tag{2.35}$$

where $H(\mathbf{P}, \mathbf{X}) = E$ is the Hamilton function which corresponds to the total system energy E and

$$\int \mathcal{D}P\mathcal{D}X \equiv C_N \prod_{n=1}^{N}\left[\iint \frac{d^3p_n d^3x_n}{(2\pi\hbar)^3}\right]. \tag{2.36}$$

The prefactor $C_N = 1/N!$ is the Boltzmann correction, which takes into account the trivial multi-counting of microstates generated by permuting identical particles – provided the particles can be exchanged at all. For a single molecular chain, where monomers are bonded and cannot change their positions within the chain, no Boltzmann correction is needed.

The factor $1/(2\pi\hbar)$ in the integral measure sets the scale of the infinitesimal phase–space volume $dp_n dx_n$ for each phase–space component. The "volume" element has the physical dimension of an action (measured in units of Js) and represents the smallest phase-space volume in which a single particle state can reside. This is a consequence of Heisenberg's uncertainty principle that even in classical statistics cannot be ignored.

If the Hamilton function is of the standard form

$$H(\mathbf{P}, \mathbf{X}) = \sum_{n=1}^{N} \frac{\mathbf{p}_n^2}{2m} + V(\mathbf{X}), \tag{2.37}$$

where the first term is the total kinetic energy (m is the particle mass, which shall be identical for all particles, for simplicity) and the second term is the potential energy

$$V(\mathbf{X}) = \sum_{n=1}^{N} V_1(\mathbf{x}_n) + \frac{1}{2} \sum_{\substack{n,m=1 \\ n \neq m}}^{N} V_2(|\mathbf{x}_n - \mathbf{x}_m|), \tag{2.38}$$

representing the respective coupling to an external field and the interactions between the particles as the sum over all pair potentials, which typically only depend on the particle–particle distance. Then, the momentum integrals factorize and can be solved exactly for each of the components, yielding in total $1/\lambda_{\text{th}}^{3N}$, where

$$\lambda_{\text{th}} = \sqrt{2\pi\hbar^2\beta/m} \tag{2.39}$$

is called the thermal wavelength. It sets the temperature-dependent length scale of the thermal fluctuations.

The resulting expression for the partition function reads

$$Z_{\text{can}} = \int \mathcal{D}X \exp\left[-\beta V(\mathbf{X})\right], \tag{2.40}$$

where the integral measure has been redefined as

$$\int \mathcal{D}X \equiv C_N \prod_{n=1}^{N}\left[\int \frac{d^3x_n}{\lambda_{\text{th}}^3}\right]. \tag{2.41}$$

Expression (2.40) is called the statistical path integral in analogy to the quantum-mechanical Feynman path integral introduced as an alternative means of quantization, where all possible particle paths, which in contrast to classical Newtonian mechanics may vary due to quantum fluctuations, are integrated over [59, 60].

Mean values of quantities O that can be parametrized with respect to the momenta and coordinates are expressed as

$$\langle O(\mathbf{P}, \mathbf{X})\rangle = \frac{1}{Z_{\text{can}}}\int \mathcal{D}P\mathcal{D}X\, O(\mathbf{P}, \mathbf{X}) \exp\left[-\beta H(\mathbf{P}, \mathbf{X})\right]. \tag{2.42}$$

A prominent example is the total system energy $E = H(\mathbf{P}, \mathbf{X})$. Its mean value can with Eq. (2.42) be written as

$$\langle E\rangle = -\frac{\partial}{\partial\beta}\ln Z_{\text{can}}, \tag{2.43}$$

in agreement with Eq. (2.27). It should be noted that the contribution for the kinetic energy can easily be separated. Since

$$\frac{1}{2m}\sum_{n=1}^{N}\langle \mathbf{p}_n^2\rangle = \frac{C_N}{Z_{\text{can}}}\prod_{n=1}^{N}\left[\int \frac{d^3x_n}{(2\pi\hbar)^3}\right]e^{-\beta V(\mathbf{X})}\left(-\frac{\partial}{\partial\beta}\right)\prod_{n=1}^{N}\left[\int d^3p_n\, e^{-\beta\mathbf{p}_n^2/2m}\right]$$
$$= \frac{3}{2}Nk_{\text{B}}T, \tag{2.44}$$

the mean total energy is

$$\langle E\rangle = \frac{3}{2}Nk_{\text{B}}T + \langle V(\mathbf{X})\rangle. \tag{2.45}$$

For practical purposes as, for example, in Monte Carlo simulations, it is therefore completely sufficient to calculate the mean potential energy $\langle V(\mathbf{X})\rangle$. Thus, in the canonical ensemble, mean energy differences are identical to mean potential energy differences. The contribution of the kinetic energy fluctuations to the heat capacity is thus also constant:

$$C_V(T) = \frac{\partial \langle E \rangle (T)}{dT} = \frac{3}{2} N k_B + \frac{1}{k_B T^2} \left[\langle V^2(\mathbf{X}) \rangle (T) - \langle V(\mathbf{X}) \rangle^2 (T) \right]. \qquad (2.46)$$

Not surprisingly, the kinetic energy contributions in Eqs. (2.45) and (2.46) are identical to the thermodynamic results for the ideal gas, the idealized model for noninteracting particles $[V(\mathbf{X}) = 0]$. In the applications presented in the following chapters, we will omit these contributions, but we have to keep in mind that they must be added to make any statistical analysis of a Hamiltonian system quantitatively correct.

2.4 Phase and pseudophase transitions

The set of microstates that dominates under given external conditions, like the equilibrium temperature, forms the macrostate of the system. If in a certain range of the external parameters, such as the temperature, the corresponding macrostates exhibit significant similarities, they are said to belong to the same phase. A system experiences a phase transition, if a small change of an external parameter leads to a dramatic change of the macrostate properties making it belong to another phase.

The quantitative analysis of phase transitions makes explicit use of the thermodynamic limit and is thus not directly transferable to small systems. Usually, two types of phase transitions are distinguished. As an example, let us consider temperature-driven phase transitions. Discontinuous or first-order transitions refer to the discontinuity of the entropy as a function of temperature:

$$\left(\frac{\partial F}{\partial T} \right)_{N,V} = -S(T, V, N). \qquad (2.47)$$

First-order transitions are characterized by the coexistence of two phases I and II at the transition temperature T_0. The difference of the respective entropies near the transition point, multiplied by the transition temperature, defines the latent heat:

$$\Delta Q_{\text{lat}} = \lim_{t \to 0} T_0 [S_{\text{II}}(T_0 + t) - S_{\text{I}}(T_0 - t)]. \qquad (2.48)$$

A phase transition is of first order if $\Delta Q_{\text{lat}} > 0$.

Consequently, a transition is continuous or of second order if $\Delta Q_{\text{lat}} = 0$. In this case, the entropy is continuous at the transition temperature. However, the second derivative of F with respect to T,

$$\left(\frac{\partial^2 F}{\partial T^2} \right)_{N,V} = -\left(\frac{\partial S}{\partial T} \right)_{N,V} = -\frac{1}{T} C_V(T), \qquad (2.49)$$

is discontinuous and thus so is the heat capacity at the "critical" transition temperature T_C. Defining the dimensionless parameter $\tau = (T - T_C)/T_C$, the heat capacity follows near the transition point a power law, $C_V(\tau) \sim |\tau|^{-\alpha}$, where α is the critical exponent associated to the heat capacity. Other fluctuation quantities such as the compressibility of a gas or the susceptibility of a magnetic system – both response quantities to external fields pressure and magnetic field, respectively – and correlation lengths also exhibit power law monotony near the transition temperature, but possess different characteristic critical exponents (γ for compressibility and susceptibility, ν for the correlation length). For the identification of phases and the type of phase transitions (e.g., the transition between the ferro- and the paramagnetic phase in a magnetic system), it is often very useful to define an order parameter

$$O \begin{cases} \neq 0, & T < T_C, \\ = 0, & T > T_C, \end{cases} \tag{2.50}$$

which thus ideally allows for the unique separation of phases. Near the transition temperature, the order parameter also follows a power law in the ordered phase: $O \sim (-\tau)^\beta$ with the critical exponent β (not to be confused with the thermal energy). For a magnetic system, it is convenient to choose the mean value of the spontaneous magnetization as the order parameter, since it is zero in the disordered phase (paramagnetism) and nonzero in the ordered phase (ferromagnetism). In other systems, the definition of a suitable order parameter is much less obvious.

One of the most striking advances in physics was the discovery of universality: physically completely different systems that share the same values of critical exponents belong to the same universality class. Thus, physical systems can be classified with respect to their transition behavior.

These powerful fundamental concepts, provided only by statistical physics, are idealized in a way that they can in this form only be applied to "large" systems, where the thermodynamic limit conditions are satisfied. Unfortunately, the transfer to systems on mesoscopic scales, like the molecular systems discussed in this book, is not straightforward. The "collapse" of the fluctuation and correlation quantities at a single transition temperature T_C does not occur for finite systems. Furthermore, a suitably defined parameter that could serve as an order parameter will not behave exactly in the way that expression (2.50) suggests. Finally, most relevantly, fluctuations of thermodynamic quantities do not exhibit power-law behavior near the transition points. Thus, a classification of different finite systems by means of sets of critical exponents is impossible. Nonetheless, as we will show in subsequent chapters, different conformational transitions can exhibit similarities and a classification of molecular systems is highly desirable. This regards, in particular, the classification of proteins with respect to their folding behavior. Another point is the difficult classification of the transitions into first- and second-order transitions. Actually, even from "real" phase transitions, which are of second order in the thermodynamic limit, it is known that they can exhibit first-order features in the finite system. In order to discriminate among transitions of finite systems and systems in the thermodynamic limit, we will denote noticeable structural changes as *pseudophase transitions*. The associated dominating macrostates shall be denoted *pseudophases* rather than phases.

It is necessary to further extend and to generalize the theory of phase transitions to make it applicable to finite systems as well. Thus, it is useful to return to the initial idea of cooperativity, which is the basis of both phase and pseudophase transitions.

2.5 Relevant degrees of freedom

When a system behaves cooperatively, its degrees of freedom do not act independently. This is particularly apparent near transitions from disordered to ordered phases. In the disordered phase, the correlation length is very small, i.e., a local system change is not necessarily felt by distant parts of the system. A collective response to the excitation does not occur. If the system is in a macrostate near the transition point, it can happen that a system change due to a small change of the environmental parameters initiates a spontaneous ordering effect.

An example is the behavior of a flexible polymer at the so-called Θ transition, where it "collapses" from large, extended random-coil structures to compact, densely packed globular conformations. A suitable model to describe the behavior of polymers in the random-coil phase are random-walk or self-avoiding random-walk models (if volume exclusion effects must be considered). The energetic interactions between the monomers are completely irrelevant; only "hard" constraints (connectivity, volume exclusion, chain length) do matter. In the globular phase, however, the attractive van der Waals forces between many atoms of the monomers, that stabilize the globular shapes, are necessarily to be taken into account in order to describe the formation of compact structures correctly. Since this structure formation process is highly cooperative, i.e., many atoms are involved, it is sufficient to introduce a general "hydrophobic" interaction between abstracted "united atoms" which comprise subsets of atoms. Thus, the effect can be described qualitatively and, for large systems, even quantitatively by a simplified, coarse-grained model that in addition to the system constraints only includes an effective interaction among nonbonded monomers. Compared to an atomic model, the number of degrees of freedom necessary for the understanding of the thermodynamics of the collapsed phase can be drastically reduced. This is not only very helpful for the efficiency of computer simulations of such systems, it also helps understand structure formation processes as effects of cooperativity.

One of the most contemporary challenges in this field of research is to find the minimal set of degrees of freedom that allows for the description of structural transitions. On a very abstract level, this could be an order parameter or, in analogy to chemical reactions, a reaction coordinate. The reason why this set of degrees of freedom should be small is the reconstruction of the free-energy landscape in an intuitive way, i.e., to express the free energy of the system not only as a function of the external macroscopic thermodynamic parameters such as the temperature, but also as a function of the set of relevant degrees of freedom. As such, "paths" from a macrostate A to a state B (for example from an unfolded protein structure to the native state) can be parametrized and different paths can be assigned statistical weights. This helps to reconstruct the most likely paths the system can follow under thermal conditions.

2.5.1 Coarse-grained modeling on mesoscopic scales

In Fig. 1.7, the general idea of coarse-graining an atomic model for a protein by introducing "united atoms" has already been depicted. The reduced set of coordinates of the monomers (i.e., the new interaction sites) represents the coarse-grained degrees of freedom. If it is justified that the main properties of structural behavior in a cooperative structure formation process can be described by these new monomer positions only, these can be considered as the *relevant* effective degrees of freedom. Other monomer properties, such as hydrophobicity or charges, are then also encoded in effective parameters. In the following, we will develop the formalism to correctly derive the effective model which only depends on the relevant degrees of freedom.

Let us represent the set of relevant mesoscopic degrees of freedom by the L-dimensional vector

$$\mathbf{Q}(\mathbf{X}) = (q_1(\mathbf{X}), q_2(\mathbf{X}), \ldots, q_l(\mathbf{X}), \ldots, q_L(\mathbf{X})). \tag{2.51}$$

Then, the path integral (2.40) can be decomposed into the new and the old coordinates by writing

$$Z_{\text{can}} = \frac{\lambda_{\text{th}}^L}{C_L} \int \mathcal{D}Q \int \mathcal{D}X \prod_{l=1}^{L} \left[\delta(q_l - q_l^{\text{s}}(\mathbf{X}))\right] e^{-\beta V(\mathbf{X})}, \tag{2.52}$$

where $q_l^{\text{s}}(\mathbf{X})$ represents the mapping from the $\mathbf{X}$ into the reduced $\mathbf{Q}$ space; $\delta(x)$ is the Dirac δ distribution. Thus, the partition function can now be expressed as a path integral in $\mathbf{Q}$ space,

$$Z_{\text{can}} = \int \mathcal{D}Q e^{-\beta \tilde{V}(\mathbf{Q})}, \tag{2.53}$$

with $\int \mathcal{D}Q \equiv C_L \prod_{l=1}^{L} \left[\int dq_l / \lambda_{\text{th}}\right]$. The *coarse-grained model* of the original system is therefore governed by the effective potential

$$\tilde{V}(\mathbf{Q}) = -k_{\text{B}} T \ln Z_{\text{can}}^{\text{res}}(\mathbf{Q}), \tag{2.54}$$

where we have introduced the restricted partition function

$$Z_{\text{can}}^{\text{res}}(\mathbf{Q}) = \frac{\lambda_{\text{th}}^L}{C_L} \int \mathcal{D}X \prod_{l=1}^{L} \left[\delta(q_l - q_l^{\text{s}}(\mathbf{X}))\right] e^{-\beta V(\mathbf{X})}. \tag{2.55}$$

The coarse-grained potential (2.54) is the most general variant of an effective potential that would even allow for a correct quantitative description of the thermodynamic system behavior. However, typically, coarse-grained models are still drastically simplified such that models like the HP and the AB models for proteins (see Section 1.5) only represent approximate and sometimes crude versions of coarse-grained models. This further reduction of complexity is justified as long as the main features of system-relevant thermodynamic processes are reproduced, at least qualitatively.

2.5.2 Macroscopic relevant degrees of freedom: The free-energy landscape

Comparing the formal expressions of Eqs. (2.21) and (2.54), we can immediately identify the effective potential as the free energy of the system parametrized as a function of the relevant degrees of freedom. It therefore represents the free-energy landscape of the system under given external parameters. However, although the complexity of the original system has already been reduced, the dimension L of the $\mathbf{Q}$ space often remains large. For this reason, it does not help the intuition to "visualize" the macrostate of a system by means of a high-dimensional free-energy landscape. Rather it appears necessary to introduce only very few components q_l. For this, we free ourselves from the concept of coarse-graining the true degrees of freedom and introduce macroscopic parameters $\mathbf{Q}$ similar to what is generally done in thermodynamics. The few parameters take over the role of order parameters and shall as such represent the macrostate of the system. A simple and popular example from polymer science is the end-to-end distance of a polymer. It is very large in the disordered, random-coil phase and very small in the phase of collapsed conformations. It therefore exhibits a sharp change while passing the collapse transition point and is perfectly suited to discriminate between these two phases and to identify the transition point. However, it is much less useful to signal the transition from the globular to the crystalline phase. This is a disadvantage of order parameters. In contrast to the general coarse-grained model (2.54) based on mesoscopic degrees of freedom, the introduction of macroscopic relevant degrees of freedom is far less general and often only useful to discriminate parts of a phase diagram.

With the vector (2.51) now representing the set of macroscopic relevant degrees of freedom, the statistical expectation value of each component is in the canonical ensemble given by

$$\langle q_l(\mathbf{X}) \rangle = \frac{1}{Z_{\text{can}}} \int \mathcal{D}X \, q_l(\mathbf{X}) e^{-\beta V(\mathbf{X})} \qquad (2.56)$$

and the fluctuation width about this average value is obtained by the derivative with respect to the temperature

$$\frac{d\langle q_l \rangle}{dT} = \frac{1}{k_B T^2} \left(\langle q_l E \rangle - \langle q_l \rangle \langle E \rangle \right), \qquad (2.57)$$

in accordance with Eqs. (2.42) and (2.34), respectively. The probability (density) to find a system state that is represented by a certain vector of relevant degrees of freedom $\mathbf{Q}$ is

$$p(\mathbf{Q}) = \frac{Z_{\text{can}}^{\text{res}}(\mathbf{Q})}{Z_{\text{can}}}, \qquad (2.58)$$

where the restricted partition function is given by Eq. (2.55). According to Eq. (2.21), we can relate this restricted partition function to the free energy of the system in this macrostate,

$$F^{\text{res}}(T, V, N; \mathbf{Q}) = -k_B T \ln Z_{\text{can}}^{\text{res}}(T, V, N; \mathbf{Q}). \qquad (2.59)$$

This function represents for given system parameters T, V, and N the free-energy landscape in dependence of the components of the vector of relevant degrees of freedom $\mathbf{Q}$. Minima in this landscape correspond to locally stable (metastable) equilibrium system states. Peaks in this landscape represent free-energy barriers. A structural transition requires the system to circumvent the barrier or to overcome it by a fluctuation with thermal energy that exceeds the barrier height.

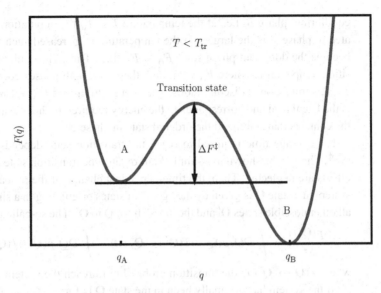

Fig. 2.1 Sketch of a free-energy landscape parametrized by the single relevant degree of freedom q at the externally fixed temperature $T < T_{tr}$, where T_{tr} is the transition temperature. The two minima of $F(q)$ at q_A and q_B correspond to equilibrium states A and B, respectively (e.g., unfolded and folded protein states).

2.6 Kinetic free-energy barrier and transition state

Figure 2.1 show a typical sketch of a free-energy landscape $F(T; q)$ at fixed temperature T. It is parametrized by the single relevant degree of freedom q. The minima $F_A = F(T; q_A)$ and $F_B = F(T; q_B)$ in the free-energy landscape correspond to the equilibrium states A and B, respectively. We assume that q is getting larger the more ordered the state is. For a protein with two-state folding characteristics, A is associated to the unfolded state (i.e., random, unstructured conformations), whereas B represents the ensemble of native-like conformations. Since B is the global minimum, the probability for the system to reside in B is larger than in A, i.e., the protein is likely to be folded. In this example, q could be defined as the number of "native contacts" n_{native}, which is the number of correctly arranged monomer–monomer positions. This number obviously increases if the molecular conformations are getting closer to the native (functional) structure.

The minima are separated by the maximum at the transition state, $F_{ts} = F(T; q_{ts})$. If q is indeed the relevant degree of freedom and the system resides in the local minimum at A, the free-energy difference $\Delta F^{\ddagger} = F_{ts} - F_A$ is the kinetic barrier the system has to climb to reach the transition state. Only after reaching the ensemble of states belonging to the thermodynamically unstable transition macrostate, the system is with a certain probability p able to enter the ordered stable equilibrium state at B. With the probability $1 - p$ it can also happen that the system returns to A again. However, since in the example shown in Fig. 2.1 $F_B < F_A$, the system will macroscopically reside in the thermodynamic equilibrium phase B, i.e., at the temperature $T < T_{tr}$ the population of microstates associated to phase B is the largest. If the temperature is increased such that $T > T_{tr}$, A would become the dominant phase since $F_A < F_B$ then. Thus, T_{tr} is suitably defined as the transition temperature, where $F_A = F_B$ and, therefore, both phases coexist. Since $q_A \neq q_B$ at T_{tr}, this transition is discontinuous. The energy difference $U_A - U_B = T_{tr}\Delta S = \Delta Q_{lat} > 0$ is the latent heat and corresponds to the energy required in the transition toward A to break the contacts that stabilized the ordered state in phase B.

The average time required to reach the transition state depends on the barrier height $\Delta F^{\ddagger}$. This can be shown in a simple Markovian-type transition state theory, where memory effects are neglected. Then, the time-dependent change of the probability $p(\mathbf{Q})$ to find the system in a state $\mathbf{Q}$ is given by the "gain" of states originating in a single time step Δt from all other possible states $\mathbf{Q}'$ and the "loss" from $\mathbf{Q}$ to $\mathbf{Q}'$. The so-called master equation reads

$$\frac{\Delta p(\mathbf{Q}, t)}{\Delta t} = \int \mathcal{D}Q' \, p(\mathbf{Q}', t) T(\mathbf{Q}' \rightarrow \mathbf{Q}, \Delta t) - \int \mathcal{D}Q' \, p(\mathbf{Q}, t) T(\mathbf{Q} \rightarrow \mathbf{Q}', \Delta t), \quad (2.60)$$

where $T(\mathbf{Q} \rightarrow \mathbf{Q}')$ is the transition probability to reach the system state $\mathbf{Q}'$ in a time step Δt if the system has originally been in the state $\mathbf{Q}$ before.

In equilibrium, the probability distribution $p(\mathbf{Q}', t)$ is constant in time, $\Delta p(\mathbf{Q}, t)/\Delta t = 0$. Since all states $\mathbf{Q}'$ can be considered to be independent of each other, this leads to the detailed-balance equilibrium condition

$$p(\mathbf{Q})T(\mathbf{Q} \rightarrow \mathbf{Q}') = p(\mathbf{Q}')T(\mathbf{Q}' \rightarrow \mathbf{Q}). \quad (2.61)$$

Inserting expression (2.58) for the probability density and Eq. (2.59) to relate the restricted partition function to the $\mathbf{Q}$-dependent restricted free energy yields the transition rate

$$k_{\mathbf{Q} \rightarrow \mathbf{Q}'} = \frac{T(\mathbf{Q} \rightarrow \mathbf{Q}')}{T(\mathbf{Q}' \rightarrow \mathbf{Q})} = e^{-\Delta F_{\mathbf{Q} \rightarrow \mathbf{Q}'}/k_B T}, \quad (2.62)$$

with $\Delta F_{\mathbf{Q} \rightarrow \mathbf{Q}'} = F^{res}(T; \mathbf{Q}') - F^{res}(T; \mathbf{Q})$. The rate $k_{\mathbf{Q} \rightarrow \mathbf{Q}'}$ is smaller than unity if $F^{res}(T; \mathbf{Q}') > F^{res}(T; \mathbf{Q})$, as expected. The transition rate can be related to the average time it needs to reach $\mathbf{Q}'$:

$$\tau_{\mathbf{Q} \rightarrow \mathbf{Q}'} \sim k_{\mathbf{Q} \rightarrow \mathbf{Q}'}^{-1} = e^{\Delta F_{\mathbf{Q} \rightarrow \mathbf{Q}'}/k_B T}. \quad (2.63)$$

The transition time thus grows at constant temperature exponentially with the free-energy difference.

Carrying over these results to the exemplified free-energy landscape shown in Fig. 2.1, we find that the time to reach the transition state from macrostate A is

$$\tau_{A \rightarrow ts} \sim e^{\Delta F^{\ddagger}/k_B T} \quad (2.64)$$

and grows exponentially with the barrier height. On the other hand, the time $\tau_{ts \rightarrow B}$ needed to reach the ordered state B from the transition state decays exponentially with the free-energy difference $|F^{res}(T; q_B) - F^{res}(T; q_{ts})|$. Note that in this theory the total average time to reach the ordered state B from the disordered state A by passing the transition state is $\tau_{A \rightarrow B} = \tau_{A \rightarrow ts} + \tau_{ts \rightarrow B}$.

This exemplified behavior generally applies to systems of rather little cooperativity, where many local conformational changes are required to perform a macroscopic change of the system state. "Climbing the hill" toward the transition state is thus a necessity to reach another (meta)stable macrostate. In other words: to reach B from A in a single fluctuation without the intermediate residence in the transition state is extremely unlikely. Thus, since such a transition is a time-dependent process influenced and activated by thermal fluctuations, it is frequently called a kinetic transition with $\Delta F^{\ddagger}$ being the kinetic barrier.

In systems of high cooperativity, a cooperative transition such as a helical alignment move in a protein during a tertiary folding process can in principle occur within a single time step. This is similar to tunneling in quantum mechanics. However, effects of thermal fluctuations in mesoscopic systems shall generally not be confused with quantum fluctuations in nanoscopic systems, where tunneling events are of much more relevance.

In this context, "cooperativity" does not necessarily mean that different parts of the system depend on and need to interact with each other to change the macrostate (or the phase). Rather, local parts of the system can react individually in the same way upon a weak change of the environmental conditions. In the freezing transition of water, nucleation cores form independently and attract other molecules in the local environment of each nucleus to join. This leads to macroscopic crystalline structures which finally bind to each other in order to reduce instabilities due to surface effects. However, the individual growth of the nucleation centers also causes dislocations that typically appear at the boundaries of these crystalline substructures.

A molecular folding process is much more complex and generally requires interactive cooperativity, i.e., the formation of a unique functional protein fold can take comparatively long (up to the order of seconds) and include weakly stable intermediate states. The collective folding is also necessary to avoid dislocations. Hence, the free-energy contour depicted in Fig. 2.1 will only apply to a subclass of proteins, so-called two-state folders.

The transition state theory and the assumption of "complex and rugged" free-energy landscapes are still under debate. One reason is the inherent difficulty to identify the "true" relevant degrees of freedom $\mathbf{Q}$ which are typically highly system-specific. The problem is that the kinetic barrier $\Delta F^{\ddagger}$ frequently depends on the choice of $\mathbf{Q}$, which makes an experimental verification of a theoretically proposed free-energy landscape more difficult. Nonetheless, the free-energy landscape concept is helpful in understanding details of phase transitions (such as, e.g., the occurrence of barriers that slow down the transition process) and to quantify these.

Throughout this book we will frequently discuss conformational transitions by means of interpretations of corresponding free-energy landscapes.

2.7 Microcanonical statistical analysis

There are mainly historical reasons why the canonical approach, as discussed in the previous section, has been used most frequently for the analysis of structural and phase transitions. This is because the exponential form of the canonical microstate probability (2.22) is suitable for the development of approximation methods and field-theoretical formulations which enable analytic calculations of thermodynamic quantities.

Before computers and efficient simulation methods became available, the canonical approach was the only way for a theoretical discussion of thermodynamic phenomena. On the experimental side, it was appealing to use the heat bath temperature as an external control parameter. Because of this, in equilibrium, "temperature" seemed to be an easily accessible parameter to control the macrostate of the system, and transition points of thermally driven phase transitions are typically defined by transition temperatures. The canonical analysis of phase transitions has been extremely successful and it enabled the introduction of fundamental physical concepts such as universality.

However, the uniqueness of the canonical approach can only be maintained as long as the investigated system fulfills the requirement of the thermodynamic limit, i.e., if finiteness, in particular surface effects, do not matter. Thus, the rapidly grown interest in the structural behavior of notoriously finite systems such as biomolecules confronted the thermodynamic analysis with the problem, to what extent a canonical analysis is still appropriate for the discussion of cooperative thermal behavior associated with structure formation processes such as protein folding or aggregation. Because of its simplicity, it is still popular to also apply the conventional canonical approach to such systems. However, the obvious violation of the thermodynamic limit condition leads to conceptual problems. Even the imagination of what temperature is, is strongly affected and requires a careful consideration.

2.7.1 Temperature as a derived quantity

It is useful to start from the basic statistical point of view again and consider the system quantities that are fundamental for its thermodynamic behavior: the energy E representing the system's microstate and the number (or density) of possibilities to create a microstate with this energy, $g(E)$. Note that E is the potential energy of the system configuration (or polymer conformation) and does not include kinetic energetic components due to motion. Therefore, $g(E)$ is the configurational (or conformational) density of states.

We remember that in the canonical statistical analysis, the canonical temperature corresponds to the canonical average of the kinetic energy per degree of freedom e_{kin}, which is simply $\langle e_{kin} \rangle = k_B T_{can}/2$. This means that the canonical temperature is a system-independent quantity and, therefore, in equilibrium, it is considered as an external parameter associated with the environment surrounding the system – the hypothetic heat bath coupled to the system. This is a satisfying approach as long as the distribution of energetic microstates at T_{can} fixed is so sharp that in a stable phase fluctuations about the mean are very small if compared with the fluctuations at phase transition points. This is the case only if the number of degrees of freedom is very large, effectively infinite. This is why phase transitions can only be described in the thermodynamic limit.

We have already mentioned that for small systems with a finite number of degrees of freedom, the relative fluctuations have much more impact and cause broader distributions. Calculating canonical averages cancels out relevant information about the system's thermodynamic behavior. Therefore, it might be useful to decouple the temperature from the heat bath and introduce it as a structural property of the system itself. The idea is that a conformational "phase" and transitions between different such "phases" can then be described entirely by the change of the configurational/conformational entropy S upon a small variation of the macrostate energy E. The macrostate with energy E contains all conformations (microstates) with potential energy E.

As has already been discussed in Section 2.2.3, the microcanonical entropy is given by $S(E) = k_B \ln g(E)$.[5] In this context, the (microcanonical) temperature $T(E)$ becomes a *defined* quantity, which is associated to the entropic change caused by a variation of macrostate energy. We strictly introduce it via

$$\beta(E) = \frac{dS(E)}{dE} \qquad (2.65)$$

as $T(E) = \beta^{-1}(E)$.[6] Note that, for a finite system, the temperature can be negative! Negative temperatures occur if the entropy decays with increasing energy.[7] This is not an unusual or even unphysical phenomenon, although it clearly contradicts our perception of what temperature is supposed to be. What is counterintuitive and questions the typical interpretation of the temperature (but not β!) is that negative temperatures refer to *hotter* macrostates than positive temperatures, i.e., the temperature scale reads $T \in \lim_{\epsilon \to 0}[+\epsilon, \ldots, +\infty, -\infty, \ldots, -\epsilon]$. For this reason and another one that will be discussed later on in connection with the identification of transitions by investigating its monotonic behavior, it is useful to give up the temperature in favor of the inverse temperature $\beta(E) \in [+\infty, \ldots, 0, \ldots, -\infty]$, as defined by Eq. (2.65), which does not exhibit a jump while turning from positive to negative values.

What is now the advantage of the inverse microcanonical temperature compared to its canonical counterpart? Since $\beta(E)$ is directly derived from the fundamental system quantities S and E, its curvature should contain all information about the system behavior if it undergoes a macroscopic change such as a cooperative transition. It will turn out that transitions occur if $\beta(E)$ responds least sensitively to changes in system energy. By means of this least-sensitivity principle, it is possible to identify transitions uniquely, even in small systems. This also means that a transition point can be uniquely assigned a single transition temperature – this is typically not possible in the canonical formalism, where the transition point depends on the fluctuation extremum of the chosen order parameter. In contrast to the canonical counterpart, the microcanonical temperature is a system property and not an external parameter that can be controlled.

2.7.2 Identification of first-order transitions by Maxwell construction

Maxwell construction in the thermodynamic limit

Traditionally, transitions with phase coexistence are analyzed by means of Maxwell's construction. It was introduced to "cure" and interpret the unphysical backbending branch that occurs for real gases described by the van der Waals equation in the pressure (p) versus

[5] Sometimes, the microcanonical entropy is defined as $S(E) = k_B \ln G(E)$, where $G(E) = \int_{E_{min}}^{E} dE' \, g(E')$ is the integrated density of states. Although this definition also has advantages, it lacks an intuitive physical meaning.

[6] Note that in the canonical representation, β typically is defined as the inverse thermal energy, $\beta = 1/k_B T$.

[7] This is not as exotic as it sounds. Even the simple Ising model, used for the qualitative description of magnetic phase transitions, exhibits this feature.

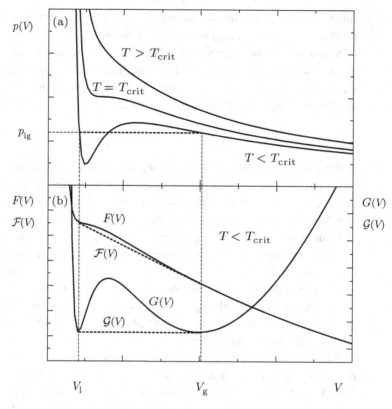

Fig. 2.2 (a) Isotherms of the van der Waals gas $p(V)$ for temperatures T below, at, and above the critical temperature T_{crit}, and Maxwell line at $p = p_{\mathrm{lg}}$. (b) Free energy $F(V)$, Gibbs construction for the free energy $\mathcal{F}(V)$ and Gibbs free energy $G(V)$ and its Gibbs hull $\mathcal{G}(V)$ at pressure $p = p_{\mathrm{lg}}$ for the isotherm in (a) at $T < T_{\mathrm{crit}}$.

volume (V) phase diagram for isotherms below the critical temperature T_{crit}. In this region, $(\partial p/\partial V)_{T<T_{\mathrm{crit}}} > 0$, which in consequence would cause a negative compressibility. Matter, although in an equilibrium state, would be unstable. Figure 2.2(a) shows three van der Waals isotherms at temperature above, at, and below the critical temperature.

Maxwell consequently utilized the fact that in equilibrium the Gibbs free energies of the two coexisting phases (liquid, gas) must be identical for each macrostate in the transition region, $G_{\mathrm{lg}} = G_{\mathrm{l}} = G_{\mathrm{g}}$, because the unphysical behavior of the van der Waals equation can only be avoided if the pressure in the transition region is assumed to be constant. The value of this pressure, p_{lg}, is determined by the following consideration. Along the isotherm ($dT = 0$), $dG = Vdp$. Hence, if the Gibbs free energies in both phases were not identical, the gradient ($V = (\partial G/\partial p)_T$) would spontaneously drive the system into one of the phases. This apparently does not happen in a real gas-liquid mixture. The stable equilibrium liquid-gas phase coexistence can be observed. Thus, it requires additional work, given by the change of free energy $dF = -pdV$, to change the macrostate (actually only V, because p and T are constant) within the transition region. Therefore, the total transition free energy released in the evaporation process is given by

$$F_{\mathrm{lg}} \equiv -\int_{V_{\mathrm{l}}}^{V_{\mathrm{g}}} p(V)dV = -p_{\mathrm{lg}}(V_{\mathrm{g}} - V_{\mathrm{l}}), \qquad (2.66)$$

where V_1 and V_g denote the volumes at the respective phase boundaries, satisfying

$$\left(\frac{\partial F}{\partial V}\right)_{V_1} = \left(\frac{\partial F}{\partial V}\right)_{V_g} = -p_{lg} \tag{2.67}$$

and $p = p_{lg} = \text{const}$ in the interval $V \in [V_1, V_g]$. Thus, the area under the original curve $p(V)$ in this volume interval and the area bounded by p_{lg} are supposed to be identical. The free energy as a function of the volume, $F(V)$, is shown in Fig. 2.2(b). Note the concavity $0 > (\partial^2 F/\partial V^2)_{T < T_{crit}} = -(\partial p/\partial V)_{T < T_{crit}} = 1/V\kappa$ of this function in the transition interval. Since neither V nor the compressibility κ can be negative, this behavior of F is unphysical.

The Maxwell construction for the subcritical isotherm corresponds to a Gibbs construction for the free energy: $\mathcal{F}(V) = F(V_1) - p_{lg}(V - V_1)$, i.e., the slope of this double tangent of $F(V)$ is $-p_{lg}$. Since $\mathcal{F}(V) < F(V) \; \forall V \in [V_1, V_g]$, the macrostates $(p_{lg}, V \in [V_1, V_g], T_{lg} < T_{crit})$ then indeed represent the stable and physical equilibrium states in the liquid–gas transition region. The hull of the Gibbs free energy, $\mathcal{G}(V) = \mathcal{F}(V) + p_{lg}V$, is constant and minimal in the coexistence regime, as required. Thus, it replaces $G(V) = F(V) + p_{lg}V$ as the appropriate thermodynamic potential for the van der Waals gas in the first-order transition region [see Fig. 2.2(b)]. To summarize, the Maxwell construction is necessary for a physically consistent thermodynamic description of this model.

Microcanonical Maxwell construction for finite systems

In the context of microcanonical statistical analyses of finite systems, Maxwell constructions can also be beneficial. In these cases, however, they are used to help identify and define transition points of first-order-like transitions, but are not required for thermodynamic consistency. As discussed earlier in this chapter, the microcanonical entropy $S(E) = k_B \ln g(E)$ is considered to be the central quantity of any microcanonical analysis. Apart from energetic regions that we will later connect to first-order-like transitions, $S(E)$ is a strictly concave function. Since $T(E) = (\partial S(E)/\partial E)^{-1}$ is the microcanonical temperature, the microcanonical heat capacity $C_V(E) = (\partial T(E)/\partial E)^{-1} = -(\partial S/\partial E)^2/(\partial^2 S/\partial E^2)$ is positive. It only can become negative in regions where $S(E)$ is convex. This can be the case if an energetic zone of a certain width exists, in which the number of energetic microstates is suppressed to such an extent that the entropy curve loses its concave monotony. In this entropic depletion zone $T(E)$ fluctuates and the microcanonical heat capacity is negative. At this point, one should not be confused with the never observed negativity of the *canonical heat capacity*, which is a function of the canonical temperature. Thermodynamic stability of matter requires this function to be strictly positive. The microcanonical heat capacity does not have such constraints and negative values are not unphysical.

Since the depletion region of the entropy has an energetic width $\Delta Q > 0$ and connects energetic spaces which are associated with different phases of reduced thermodynamic activity, the extended region in between accommodates macrostates, in which two phases coexist. It is therefore common to interpret ΔQ as the latent heat and to associate the entropic suppression in this region with a first-order transition, in analogy to thermodynamic first-order phase transitions in the thermodynamic limit.

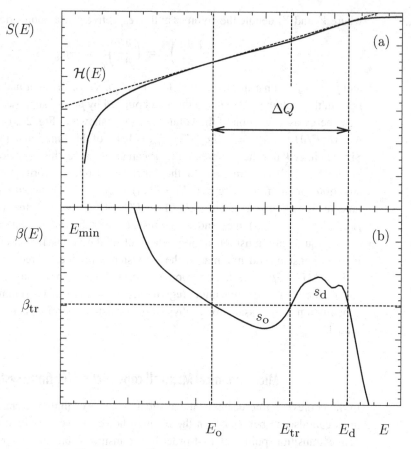

Fig. 2.3 Microcanonical analysis by Maxwell construction: (a) Microcanonical entropy $S(E)$ and Gibbs construction $\mathcal{H}(E)$ as functions of energy E; (b) Inverse thermal energy $\beta(E)$ and Maxwell line at the first-order transition point β_{tr}. The first-order-like transition behavior is characterized by the "backbending" effect between E_0 and E_d. The Maxwell line intersects $\beta(E)$ at energies E_0, E_{tr}, and E_d. It is defined by the equality of the areas s_0 and s_d, which correspond to the surface entropy. These results were obtained by simulations of heteropolymer aggregation. Note the slight deformation of $\beta(E)$ right below E_d that indicates a subphase transition within the Maxwell regime, which cannot be considered separately in a microcanonical analysis based on Maxwell's construction.

Figure 2.3(a) shows a typical example for an entropic curve with a convex region. The depletion zone is bounded by the limiting energies of the ordered phase, E_0, and the disordered phase, E_d. The width of the entropic depletion zone is $\Delta Q = E_d - E_0$. Also shown is the double-tangent, i.e., the Gibbs hull. This corresponds to a Maxwell construction in the (inverse) temperature curve $\beta(E) = 1/T(E)$, as plotted in Figure 2.3(b). The advantage of this construction is that it enables the definition of a unique transition temperature β_{tr} – a notorious problem for finite systems. The boundaries E_0 and E_d are defined by the points where the Gibbs hull

$$\mathcal{H}(E) = S(E_0) + \beta_{tr}(E - E_0) = S(E_d) + \beta_{tr}(E - E_d) \qquad (2.68)$$

touches $S(E)$: $\mathcal{H}(E_{o,d}) = S(E_{o,d})$. The total entropy change during the transition $\Delta S_{tr} = S(E_d) - S(E_o) = \mathcal{H}(E_d) - \mathcal{H}(E_o)$ can then be expressed in two ways:

$$\Delta S_{tr} = \int_{E_o}^{E_d} dE\, \beta(E) = \beta_{tr}(E_d - E_o). \tag{2.69}$$

As Fig. 2.3(b) also shows, the β curve resembles a third-order polynomial. For that reason, the Maxwell line β_{tr} possesses a third intersection point with $\beta(E)$, denoted by E_{tr}. It can be considered as the separation point, where the entropic weight of each of the two phases in the coexistence region is identical. In this formalism, it defines the transition state. The total entropy change in the order-dominated coexistence regime is:

$$\Delta S_o = \int_{E_o}^{E_{tr}} dE\, \beta(E) = S(E_{tr}) - S(E_o) = \beta_{tr}(E_{tr} - E_o) - s_o, \tag{2.70}$$

and analogously follows for the disordered part:

$$\Delta S_d = \int_{E_{tr}}^{E_d} dE\, \beta(E) = S(E_d) - S(E_{tr}) = \beta_{tr}(E_d - E_{tr}) + s_d, \tag{2.71}$$

where the right-hand side expressions are simply obtained by adding the areas according to the definitions in Fig. 2.3(b). What remains to be done is to resolve the meaning of s_o and s_d, and their relationship in correspondence with the Maxwell construction. For this purpose, let's consider the difference of the value of the Gibbs hull and the entropy,

$$\Delta S(E) = \mathcal{H}(E) - S(E), \tag{2.72}$$

at the separation point E_{tr}:

$$\Delta S(E_{tr}) = S(E_o) + \beta_{tr}(E_{tr} - E_o) - S(E_{tr}) = S(E_d) + \beta_{tr}(E_{tr} - E_d) - S(E_{tr}), \tag{2.73}$$

where we made use of the two parametrizations of the Gibbs tangent given in Eq. (2.68). Since $\Delta S(E_{tr}) \neq 0$ is a very characteristic feature of first-order-like transitions in finite systems, where surface effects are not negligible, it is often called surface entropy, $\Delta S_{surf} \equiv \Delta S(E_{tr})$. Compared to the bulk of topological dimension D, the conformational entropic freedom at the surface of such systems is smaller, because of the reduced dimension $D-1$. This has the effect that particles prefer to be part of the interior volume and not of the surface layer of the system. This results in shapes with minimal surfaces (e.g., spherical or globular conformations in an isotropic medium). The "surface entropy" ΔS_{surf} is actually the amount of entropy suppression by the formation of surfaces. If the effective attractive interactions between the particles have a finite range, surface effects decrease with increasing system size (number of particles N) and so does ΔS_{surf}. For such systems, in the thermodynamic limit, $\lim_{N \to \infty} \Delta S_{surf} = 0$. Thus, in this limit, the entropy cannot be convex and thermodynamic first-order phase transitions do not exhibit this "backbending" feature that we discuss here in this section.

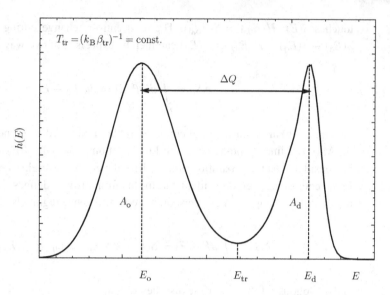

$T_{tr} = (k_B \beta_{tr})^{-1} = \text{const.}$

ΔQ

$h(E)$

A_o A_d

E_o E_{tr} E_d E

Fig. 2.4   Canonical energy histogram at $T_{tr} = (k_B \beta_{tr})^{-1}$ of the same transition as identified in Fig. 2.3.

Eliminating $S(E_{tr})$ in Eq. (2.73) by employing Eqs. (2.70) and (2.71), we eventually find that

$$\Delta S_{surf} = s_o = s_d. \tag{2.74}$$

Thus, the Gibbs construction $\mathcal{H}(E)$, entailing Maxwell's line at β_{tr}, automatically leads to identical areas s_o and s_d and requires them to be the same as the surface entropy ΔS_{surf}. This is a remarkable result, as it shows that the surface entropy corresponds to the maximum deviation of the Gibbs tangent from the original $S(E)$ curve, located at E_{tr}. The lesson to learn is that the transition point between the ordered and the disordered phase is well-defined by β_{tr}. Thus, the microcanonical analysis by Maxwell construction does not possess any ambiguity in the definition of transition points for finite systems, in contrast to the canonical analysis of peak positions and "shoulders" of fluctuations of typically not-well-defined order parameters.

In the context used here, we actually consider the energy itself as an order parameter. It is not very specific (we do not learn much about the physical properties of the transition), but it allows us to clearly separate the phases. This is particularly apparent when falling back to the canonical interpretation. Figure 2.4 shows the canonical histogram (unnormalized probability density distribution) $h(E) = g(E)e^{-E/k_B T_{can}}$ at the transition temperature $T_{can} = T_{tr}$ identified by Maxwell construction. This distribution has a characteristic bimodal form, with peaks at energies E_o and E_d, separated by the latent heat ΔQ. The minimum between the peaks is located at the separation point E_{tr}. Note that the theory we have introduced so far is based on the so-called equal-height criterion, i.e., the transition point is defined by the temperature T_{tr}, where $h(E_o) = h(E_d)$. This is, as we have seen before, a direct consequence of the microcanonical analysis. However, as is already obvious from the example

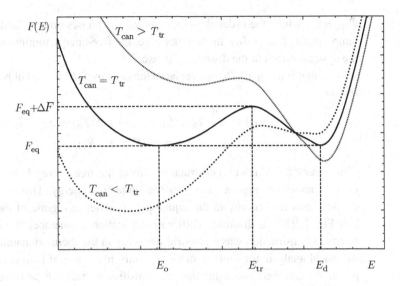

Fig. 2.5 Free-energy profiles $F(E)$ for temperatures below, at, and above the transition point $T_{can} = T_{tr}$.

shown in Fig. 2.4, the areas under the peaks in the ordered and in the coexisting disordered phase, A_o and A_d, respectively, are not necessarily identical.

In complete analogy to the general discussion of the transition state theory, $h(E)$ already represents the restricted partition sum (2.55), if the energy is understood as an order parameter

$$h(E) \equiv Z_{can}^{res}(E) = \int dE' \delta(E - E') g(E') e^{-E'/k_B T_{can}} \qquad (2.75)$$

and T_{can} is a constant parameter representing the canonical temperature. Then, with Eq. (2.59), the free-energy landscape in energy space is parametrized by

$$F(E) = -k_B T_{can} \ln h(E) = E - T_{can} S(E). \qquad (2.76)$$

Thermodynamically, only global minima in the free-energy landscape represent stable equilibrium states. The extremum condition $(dF/dE)_{E=E_{ext}} = 0$ thus yields

$$T(E_{ext}) = T_{can} \qquad (2.77)$$

and connects the microcanonical and canonical pictures. Extrema in $h(E)$ at T_{can} are thus mirrored by extremal points in the $F(E)$ landscape, too. Because of the Maxwell construction, the transition point is defined by $T(E) = T_{tr}$. This is only satisfied at energies E_o, E_{tr}, and E_d, which, according to the condition (2.77) are also the locations of extremal points in the free-energy landscape at the transition point. In this interpretation, $F(E_o) = F_{eq}$ and $F(E_d) = F_{eq}$ represent the degenerate macrostates of the coexisting phases, and $F(E_{tr}) = F_{eq} + \Delta F$ is the free energy of the transition barrier. The height of the barrier, ΔF, can be interpreted kinetically, according to the discussion in Section 2.6. Figure 2.5 shows free-energy profiles for temperatures below the transition point $T_{can} < T_{tr}$, where the system is in the ordered phase, at the microcanonically defined transition point

$T_{can} = T_{tr}$, where the ordered and the disordered phases coexist, and above the transition temperature ($T_{can} > T_{tr}$). In the latter case, the free-energy minimum lies beyond E_d, i.e., the system resides in the disordered phase.

Note that if in Eq. (2.76) one replaces the entropy $S(E)$ by its Gibbs hull in the transition regime,

$$F(E) = E - T_{tr}\mathcal{H}(E) = E_o - T_{tr}S(E_o) = E_d - T_{tr}S(E_d) \tag{2.78}$$

$$= F_{eq} = \text{const.} \quad \forall E \in [E_o, E_d]. \tag{2.79}$$

This is how the Maxwell construction affects the free-energy behavior: each macrostate in the coexistence regime possesses the same free energy. This resembles the behavior of the Gibbs free energy in the liquid-gas coexistence regime of the van der Waals gas [see Fig. 2.2(b)], as discussed earlier in this section. Remember that in the latter case, the Maxwell construction was a physical necessity in the thermodynamic limit. In the microcanonical analysis for finite systems, we only made use of it to determine the transition point. We can therefore argue that the transition barrier $\Delta F > 0$ for the finite system is physical reality, but it disappears in the thermodynamic limit and the $F(E)$ curve converges to the Maxwell line. It is worth noting at this point that the kinetics of processes such as protein folding is strongly affected by the existence of a transition barrier. Nonetheless, it is only a finite-size effect because of its surface-entropic origin.

The general disadvantage of the microcanonical analysis based on Gibbs/Maxwell construction is that it is only applicable to transitions with a clear, single signature, the "backbending effect." Often, this is actually not the case. As we will investigate in more detail later, in the discussion of aggregation transitions, first-order transitions typically exhibit hierarchies of individual first-order-like subphase transitions. A first indication can already be noticed in the example shown in Fig. 2.3(b), where an additional mini-backbending effect can be observed between E_{tr} and E_d. In a systematic microcanonical analysis, such signals shall not be ignored. However, since this transition lies within the Maxwell coexistence regime, it cannot be resolved by this kind of analysis. Transitions without backbending feature, which we will later define as transitions of higher than first order, cannot be analyzed by Maxwell construction as well. In order to enable a thorough and systematic microcanonical analysis that includes hierarchies in first-order transitions and the identification of higher-order transitions, a revised theory is needed. A simple approach will be discussed in the following.

2.7.3 Systematic classification of transitions by inflection-point analysis

We have seen that the Maxwell construction can be employed successfully for the analysis of strong first-order transitions. Geometrically, the backbending region in the transition regime of the energetic temperature curve in Fig. 2.3 is replaced by an entirely flat segment. We also know that the backbending effect is due to surface effects that become negligible for very large systems, where volume properties are dominant. Thus, this flattening of the $\beta(E)$ curve is a physical property of this transition. It means that, in the transition region, the (volume part of the) system is not very sensitive to energetic changes. The system relaxes; macrostate changes require more energy (latent heat) while temperature changes decrease. The kinetics of macrostate changes is affected by the transition barrier. However, the Maxwell construction only applies to single transitions of first order.

To make it applicable to other transition types as well, it is attractive to extend this "flatness" idea by replacing the Maxwell construction with a more general principle, the *principle of least sensitivity*. This is a weaker condition, but it allows us to investigate first- and higher-order transitions by a microcanonical analysis more systematically and in much more detail [61].

This approach takes into account that all qualitative changes in the interplay of entropy and energy as signaled by alterations in the curvature of the microcanonical entropy $S(E)$, are indicators of cooperative behavior in the system. The strength of these changes can then also be considered as quantitative measures for the significance of cooperativity of the associated transitions. The slope of a tangent at each point of the curve $S(E)$ is unique, and we now strictly define the reciprocal microcanonical temperature via the caloric derivative of S,

$$\beta(E) \equiv T^{-1}(E) = (dS/dE)_{N,V}, \qquad (2.80)$$

where system size N and volume V are kept constant. Since the complete phase behavior is already encoded in $S(E)$, $\beta(E)$ is considered to be the only unique parameter to identify transition points. The analysis of the monotonic behavior of $\beta(E)$, expressed by its derivative with respect to energy,

$$\gamma(E) = d\beta(E)/dE = d^2 S/dE^2, \qquad (2.81)$$

then allows for the introduction of a systematic classification scheme of transitions in finite systems [61], but it can also be used for scaling analyses toward the thermodynamic limit. The function $\gamma(E)$ describes the variation of the inverse temperature with respect to energy at a given energy value E. As such it is related with the microcanonical heat capacity via $C_V(E) = [dT(E)/dE]^{-1} = -\beta^2(E)/\gamma(E)$.

In this scheme, a transition between phases is defined to be of *first order* if the slope of the corresponding inflection point of $\beta(E)$ at $E = E_{\mathrm{tr}}$ is positive,

$$\gamma_{\mathrm{tr}} = \gamma(E_{\mathrm{tr}}) > 0 : \text{first-order transition}. \qquad (2.82)$$

Only in this case is the temperature curve non-monotonic and there is no unique mapping between β and E. As has been discussed in the context of the Maxwell construction, this is the regime where both phases coexist. The overall energetic width of the undercooling, backbending, and overheating regions is identical to the latent heat. Thus, for a first-order transition, $\Delta Q > 0$. Note that the inflection point is not necessarily identical to the separation point obtained by Maxwell construction. For this reason, the latent heats obtained by Maxwell construction and by inflection-point analysis can differ slightly.

In the case where the inflection point has a negative slope, the phases cannot coexist and the latent heat is zero, $\Delta Q = 0$. In complete analogy to phase transitions in the thermodynamic limit, such transitions are classified as of *second order*:

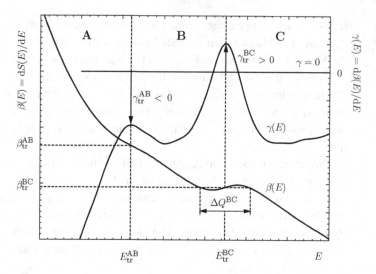

Fig. 2.6 Inflection-point analysis of the inverse temperature $\beta(E)$ and its derivative $\gamma(E)$ as functions of energy E for a system exhibiting a first- and a second-order transition. The maxima of $\gamma(E)$ indicate transitions between the phases A and B at $E_{\mathrm{tr}}^{\mathrm{AB}}$ and B and C at $E_{\mathrm{tr}}^{\mathrm{BC}}$. The associated points $\beta(E_{\mathrm{tr}}^{\mathrm{AB}}) = \beta_{\mathrm{tr}}^{\mathrm{AB}}$ and $\beta(E_{\mathrm{tr}}^{\mathrm{BC}}) = \beta_{\mathrm{tr}}^{\mathrm{BC}}$ define the transition temperatures $T_{\mathrm{tr}}^{\mathrm{AB}} = (\beta_{\mathrm{tr}}^{\mathrm{AB}})^{-1}$ and $T_{\mathrm{tr}}^{\mathrm{BC}} = (\beta_{\mathrm{tr}}^{\mathrm{BC}})^{-1}$. According to the classification scheme of inflection-point analysis, the transition between A and B is of second order, since the slope of the inflection point is negative. On the other hand, B$\leftrightarrow$C is a first-order transition as the respective slope at $\beta(E_{\mathrm{tr}}^{\mathrm{BC}})$ is positive. The non-monotonic "backbending" is a characteristic signal of phase coexistence. The latent heat ΔQ^{BC} is defined as the energetic width of this transition region.

$$\gamma_{\mathrm{tr}} = \gamma(E_{\mathrm{tr}}) < 0 : \text{ second-order transition.} \tag{2.83}$$

Since the inflection points of $\beta(E)$ correspond to maxima in $\gamma(E)$, it is therefore sufficient to analyze the peak structure of $\gamma(E)$ in order to identify the transition energies and temperatures. The sign of the peak values classifies the transition. This very simple and general classification scheme applies to all physical systems.

Figure 2.6 illustrates the procedure for the identification of the transitions by means of inflection-point analysis. Plotted are the inverse temperature $\beta(E)$ and its energetic derivative $\gamma(E)$. There are two regions where the weak-sensitivity condition applies to $\beta(E)$. One is located around the inflection point at $E_{\mathrm{tr}}^{\mathrm{AB}}$ and the other is the backbending regime surrounding the central inflection point at $E_{\mathrm{tr}}^{\mathrm{BC}}$. The latter exhibits the already well-described features of a first-order transition: $\gamma_{\mathrm{tr}}^{\mathrm{BC}}$ is positive and the intersection points of the inverse transition temperature $\beta_{\mathrm{tr}}^{\mathrm{BC}}$ with the $\beta(E)$ curve defining the coexistence region. The width is interpreted as the latent heat, which is obviously nonzero: $\Delta Q^{\mathrm{BC}} > 0$. The behavior is qualitatively different at $E_{\mathrm{tr}}^{\mathrm{AB}}$, where $\gamma_{\mathrm{tr}}^{\mathrm{AB}} < 0$. There is no phase coexistence so that the latent heat is zero. The A$\leftrightarrow$B transition is consequently classified as a second-order transition. The transition point is given by the inverse temperature $\beta_{\mathrm{tr}}^{\mathrm{AB}}$.

In cases where it is difficult to find first- or second-order traces of "hidden" finite-size effects in $\beta(E)$, the weak-sensitivity criterion can be extended by allowing higher-than-second-order transitions. This can be done by analyzing the inflection points of $\gamma(E)$, or if necessary, of even higher derivatives of $\beta(E)$.

This type of microcanonical analysis is very similar to Ehrenfest's classification scheme for phase transitions in the thermodynamic limit. In this scheme, the order of the transition is fixed by the smallest value of n, at which the nth-order derivative of the free energy with respect to an independent thermodynamic variable, e.g., $(\partial^{(n)} F(T, V, N)/\partial T^{(n)})_{V,N}$, becomes discontinuous at any point. Obviously, first-order transitions are characterized by a discontinuity in the entropy as a function of temperature, $S(T) = -(\partial F(T, V, N)/\partial T)_{V,N}$, at the transition temperature T_{tr}. The discontinuity at the transition point ΔS corresponds to a non-vanishing of the latent heat $T_{\text{tr}} \Delta S = \Delta Q > 0$. In a second-order phase transition, the entropy is continuous, but the second-order derivative, which is related to the heat capacity, $(\partial^2 F(T, V, N)/\partial T^2)_{V,N} \sim C_V(T)$, is not. The heat capacity (or better the specific heat $c_V = C_V/N$) possesses a discontinuity (often a divergence) at the critical temperature T_{cr}. Although higher-order phase transitions are rather rare, Ehrenfest's scheme accommodates these transitions as well.

Although the analysis of discontinuities is not useful for finite systems at all (there are none), the inflection-point analysis in its most general form can be interpreted in a similar way. Cooperative behavior is encoded in the curvature of an appropriate "potential." In the thermodynamic limit, one might want to choose the free energy; in the microcanonical analysis of small systems, it is the entropy as a function of energy. In a hierarchical inflection-point analysis, the variational principle standing behind the weak-sensitivity criterion tells us that first-order transitions are visible in the first derivative of $S(E)$ (backbending effect or Maxwell construction), second-order transitions are characterized by specific inflection points of $S(E)$, i.e., second-order derivatives are needed for their characterization, and higher-order transitions require higher-order derivatives to be identified.

There is another issue worth mentioning. The consistency of the hierarchy $S \rightarrow \beta = dS/dE \rightarrow \gamma = d^2/dE^2 \rightarrow \cdots$ tells us that, in principle, the *inverse* temperature β is more fundamental than the temperature T. This is not a trivial statement. Second-order transition points cannot easily be identified as inflection points of $T(E)$. The introduction of the temperature T (or, more precisely, $k_B T$) as a "measure" for thermal energy had merely happened as a historical accident, but there were no substantial physical reasons to favor it. Microcanonical analysis based on the close relationship between entropy and *inverse* temperature suggests that β is the more appropriate thermodynamic variable.

The complexity of minimalistic lattice models for protein folding

3.1 Evolutionary aspects

The number of different functional proteins encoded in human DNA is of the order of about 100 000, which is an extremely small number compared to the total number of possibilities: Recalling that 20 types of amino acids occur in natural proteins and typical proteins consist of $N \sim \mathcal{O}(10^2 - 10^3)$ amino acid residues, the number of possible primary structures, 20^N, lies somewhere far, far above $20^{100} \sim 10^{130}$. Assuming all proteins were of size $N = 100$ and a single folding event would take 1 ms, a sequential enumeration process would need about 10^{119} years to generate structures of all sequences, irrespective of the decision about their "fitness," i.e., the functionality and ability to efficiently cooperate with other proteins in a biological system. Of course, one might argue that the evolution is a highly parallelized process that drastically increases the generation rate. So, we can ask the question, how many processes can maximally run in parallel. The visible universe contains of the order of 10^{80} protons. Assuming that an average amino acid consists of at least 50 protons, a chain with $N = 100$ amino acids has of the order $\mathcal{O}(10^3)$ protons, i.e., 10^{77} sequences could be generated in each millisecond (forgetting for the moment that some proton-containing machinery is necessary for the generation process and only a small fraction of protons is assembled in Earth-bound organic matter). The age of our universe is about 10^{10} years (we also forget that Earth is even about one order of magnitude younger) or 10^{21} ms. Hence, about 10^{98} sequences could have been tested to date, if our drastic simplifications were right. But even this yet much too optimistic estimate is still noticeably smaller than the abovementioned reference number of 10^{130} possible sequences for a 100-mer.

At least two conclusions can be drawn from this crude analysis. One is that the evolutionary process of generating and selecting sequences is ongoing, as it is likely that only a small fraction of functional proteins has been identified yet by nature. On the other hand, the existence of complex biological systems, where hundreds of thousands of different types of macromolecules interact efficiently, can only be explained by means of efficient evolutionary strategies of adaptation to environmental conditions on Earth which dramatically changed through billions of years. Furthermore, the development from primitive to complex biological systems leads to the conclusion that within the evolutionary process of protein design, particular patterns in the genetic code have survived over generations, while others were improved (or deselected) by recombinations, selections, and mutations. But the sequence question is only one side. Another regards the geometric structure of proteins that is directly connected to biological functionality. The conformational similarity among human functional proteins is also quite surprising; only of the order of 1000 qualitatively significantly different "folds" were identified [13].

Since the conformation space is infinitely large because of the continuous degrees of freedom and the sequence space is also giant, the protein folding problem is typically attacked from two sides: the *direct folding problem*, where the amino acid sequence is given and the associated native, functional conformation has to be identified, and the *inverse folding problem*, where one is interested in all sequences that fold into a given target conformation. With these two approaches, it is, however, virtually impossible to unravel evolutionary factors that led to the set of present functional proteins. Only for small, discrete protein models, a complete and exact statistical analysis of the entire sequence and conformations space is possible. Such an analysis [19, 20] will be performed in the following by employing hydrophobic–polar (HP) lattice models [12, 13], as introduced in Section 1.5.2.

3.2 Self-avoiding walks and contact matrices

Flexible lattice polymers are typically modeled by self-avoiding walks (SAW). The total number of conformations for a chain with N monomers is not known exactly. For $N \to \infty$ it is widely believed that in leading order the scaling law [62, 63]

$$C_n = A\mu_C^n n^{\gamma-1} \tag{3.1}$$

holds, where $n = N - 1$ is the number of self-avoiding steps. In this expression, μ_C is the effective coordination number of the lattice, γ is a universal exponent, and A is a non-universal amplitude. In Table 3.1, the exactly enumerated number for self-avoiding conformations is listed for chains with up to $N = n + 1 = 19$ monomers. By extrapolating these data, e.g., by using ratio methods [62, 64], the estimates $\mu_C \approx 4.684$ and $\gamma \approx 1.16$ [19] are obtained for the simple-cubic lattice. Despite the short chain lengths, these values are already very close to results obtained in extended enumeration studies [65–67] and Monte Carlo simulations [68], and to field-theoretic estimates [69] for γ. However, here we are not only interested in the structural dimension; rather we have to scan the combined space of HP sequences and conformations, which contains for chains of $N = 19$ monomers $2^{19} C_{19} \approx 1.17 \times 10^{18}$ possible combinations.

In models with the general form (1.7), where the calculation of the energy reduces to the summation over contacts (i.e., pairs of monomers being nearest neighbors on the lattice but nonadjacent along the chain) of a given conformation, the number of conformations that must necessarily be enumerated can be decreased drastically by considering only classes of conformations, so-called contact sets [17, 70]. A contact set is uniquely characterized by a corresponding contact map (or contact matrix), but a single conformation is not. Thus, for determining energetic quantities of different sequences, it is sufficient to carry out enumerations over contact sets. In a first step, however, the contact sets and their degeneracy, i.e., the number of conformations belonging to each set, must be determined and stored. Then, the loop over all nonredundant sequences is performed for all contact sets instead of

Table 3.1 Number of conformations C_N and contact matrices M_N for chains with N monomers (or, equivalently, self-avoiding walks with $n = N - 1$ steps).

N	n	$\frac{1}{6}C_N$	M_N	$\frac{1}{6}C_N/M_N$
4	3	25	2	13
5	4	121	3	40
6	5	589	9	65
7	6	2 821	20	141
8	7	13 565	66	206
9	8	64 661	188	344
10	9	308 981	699	442
11	10	1 468 313	2 180	674
12	11	6 989 025	8 738	800
13	12	33 140 457	29 779	1 113
14	13	157 329 085	121 872	1 290
15	14	744 818 613	434 313	1 715
16	15	3 529 191 009	1 806 495	1 954
17	16	16 686 979 329	6 601 370	2 528
18	17	78 955 042 017	27 519 000	2 869
19	18	372 953 947 349	102 111 542	3 652

conformations. This can be done by efficient parallelized exact enumeration of the whole space of contact sets and nonredundant sequences [20].

In Table 3.1, the resulting numbers of contact sets M_N are summarized and, although also growing exponentially [see Figs. 3.1(a) and (b)], the gain of efficiency by enumerating contact sets is documented by the ratio between C_N and M_N in the last column. Assuming that the number of contact sets M_n follows a scaling law similar to Eq. (3.1), the effective coordination number can be estimated, yielding approximately $\mu_M \approx 4.38$. The ratios of numbers of contact sets for even and odd numbers of walks oscillate more strongly than for the number of conformations, as is shown in Fig. 3.1(b). This renders an accurate scaling analysis (in particular for the exponent γ) based on the data for the relatively small number of steps much more difficult than for self-avoiding walks, but in the context of protein folding studies, scaling properties are of more secondary interest.

3.3 Exact statistical analysis of designing sequences

In this section, we will discuss statistical properties of the complete sets $\mathbf{S}_N$ of designing sequences for HP proteins for given numbers of residues $N \leq 19$. A sequence σ is called *designing* if there is only one conformation associated with the native ground state, not counting rotation, translation, and reflection symmetries that altogether contribute on a

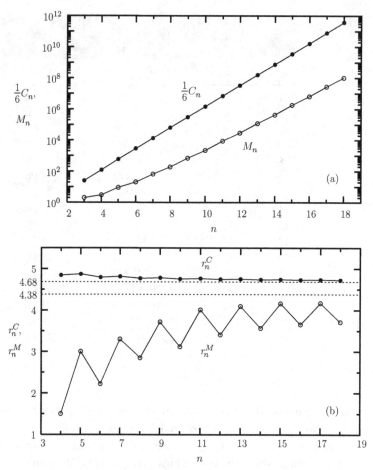

Fig. 3.1 (a) Numbers of self-avoiding walks C_n and contact matrices M_n vs. the number of steps $n = N - 1$. (b) Ratios of numbers of self-avoiding walks $r_n^C = C_n/C_{n-1}$ and contact matrices $r_n^M = M_n/M_{n-1}$. The dotted lines indicate the values the respective series converge to, $r_\infty^C = \mu_C \approx 4.68$ and $r_\infty^M = \mu_M \approx 4.38$, respectively. From [19].

simple cubic lattice a symmetry factor 6 for linear, 24 for planar, and 48 for conformations spreading into all three spatial directions. In Table 3.2 the numbers of designing sequences S_N are listed for the original HP model (1.10) and for the "mixed" HP model (1.12). In contrast to two-dimensional lattice proteins (e.g., on the square lattice [17]), the number of designing sequences obtained with the pure HP model is extremely small on the simple cubic lattice. This does not allow for a reasonable statistical study of general properties of designing sequences, at least for very short chains. The situation is much better using the more adequate MHP model.

The first quantity we want to investigate is the hydrophobicity of a sequence σ, i.e., the number of hydrophobic monomers N_H, normalized with respect to the total number of residues:

Table 3.2 Number of designing sequences S_N (only relevant sequences, see text) in the HP and MHP models.

N	4	5	6	7	8	9	10	11	12	13	14	15	16	17	18	19
S_N^{HP}	3	0	0	0	2	0	0	0	2	0	1	1	1	8	29	47
S_N^{MHP}	7	0	0	6	13	0	11	8	124	14	66	97	486	2196	9491	4885

$$m(\boldsymbol{\sigma}) = \frac{N_H}{N} = \frac{1}{N} \sum_{i=1}^{N} \sigma_i. \tag{3.2}$$

The average hydrophobicity over a set of designing sequences of given length N can then be introduced as

$$\langle m \rangle_N = \frac{1}{S_N} \sum_{\boldsymbol{\sigma} \in S_N} m(\boldsymbol{\sigma}). \tag{3.3}$$

The hydrophobicity distribution for all sequences is not binomial since we here distinguish only *relevant* sequences, i.e., two sequences that are symmetric under reversal of their residues enter only once into the statistics. There are, for example, only 10 relevant sequences with length $N = 4$ instead of $2^4 = 16$. Taking into account *all* 2^N sequences would obviously lead to a binomial distribution for N_H, since there are then exactly

$$\binom{N}{N_H} \tag{3.4}$$

sequences with N_H hydrophobic monomers.

In Fig. 3.2(a), the distribution of hydrophobicity is plotted for the *designing sequences* with $N = 18$ monomers in the MHP model and, for comparison, for *all sequences* with $N = 18$. For this example, we see that the width of the hydrophobicity distribution for the designing sequences, which has its peak at $\langle m \rangle_{18}^{\mathrm{MHP}} \approx 0.537 > 0.5$, is smaller than that of the distribution over all sequences. In order to gain more insight into how the hydrophobicity distributions differ, we compare the widths of both distributions in their dependence on the chain length $N \leq 19$. This is shown in Fig. 3.2(b). It seems that for $N \to \infty$ the widths of the hydrophobicity distributions for the designing sequences asymptotically approach the curve of the widths of the hydrophobicity distributions of all sequences.

After having discussed sequential properties of designing sequences, we now analyze the properties of their unique ground-state structures, the native conformations. From Table 3.3 we see that the number of *different* native conformations D_N is usually much smaller than the number of designing sequences, i.e., several designing sequences share the same ground-state conformation. The number of designing sequences that fold into a certain given target conformation $\mathbf{X}^{(0)}$ (or conformations being trivially symmetric to this by translations, rotations, and reflections) is called *designability* [71]:

$$F_N(\mathbf{X}^{(0)}) = \sum_{\boldsymbol{\sigma} \in S_N} \Delta \left(\mathbf{X}_{\mathrm{gs}}(\boldsymbol{\sigma}) - \mathbf{X}^{(0)} \right), \tag{3.5}$$

Table 3.3 Number of designable conformations D_N in both models.

N	4	5	6	7	8	9	10	11	12	13	14	15	16	17	18	19
D_N^{HP}	1	0	0	0	2	0	0	0	2	0	1	1	1	8	28	42
D_N^{MHP}	1	0	0	2	2	0	5	6	30	8	31	58	258	708	1447	1623

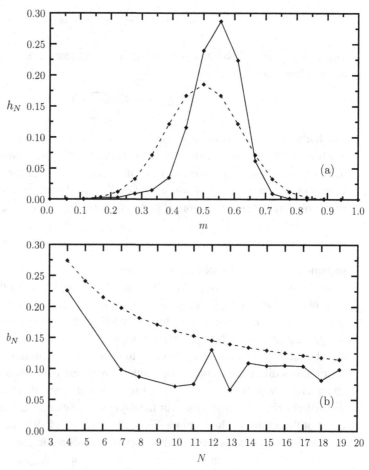

Fig. 3.2 (a) Distribution of hydrophobicity h_N of all designing sequences with $N = 18$ monomers (solid line) compared with the distribution of hydrophobicity of all sequences of this length (dashed line) for the MHP model. (b) Widths of the hydrophobicity distribution of the designing sequences, b_N, depending on the chain length N (solid line) compared with the widths of the hydrophobicity distribution of all sequences (dashed line) for the MHP model. From [19].

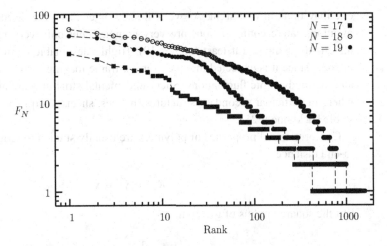

Fig. 3.3 Designability F_N of native conformations in the MHP model for $N = 17, 18$, and 19. The abscissa is the rank obtained by ordering all designable conformations according to their designability. From [19].

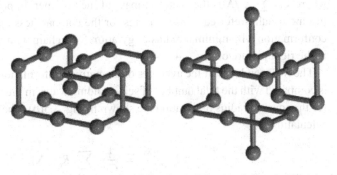

Fig. 3.4 Structure ($N = 18$) with the highest designability of all native conformations (left) and with minimal radius of gyration (right). From [19].

where $\mathbf{X}_{gs}(\boldsymbol{\sigma})$ is the native (ground-state) conformation of the designing sequence $\boldsymbol{\sigma}$. The function $\Delta(\mathbf{Z})$ is the generalization of Eq. (1.8) to $3N$-dimensional vectors. It is unity for $\mathbf{Z} = \mathbf{0}$ and zero otherwise.

The designability is plotted in Fig. 3.3 for all native conformations that HP proteins with $N = 17, 18$, and 19 monomers can form in the MHP model. In this figure, the abscissa is the rank of the conformations, ordered according to their designability. The conformation with the lowest rank is therefore the most designable structure and we see that a majority of the designing sequences fold into a small number of highly designable conformations, while only a small number of designing sequences possesses a native conformation with low designability (note that the plot is logarithmic). The left picture in Fig. 3.4 shows the conformation with the lowest rank (or highest designability) with $N = 18$ monomers. The designability distribution is similar to what is known from lattice proteins which are restricted to exist on cuboid lattices [72]. However, the characteristic distribution of the designing sequences is not restricted to cuboid lattices only. This result is less trivial than one may think at first sight. As we will show later on in the discussion of the radius of gyration, native conformations are very compact, but only very few conformations are maximally compact (at least for $N \leq 19$). Highly designable conformations are of great interest, since it is expected that they form a frame making them stable against mutations and thermodynamic fluctuations. Such fundamental structures are also relevant in nature, where in particular secondary structures (helices, sheets, hairpins) supply proteins with a stable backbone [72].

Conformational properties of polymers are usually studied in terms of the squared end-to-end distance

$$R_e^2 = (\mathbf{x}_N - \mathbf{x}_1)^2 \tag{3.6}$$

and the squared radius of gyration

$$R_g^2 = \frac{1}{N} \sum_{i=1}^{N} (\mathbf{x}_i - \bar{\mathbf{x}})^2, \tag{3.7}$$

where $\bar{\mathbf{x}} = \sum_i \mathbf{x}_i / N$ is the center of mass of the polymer. In polymer physics both quantities are usually referred to as measures for the compactness of a conformation. A typical conformation with minimal radius of gyration for a chain with $N = 18$ monomers is shown in the right picture of Fig. 3.4.

The N-dependence of the averages of the native conformations found in the MHP model is compared with the total number of self-avoiding walks in Fig. 3.5(a). The same quantities for the squared radius of gyration are shown in Fig. 3.5(b). The averages were obtained by calculating

$$\langle R_{e,g}^2 \rangle^{\text{SAW}} = \frac{1}{C_N} \sum_{\mathbf{X} \in \mathbf{C}_N} R_{e,g}^2(\mathbf{X}), \tag{3.8}$$

$$\langle R_{e,g}^2 \rangle^{\text{MHP}} = \frac{1}{S_N} \sum_{\sigma \in \mathbf{S}_N} R_{e,g}^2(\mathbf{X}_{\text{gs}}(\sigma)), \tag{3.9}$$

where $\mathbf{C}_N$ is the set of all self-avoiding conformations on an sc lattice. Figure 3.5(a) shows for the mixed HP model that, compared to $\langle R_{e,g}^2 \rangle^{\text{SAW}} \sim n^{2\nu}$ with the famous value $\nu \approx 0.59$ for self-avoiding walks, the average end-to-end distance $\langle R_{e,g}^2 \rangle^{\text{MHP}}$ of the native conformations only is much smaller. For an even number of monomers, the ends of an HP protein can form contacts with each other on the sc lattice. Accordingly, the values of $\langle R_{e,g}^2 \rangle^{\text{MHP}}$ are smaller for N being even and the even-odd oscillations are very pronounced. The widths (or standard deviations) $b_{R_e^2}$ of the distributions of the squared end-to-end distances are also very small. Even for heteropolymers with $N = 19$ monomers in total, there are virtually no native conformations, where the distance between the ends is larger than three lattice sites. This is the same for the standard HP model. Since the number of native conformations is very small in this model, these results are not included in the figure. Depicting the average squared radius of gyration $\langle R_g^2 \rangle$ and the widths of the corresponding distribution of the radius of gyration in Fig. 3.5(b) for all self-avoiding conformations as well as for the native ones, we see that these plots confirm the above remarks. As the average end-to-end

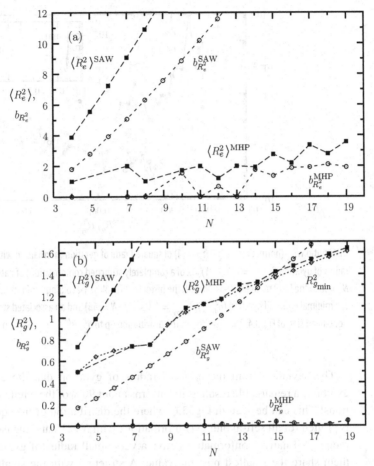

(a) Average squared end-to-end distances $\langle R_e^2 \rangle$ of native conformations in the MHP model compared with those of all self-avoiding walks (SAW). We have also inserted the widths $b_{R_e^2}$ of the corresponding distributions of end-to-end distances. (b) The same for the average squared radius of gyration $\langle R_g^2 \rangle$. Since the radius of gyration is an appropriate measure for the compactness of a conformation, we have also plotted $R_{g_{\min}}^2$ for the conformations with the minimal radius of gyration (or, equivalently, maximal compactness). From [19].

distances of native conformations are much smaller than those for the bulk of all conformations, the same trend can be observed for the mean squared radii of gyration $\langle R_g^2 \rangle^{\mathrm{MHP}}$ and $\langle R_g^2 \rangle^{\mathrm{SAW}}$ and the widths $b_{R_g^2}^{\mathrm{MHP}}$ and $b_{R_g^2}^{\mathrm{SAW}}$ as well. In particular, the width $b_{R_g^2}^{\mathrm{MHP}}$ is so small that virtually all native conformations possess the same radius of gyration. For this reason, it is useful to search for the conformations having the smallest radius of gyration $R_{g_{\min}}^2$ (these conformations are not necessarily native as we will see!) and to insert these values into this figure, too. We observe that these values differ only slightly from $\langle R_g^2 \rangle^{\mathrm{MHP}}$. Thus we may conclude that native conformations are very compact, but not necessarily maximally compact.

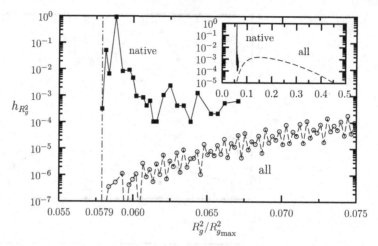

Fig. 3.6 Distribution $h_{R_g^2}$ (normalized to $\sum h_{R_g^2} = 1$) of squared radii of gyration (normalized with respect to the maximal radius of gyration $R_{g_{max}}^2 = (N^2 - 1)/12$ of a completely stretched conformation) of native conformations with $N = 18$ in the MHP model, compared with the histogram for all self-avoiding conformations. The vertical line refers to the minimal radius of gyration ($R_{g_{min}}^2/R_{g_{max}}^2 = 0.0579$ for $N = 18$) and an associated structure is shown on the right-hand side of Fig. 3.4. The inset shows the distribution up to $R_g^2/R_{g_{max}}^2 = 0.5$. From [19].

The deviation from the minimal radius of gyration that those native conformations exhibit is a remarkable result as it concerns about 90% of the whole set of native conformations! This can be seen in Fig. 3.6, where the distribution of the squared radii of gyration is plotted for all self-avoiding conformations with $N = 18$ and the native states in the MHP model. All native conformations have a very small radius of gyration, but only a few of them share the smallest possible value. A structure with the smallest radius of gyration is shown in the right-hand side of Fig. 3.4. It obviously differs from the most-designable conformation depicted in the left-hand panel of the same figure.

3.4 Exact density of states and thermodynamics

Returning to the simpler HP model (1.10), we now discuss thermodynamic properties of designing and nondesigning sequences. It has been conjectured [73] for exemplified sequences of comparable 14-mers, one of them being designing, that designing sequences in the HP model seem to show up a much more pronounced low-temperature peak in the specific heat than the nondesigning examples. This peak may be interpreted as kind of a conformational transition between structures with compact hydrophobic cores (ground states) and states where the whole conformation is highly compact (globules) [39, 40]. Another peak in the specific heat at higher temperatures, which is exhibited by all lattice proteins, is an indication for the usual globule–coil transition between compact and untan-

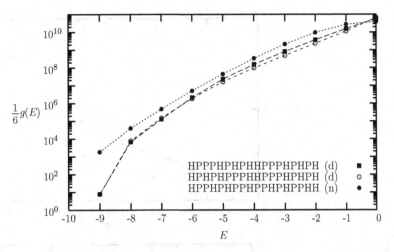

Fig. 3.7   Density of states $g(E)$ for two designing sequences (d) with $N = 18$, $m_H = 8$, and $E_{min} = -9$ in the HP model. We have divided out the symmetry factor 6 that is common to all conformations. Three-dimensional conformations have an additional symmetry factor 8, such that the states with minimal energy for these two curves are indeed unique and the sequences are designing. For comparison we have also plotted $g(E)$ for one exemplified nondesigning sequence (n) out of 525 having the same properties as quoted above, but different sequences. The ground-state degeneracy for this example is $g_0 = g(E_{min}) = 6 \times 1840$ (including all symmetries). From [19].

gled conformations. We will return to this point again when we discuss conformational transitions of proteins in more detail later in this book.

In order to study energetic thermodynamic quantities such as mean energy and specific heat, we can determine from the conformations of an HP protein with a given sequence the density of states $g(E)$ that conveniently enables the calculation of the partition sum $Z(T) = \sum_E g(E) \exp(-E/k_B T)$ and the moments $\langle E^k \rangle_T = \sum_E E^k g(E) \exp(-E/k_B T)/Z$, where the subscript T indicates the difference of calculating thermal mean values based on the Boltzmann probability from averages previously introduced in this section. Then, the specific heat as a function of temperature is given by the fluctuation formula $C_V(T) = (\langle E^2 \rangle_T - \langle E \rangle_T^2)/k_B T^2$.

In the HP model with pure hydrophobic interaction, the density of states exhibits a monotonic growth with increasing energy, at least for the very short chains. For a reasonable comparison of the behavior of designing and nondesigning sequences, we focus here on 18-mers having the same hydrophobicity ($m_H = 8$) and ground-state energy $E_{min} = -9$. There are in total 527 sequences with these properties, only two of which are designing. The densities of states for the two designing sequences and an example of a nondesigning sequence are plotted in Fig. 3.7. A global symmetry factor 6 (number of possible directions for the link connecting the first two monomers) that all conformations on an sc lattice have in common has already been divided out. Since the ground-state conformations of the designing sequences spread into all three dimensions, an additional symmetry factor $4 \times 2 = 8$ (4 for rotations around the first bond, 2 for a remaining independent reflection) makes

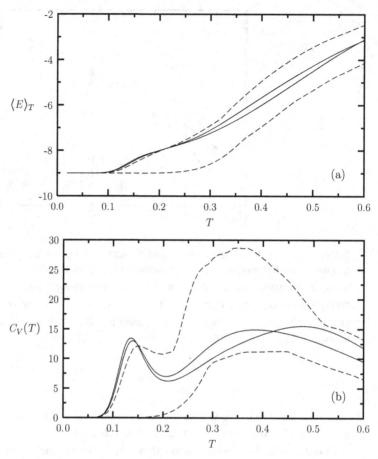

Fig. 3.8 (a) Mean energy $\langle E \rangle_T$ and (b) specific heat C_V for the two designing sequences with $N = 18$, $m_H = 8$, and $E_{min} = -9$ (solid lines) in the HP model, whose densities of states were plotted in Fig. 3.7. The curves of the same quantities for the 525 nondesigning sequences are completely included within the respective areas between the dashed lines. The low-temperature peak of the specific heat (near $T = 0.14$) is most pronounced for the two designing sequences which behave similarly for low temperatures. From [19].

a number of conformations obsolete and the ground-state degeneracy of the designing sequences is indeed unity.

Obviously this is not the case for the sequences we identified as nondesigning. In fact, the uniqueness of the ground states of designing sequences is a remarkable property as there are not less than $\sim 10^{10}$ possible conformations of HP lattice proteins with 18 monomers. As we also see in Fig. 3.7, the ratio of the density of the first excited state ($E = -8$) for the designing and the nondesigning sequences is smaller than for the ground state. This means that, at least for these short chains, the low-temperature behavior of the HP proteins in this model strongly depends on the degeneracy of the ground state. Furthermore, we expect that the low-temperature behavior of both designing sequences is very similar as their low-energy densities hardly differ. We can investigate this, once more for the 18-mers with the properties described above, by considering the mean energy $\langle E \rangle_T$ and the specific heat $C_V(T)$. The results are shown in Figs. 3.8(a) and 3.8(b), respectively. The two solid curves belong to the two designing sequences and the dashed lines are the minimum/maximum bounds of the respective quantities for the nondesigning sequences. The central conclusion is that designing and nondesigning sequences do indeed behave differently for very low temperatures. There is a characteristic, pronounced low-temperature peak in the specific heat that can be interpreted as a transition between low-energy states with hydrophobic core and very compact globules.

The upper bound of the specific heats for nondesigning sequences in Fig. 3.8(b) exposes two peaks. By analyzing all 525 nondesigning sequences we find that there are two groups: some of them experience two conformational transitions, while others do not show a characteristic low-temperature behavior. Thus, the only appearance of these two peaks is not a unique, characteristic property of designing sequences. In order to quantify this observation, we study all relevant 32 896 sequences with 16 monomers. Only one of these sequences is designing ($HP_2HP_2HPHPH_2PHPH$, with minimum energy $E_{\min} = -9$), but in total there are 593 sequences, i.e., 1.8% of the relevant sequences, corresponding to curves of specific heats with two local maxima. It should be noted that the degeneracies of the ground states associated with these sequences are comparatively small.

As we have seen in this chapter, exact data obtained in enumeration studies of conformation *and* sequence space, which are only possible for lattice peptides, are extremely helpful for a generic study of the relationship between sequence design and folding properties. Although the chains we discussed here are very short and their structural behavior is to some extent influenced by properties of the underlying discrete embedding space (we only explored the simple-cubic lattice here), we have found first remarkable insights into the structural and thermodynamic behavior of heterogeneous polymers. It should be emphasized once more that currently discrete lattice representations of simplified protein models are the only way to gather comprising information about *general* properties that are associated with effects due to sequence disorder. This problem ranks at a level of complexity similar to the spin-glass problem, where the coupling of magnetic moments can be either ferromagnetic or antiferromagnetic, which can cause unresolvable conflicts (frustration). As we have outlined in the beginning, several questions addressed in this chapter, such as the effect of mutations of the amino-acid sequence on structural properties and stability, are related to a general problem of understanding the evolutionary changes biopolymers have experienced over millions of years. Finding answers to these questions is certainly one of the most fascinating scientific problems of our time.

Enumeration is a powerful method for the investigation of some interesting features of heteropolymer folding in an evolutionary context, because it creates exact data. However, it is obviously only applicable to small lattice systems. The investigation of longer chains and more specific models requires more sophisticated stochastic approaches, which unfortunately can only generate approximate results. We will introduce some of the standard methods in the following chapter, before we return to the discussion of structural properties of polymers and proteins.

4 Monte Carlo and chain growth methods for molecular simulations

4.1 Introduction

For a system under thermal conditions in a heat bath with temperature T, the dynamics of each of the system particles is influenced by interactions with the heat-bath particles. If quantum effects are negligible (what we will assume in the following), the classical motion of any system particle looks erratic; the particle follows a stochastic path. The system can "gain" energy from the heat bath by these collisions (which are typically more generally called "thermal fluctuations") or "lose" energy by friction effects (dissipation). The total energy of the coupled system of heat bath and particles is a conserved quantity, i.e., fluctuation and dissipation refer to the energetic exchange between heat bath and system particles only. Consequently, the coupled system is represented by a *microcanonical ensemble*, whereas the particle system is in this case represented by a *canonical ensemble*: The energy of the particle system is not a constant of motion. Provided heat bath and system are in thermal equilibrium, i.e., heat-bath and system temperature are identical, fluctuations and dissipation balance each other. This is the essence of the celebrated fluctuation-dissipation theorem [74]. In equilibrium, only the statistical mean of the particle system energy is constant in time.

This *canonical* behavior of the system particles is not accounted for by standard Newtonian dynamics (where the system energy is considered to be a constant of motion). In order to perform molecular dynamics (MD) simulations of the system under the influence of thermal fluctuations, the coupling of the system to the heat bath is required. This is provided by a thermostat, i.e., by extending the equations of motion by additional heat-bath coupling degrees of freedom [75]. The introduction of thermostats into the dynamics is a notorious problem in MD and it cannot be considered to be solved satisfactorily to date [76]. In order to take into consideration the stochastic nature of any particle trajectory in the heat bath, a typical approach is to introduce random forces into the dynamics. These forces represent the collisions of system and heat-bath particles on the basis of the fluctuation-dissipation theorem.

Unfortunately, MD simulations of complex systems on microscopic and mesoscopic scales are extremely slow, even on the largest available computers. A prominent example is the folding of proteins with natural timescales of milliseconds to seconds. It is currently still impossible to simulate folding events of bioproteins under realistic conditions, since even the longest MD runs are hardly capable of generating single trajectories of more than a few microseconds. Consequently, if the intrinsic timescale of a realistic model exceeds the timescale of an MD simulation of this model, MD cannot seriously be applied in these cases.

However, many interesting questions do not require consideration of the intrinsic dynamics of the system explicitly. This regards, e.g., equilibrium thermodynamics, which includes all relevant phenomena of cooperativity – the collective original source for the occurrence of phase transitions. Stability of all matter, independently whether soft or solid, requires fundamental ordering principles. We are far away from having understood the *general physical properties* of transition processes that separate, e.g., ordered and disordered phases, crystals and liquids, glassy and globular polymers, native and intermediate protein folds, ferromagnetic and paramagnetic states of metals, Bose–Einstein condensates and bosonic gases, etc. Meanwhile, the history of research of collective or critical phenomena has already lasted for more than 100 years and the universality hypothesis has already been known for several decades [77]. Yet no complete theory exists that is capable of relating to other phenomena such as protein folding (unfolding) and freezing (melting) of solid matter. The reason is that the first process is dominated by finite-size effects, whereas the latter seems to be a macroscopic "bulk" phenomenon. However, although doubtless associated to different length scales which differ by orders of magnitude, both examples are based on cooperativity, i.e., the collective multibody interplay of a large number of atoms. Precise theoretical analyses are extremely difficult, even more so if several attractive and repulsive interactions compete with each other and if the system does not possess any obvious internal symmetries (which is particularly apparent for "glassy" heteropolymers like proteins).

On the experimental side, the situation has not been much better as the resolution of the data often did not allow an in-depth analysis of the simultaneous microscopic effects accompanying cooperative phenomena. This has dramatically improved by novel experimental techniques enabling measurement of the response of the system to local manipulations, giving insight into the mesoscopic and macroscopic multibody effects upon activation. On the other hand, a systematic understanding requires a theoretical basis. The relevant physical forces have been known for a long time, but the efficient combination of technical and algorithmic prerequisites has been missing until recently. The general understanding of cooperativity in complex systems as a statistical effect, governed by a multitude of forces acting on different energy and length scales, requires the study of the interplay of entropy and energy.

The key to this understanding is currently provided only by Monte Carlo computer simulations [78].

4.2 Conventional Markov-chain Monte Carlo sampling

4.2.1 Ergodicity and finite time series

The general idea behind all Monte Carlo methodologies is to provide an efficient stochastic sampling of the configurational or conformational phase space or parts of it with the objective to obtain reasonable approximations for statistical quantities such as expectation values, probabilities, fluctuations, correlation functions, densities of states, etc.

A given system conformation (e.g., the geometric structure of a molecule) $\mathbf{X}$ is locally or globally modified to yield a conformation $\mathbf{X}'$. This update or "move" is then accepted with the transition probability $t(\mathbf{X} \to \mathbf{X}')$. Frequently used updates for polymer models are, for example, random translational changes of single monomer positions, bond angle modifications, or rotations about covalent bond axes. More global updates consist of combined local updates, which can be necessary to satisfy constraints such as fixed bond lengths or simply to improve efficiency. It is, however, a necessary condition for correct statistical sampling that Monte Carlo moves are ergodic, i.e., the chosen set of moves must, in principle, guarantee to reach any conformation out of any other conformation. Since this is often hard to prove and an insufficient choice of move sets can result in systematic errors, great care must be dedicated to choosing appropriate moves or sets of moves. Since molecular models often contain constraints, the construction of global moves can be demanding. Therefore, reasonable and efficient moves have to be chosen in correspondence to the model of a system to be simulated. We will discuss elementary updates for polymer and protein models in Section 4.7.

A Monte Carlo update corresponds to the discrete "time step" $\Delta\tau_0$ in the simulation process. In order to reduce correlations, typically a number of updates are performed between measurements of a quantity O. This series of updates is called a "sweep" and the "time" passed in a single sweep is $\Delta\tau = N\Delta\tau_0$ if the sweep consists of N updates. Thus, if M sweeps are performed, the discrete "time series" is expressed by the vector $(O(\tau_{\mathrm{init}} + \Delta\tau), O(\tau_{\mathrm{init}} + 2\Delta\tau), \ldots, O(\tau_{\mathrm{init}} + m\Delta\tau), \ldots, O(\tau_{\mathrm{init}} + M\Delta\tau))$ and represents the Monte Carlo trajectory. The period of equilibration τ_{init} sets the starting point of the measurement. For convenience, we use the abbreviation $O_m \equiv O(\tau_{\mathrm{init}} + m\Delta\tau)$ and $\tau_m = \tau_{\mathrm{init}} + m\Delta\tau$ with $m = 1, 2, \ldots, M$ in the following. A typical example of an energetic time series in equilibrium in the random-coil phase of a polymer model is shown in Fig. 4.1.

According to the theory of ergodicity, averaging a quantity over an infinitely long time series is identical to performing the statistical ensemble average:

$$\overline{O} = \lim_{M\to\infty} \frac{1}{M} \sum_{m=1}^{M} O_m \equiv \langle O \rangle = \int \mathcal{D}X O(\mathbf{X}) p(\mathbf{X}), \qquad (4.1)$$

where $\mathcal{D}X$ represents the formal integral measure for the infinitesimal scan of the conformation space and $p(\mathbf{X})$ is the energy-dependent microstate probability of the conformation $\mathbf{X}$ in the relevant ensemble in thermodynamic equilibrium (in the canonical ensemble with temperature T, simply $p(\mathbf{X}) = \exp[-E(\mathbf{X})/k_{\mathrm{B}}T]$). This is the formal basis for Monte Carlo sampling. However, only finite time series can be simulated on a computer. For a finite number of sweeps M in a sample k, the relation (4.1) can only be satisfied approximately, $M^{-1} \sum_{m=1}^{M} O_m^{(k)} = \overline{O}^{(k)} \approx \langle O \rangle$. Note that the mean value $\overline{O}^{(k)}$ will depend on the sample k, meaning it is likely that another sample k' will yield a different value $\overline{O}^{(k')} \neq \overline{O}^{(k)}$.

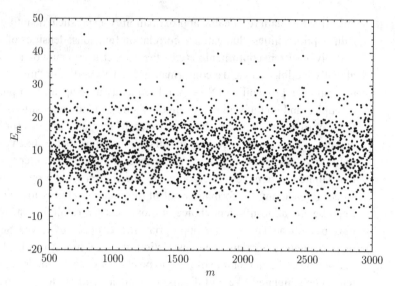

Fig. 4.1 Typical energetic time series from Metropolis Monte Carlo simulations of a polymer model in equilibrium in the random-coil phase. At fixed temperature, the fluctuation width of individual data points E_m about the mean value $\overline{E}$ is proportional to the heat capacity.

We keep this in mind in the following, omit the superscript, and redefine

$$\overline{O} := \frac{1}{M} \sum_{m=1}^{M} O_m \qquad (4.2)$$

as the symbol for the sample average of the quantity O, obtained in any finite time series.

4.2.2 Statistical error and bias

In order to define a reasonable estimator for the statistical error, it is necessary to start from the assumption that an infinite number of independent samples k has been generated. In this case, the distribution of the estimates $\overline{O}^{(k)}$ is Gaussian, according to the central limit theorem of uncorrelated samples. The exact average of the estimates is then given by $\langle \overline{O} \rangle$. It immediately follows that

$$\langle \overline{O} \rangle = \frac{1}{M} \sum_{m=1}^{M} \langle O_m \rangle = \langle O \rangle, \qquad (4.3)$$

because in statistical equilibrium the exact average cannot depend on the time of measurement. The statistical error of $\overline{O}$ is suitably defined as the standard deviation of the Gaussian:

$$\varepsilon_{\overline{O}} = \pm \sqrt{\sigma_{\overline{O}}^2} = \pm \sqrt{\left\langle \left(\overline{O} - \langle \overline{O} \rangle \right)^2 \right\rangle} = \pm \sqrt{\langle \overline{O}^2 \rangle - \langle \overline{O} \rangle^2} = \pm \sqrt{\frac{1}{M^2} \sum_{m=1}^{M} \sum_{n=1}^{M} \mathcal{A}_{mn} \sigma_O^2},$$

$$(4.4)$$

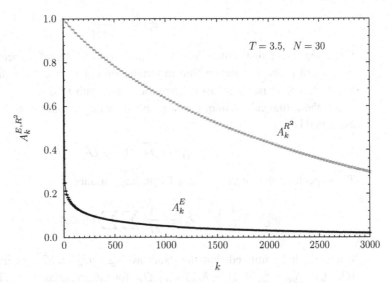

Fig. 4.2 Autocorrelation functions of energy and square radius of gyration for data obtained in Metropolis Monte Carlo simulations of a flexible polymer model in the random-coil phase [79].

where

$$\mathcal{A}_{mn} = \frac{\langle O_m O_n \rangle - \langle O_m \rangle \langle O_n \rangle}{\sigma_O^2} \tag{4.5}$$

is the normalized correlation function (normalized in the sense that $\mathcal{A}_{mm} = 1$) and

$$\sigma_O^2 = \langle O_m^2 \rangle - \langle O_m \rangle^2 \tag{4.6}$$

is the variance of the distribution of individual data O_m. In equilibrium, the variance does not depend on m. The translational invariance in equilibrium also simplifies the autocorrelation function (4.5), which can now be written as

$$A_k = \frac{\langle O_m O_{m+k} \rangle - \langle O_m \rangle^2}{\sigma_O^2}, \tag{4.7}$$

where, again, m indexes any point in the time series, but A_k is independent of time (in equilibrium only!). Figure 4.2 shows examples of autocorrelation functions for energetic and structural polymer data obtained by Metropolis Monte Carlo simulations. Apparently, the decay rate of correlations depends on the quantity measured.

If the Monte Carlo updates in each sample are performed completely randomly without memory, i.e., a new conformation is created independently of the one in the step before (which is a possible but typically very inefficient strategy), two measured values O_m and O_n are uncorrelated, if $m \neq n$. Then, the autocorrelation function simplifies to $\mathcal{A}_{mn} = \delta_{mn}$. Thus, the variances of the individual data and of the mean are related with each other by

$$\sigma_{\overline{O}}^2 = \frac{\sigma_O^2}{M} \tag{4.8}$$

and the statistical error of $\overline{O}$ is given by the celebrated relation

$$\varepsilon_{\overline{O}} = \pm \frac{\sigma_O}{\sqrt{M}}. \tag{4.9}$$

Since the exact distribution of O_m values and the "true" expectation value $\langle O \rangle$ are unchanged in the simulation (but unfortunately unknown), the standard deviation σ_O is constant, too. Thus, the statistical error decreases with $1/\sqrt{M}$.

For the actual calculation, it is a problem that σ_O^2 is unknown. However, what can be estimated is

$$\tilde{\sigma}_O^2 = \overline{(O - \overline{O})^2} = \overline{O^2} - \overline{O}^2. \tag{4.10}$$

The expected value of this so-called sampling variance is

$$\langle \tilde{\sigma}_O^2 \rangle = \frac{1}{M} \sum_{m=1}^{M} \langle O_m^2 \rangle - \frac{1}{M^2} \sum_{m=1}^{M} \sum_{n=1}^{M} \langle O_m O_n \rangle = \sigma_O^2 \left(1 - \frac{1}{M} \right), \tag{4.11}$$

where we have utilized that the exact average $\langle O_m^2 \rangle = \langle O^2 \rangle$ is independent of m and $\langle O_m O_n \rangle = \delta_{mn} \langle O_m^2 \rangle + (1 - \delta_{mn}) \langle O_m \rangle \langle O_n \rangle$ for uncorrelated data. The $1/M$ correction is the *systematic* error due to the finiteness of the time series, called bias. We define the bias-corrected estimator for the variance of individual data by

$$\sigma_{O,c}^2 = \frac{M}{M-1} \tilde{\sigma}_O^2. \tag{4.12}$$

Thus, the bias-corrected version for the statistical error of uncorrelated data [Eq. (4.9)] finally reads

$$\varepsilon_{\overline{O},c} = \pm \frac{\sigma_{O,c}}{\sqrt{M}} = \pm \sqrt{\frac{1}{M(M-1)} \sum_{m=1}^{M} (O_m - \overline{O})^2}. \tag{4.13}$$

In practice, most of the efficient Monte Carlo techniques generate correlated data. The assumption that a Gaussian approach can still be used to calculate the statistical error of correlated data is justified only if it is possible to select a set of uncorrelated data from the original data set. In the running simulation, one literally *waits* between two measurements until the correlation has sufficiently decayed. The effective statistics will be reduced by the waiting time, i.e., the autocorrelation time, which corresponds to the number of sweeps needed to let the correlations decay. If we consider the same total number of sweeps M as in the uncorrelated case, the error will be larger. We expect that it can be conveniently rewritten as

$$\varepsilon_{\overline{O}} = \pm \frac{\sigma_O}{\sqrt{M_{\text{eff}}}} \tag{4.14}$$

with the effective statistics $M_{\text{eff}} = M/\Delta\tau_{\text{ac}} \leq M$, where $\Delta\tau_{\text{ac}}$ is the autocorrelation time.

For the calculation of the autocorrelation time, we fall back to the general formula (4.4) and write it as

$$\sigma_{\overline{O}}^2 = \frac{1}{M^2} \sum_{m,n=1}^{M} (\langle O_m O_n \rangle - \langle O_m \rangle \langle O_n \rangle) = \frac{1}{M} \sigma_O^2 \left(1 + \frac{2}{M} \sum_{m=1}^{M} \sum_{n=m+1}^{M} A_{mn} \right), \tag{4.15}$$

where the diagonal part was split off and the symmetry of the correlation function $\mathcal{A}_{mn}$ was employed. Substituting $k = n - m$ in the last term yields

$$\sum_{m=1}^{M} \sum_{n=m+1}^{M} \mathcal{A}_{mn} = \sum_{m=1}^{M} \sum_{k=1}^{M-m} \mathcal{A}_{m\,m+k} = \sum_{m=1}^{M} \sum_{k=1}^{M-m} A_k = \sum_{k=1}^{M} \sum_{m=1}^{M-k} A_k = \sum_{k=1}^{M} A_k(M-k),$$

(4.16)

where we utilized first the time translation invariance of the autocorrelation function and second that the order in which the triangular summation is performed is arbitrary. The final result for the variance (4.15) reads

$$\sigma_{\overline{O}}^2 = \sigma_O^2 \frac{\Delta\tau_{ac}}{M},$$

(4.17)

with the (integrated) autocorrelation time

$$\Delta\tau_{ac} = 1 + 2 \sum_{k=1}^{M} A_k \left(1 - \frac{k}{M}\right).$$

(4.18)

Thus, the ansatz (4.14) has proven to be correct and $\varepsilon_{\overline{O}}$ and M_{eff} are completely determined. Note that for uncorrelated data $\Delta\tau_{ac} = 1$ as expected, because in this case $A_k = \delta_{0k}$.

If the data set is finite and σ_O^2 is unknown, the bias needs to be corrected and σ_O^2 has to be replaced by making use of $\langle\tilde{\sigma}_O^2\rangle = \sigma_O^2(1 - 1/M_{\text{eff}})$, in analogy to Eq. (4.11). The bias-corrected estimator of the variance of individual data now reads

$$\sigma_{O,c}^2 = \frac{M_{\text{eff}}}{M_{\text{eff}} - 1} \tilde{\sigma}_O^2$$

(4.19)

and that of the average is given by

$$\sigma_{\overline{O},c}^2 = \frac{\sigma_{O,c}^2}{M_{\text{eff}}}.$$

(4.20)

Eventually, the error of the average quantity $\overline{O}$, calculated for a finite, correlated data set, is given by

$$\varepsilon_{\overline{O},c} = \pm \frac{\sigma_{O,c}}{\sqrt{M_{\text{eff}}}} = \pm \sqrt{\frac{1}{M(M_{\text{eff}} - 1)} \sum_{m=1}^{M} (O_m - \overline{O})^2}.$$

(4.21)

Since it takes at least the time $\Delta\tau_{ac} = N_{ac}\Delta\tau_0$ to generate statistically independent conformations, a sweep can simply contain as many *updates* N_{ac} as necessary to satisfy $\Delta\tau \approx \Delta\tau_{ac}$ without losing effective statistics. In this case, the $M \approx M_{\text{eff}}$ data entering into the effective statistics are virtually uncorrelated. This is also the general idea behind advanced, computationally convenient error estimation methods such as binning and jackknife analyses described in the following section. For the correctness of the measurements, $M \approx M_{\text{eff}}$ is not a necessary condition; more sweeps with fewer updates in each sweep, i.e., periods between measurements shorter than $\Delta\tau_{ac}$, yield only redundant statistical information. This is not even wrong, but computationally inefficient as it does not improve the statistical error (4.14). This is certainly something one may keep in mind, since calculations of observables can be computationally costly.

4.2.3 Binning–jackknife error analysis

The error estimation by means of the formula (4.21) is in most cases laborious. First, the calculation of the autocorrelation time is time-consuming and tedious. Second, to estimate the statistical error of nonlinear functions of the mean $f(\overline{O})$ requires the consideration of error propagation. Fortunately, both problems can easily be solved approximately by the binning methods to be described now [80–82].

Binning method and estimation of the autocorrelation time

If the series of M correlated data points O_m ($m = 1, \ldots, M$) is cut into K pieces, each containing M^B data points such that $M = KM^B$, a set of effectively uncorrelated data subsets is generated, if $M^B \gg \Delta\tau_{ac}$. This is virtually the same as if K independent simulations (with different initializations, of course) had been performed, recording M^B data in each of them.

The average of the quantity O in each of the bins is then given by

$$\overline{O}_k^B = \frac{1}{M^B} \sum_{m^B=1}^{M^B} O_{(k-1)M^B+m^B}, \quad k = 1, \ldots, K, \tag{4.22}$$

and

$$\overline{O} = \frac{1}{K} \sum_{k=1}^{K} \overline{O}_k^B \tag{4.23}$$

obviously coincides with the average (4.2). The estimator for the variance of an individual average is in bias-corrected form:

$$\sigma_{\overline{O}^B,c}^2 = \frac{1}{K-1} \sum_{k=1}^{K} \left(\overline{O}_k^B - \overline{O}\right)^2, \tag{4.24}$$

and the variance of $\overline{O}$ is given by

$$\sigma_{\overline{O},c}^2 = \frac{1}{K}\sigma_{\overline{O}^B,c}^2. \tag{4.25}$$

Comparing this expression with Eq. (4.20), we find

$$M_{\text{eff}} = K\frac{\sigma_{\overline{O},c}^2}{\sigma_{\overline{O}^B,c}^2}. \tag{4.26}$$

Hence, the autocorrelation time can be estimated as follows:

$$\Delta\tau_{ac} = M^B \frac{\sigma_{\overline{O}^B,c}^2}{\sigma_{\overline{O},c}^2}. \tag{4.27}$$

Technically, this relation is much more convenient than (4.18), because an estimate for the autocorrelation function is not needed. For a given time series with a total of as much

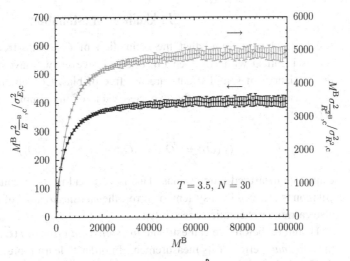

Right-hand side of Eq. (4.27) plotted as a function of the bin size M^B for time series of energy E and square radius of gyration R^2 obtained by Metropolis Monte Carlo simulations of a flexible polymer model in the random-coil phase [79]. Both curves converge to noticeably different values of integrated autocorrelation times (left scale for the energetic, right scale for the structural quantity). The data used for these plots are the same as for the autocorrelation functions in Fig. 4.2. The autocorrelation times obtained by binning analysis agree very well with the estimate obtained by integrating the autocorrelation function [according to Eq. (4.7)].

as M entries, $\sigma^2_{\overline{O},c}$ can be calculated and does not change if M^B is modified. This is also true, of course, for $\Delta\tau_{ac}$, which is an intrinsic property of the time series. Therefore, only $\sigma^2_{\overline{O}^B,c}$ depends on the bin size M^B, but in a way that $M^B\sigma^2_{\overline{O}^B,c} = \text{const}$, provided the binning condition $M^B \gg \Delta\tau_{ac}$ is satisfied. If we would trivially choose $M^B = 1$, $\sigma^2_{\overline{O}^B,c} = \sigma^2_{\overline{O},c}$, and consequently $\Delta\tau_{ac} = 1$. This is only true for uncorrelated data. For correlated data, the binning condition cannot be satisfied, i.e., too small bin sizes will underestimate the autocorrelation time. Increasing M^B reduces the number of bins K and because of this the variance $\sigma^2_{\overline{O}^B,c}$ will also decrease, but not by the same factor as M^B is increased. Therefore, we expect that the right-hand side of Eq. (4.27) converges to a constant value that is associated with $\Delta\tau_{ac}$. In practice, one plots the right-hand side of Eq. (4.27) for various values of M^B and obtains an estimate for $\Delta\tau^{est}_{ac}$ at a value M^B, where the curve apparently starts to converge. Figure 4.3 shows such curves for energetic and structural data obtained in polymer simulations. It is obvious that the values the curves converge to are different, i.e., the integrated autocorrelation time depends on the quantity considered.

With the estimated autocorrelation time, the effective number of uncorrelated data, $M^{est}_{eff} = M/\Delta\tau^{est}_{ac}$, can be estimated and eventually be used to calculate the error (4.21) of the average $\overline{O}$. This method is certainly not very accurate, but absolutely sufficient to obtain reliable error bars for $\overline{O}$ and linear functions of it ("the error of the error" is typically of no particular interest).

Jackknife error estimation

We have already found that the estimation of $\overline{O}$ is bias-free, i.e., its exact average $\langle \overline{O} \rangle$ is identical with $\langle O \rangle$ [see Eq. (4.3)]. However, we have also convinced ourselves that variances of sampled data are not free of bias. Combining Eqs. (4.10) and (4.11), we consequently find that the expected value of the simplest nonlinear function of $\overline{O}$, $f(\overline{O}) = \overline{O}^2$,

$$\langle f(\overline{O}) \rangle = \langle \overline{O}^2 \rangle = \langle O \rangle^2 + \frac{1}{M}\sigma_O^2 = f(\langle O \rangle) + \frac{1}{M}\sigma_O^2, \qquad (4.28)$$

cannot be measured without bias. This is very bad news, because not only the error calculation is affected by a systematic error; the *measured* value of a nonlinear function of $\overline{O}$ also carries a systematic uncertainty.

Therefore, before we can estimate the *statistical* error of $f(\overline{O})$, we first have to discuss the *systematic* error of its measurement. In order to learn more of the nature of the latter, we generalize the result (4.28) for the square of $\overline{O}$ to an arbitrary function $f(\overline{O})$. If we assume that the deviation $|\overline{O} - \langle O \rangle|$ is small if compared with $|\overline{O}|$, we may perform a Taylor expansion and write

$$
\begin{aligned}
\langle f(\overline{O}) \rangle - f(\langle O \rangle) &= \langle f(\langle O \rangle + \overline{O} - \langle O \rangle) \rangle - f(\langle O \rangle) \\
&= f'(\langle O \rangle)(\langle \overline{O} \rangle - \langle O \rangle) + \frac{1}{2}f''(\langle O \rangle)\langle(\overline{O} - \langle O \rangle)^2\rangle + \cdots \\
&= \frac{1}{2}f''(\langle O \rangle)\sigma_{\overline{O}}^2 + \cdots = \frac{1}{2M}f''(\langle O \rangle)\sigma_O^2 + \cdots, \qquad (4.29)
\end{aligned}
$$

where we have made use of Eqs. (4.3) and (4.8).[1] We can conclude that indeed linear functions of $\overline{O}$ can be calculated for finite time series without bias, but the estimate of any nonlinear function of $\overline{O}$ inevitably deviates systematically from the expected result.[2]

Because the values of the derivatives of $f(\langle O \rangle)$ are unknown and there are also higher-order bias terms, it is not obvious from Eq. (4.29) how to correct the bias. At least, it is reassuring to realize that it decreases with $1/M$, i.e., for sufficiently large time series, the estimate $f(\overline{O})$ is reasonable. We will see later on that one can indeed introduce an estimator that reduces the bias to the order of $1/M^2$. This is made possible by the jackknife resampling method, which will also help us estimate the statistical error.

In a sense, the jackknife method is complementary to the binning method. Instead of calculating the average of an observable O in a single bin, the jackknife uses the data of all bins except one. We therefore define the kth individual jackknife average by taking all data, reduced by those in the kth bin:

$$\overline{O}_k^{\mathrm{J}} = \frac{M\overline{O} - M^{\mathrm{B}}\overline{O}_k^{\mathrm{B}}}{M - M^{\mathrm{B}}}. \qquad (4.30)$$

---

[1] We have also utilized that $\langle f(\langle O \rangle) \rangle = f(\langle O \rangle)$, which is correct because the argument of f is already a constant such that f is not changed if replaced by $\langle f \rangle$. However, it must also be noted that generally $\langle f(O) \rangle \neq f(\langle O \rangle)$.

[2] If we would measure $\overline{f(O)}$ instead of $f(\overline{O})$, the result would be even worse. One can show that in this case $\langle \overline{f(O)} \rangle = f(\langle O \rangle) + (1/2)f''(\langle O \rangle)\sigma_O^2 + \ldots$, i.e., the bias does not even vanish for infinitely large time series ($M \to \infty$). Effectively, this means that $\overline{f(O)}$ is not an appropriate estimator for $f(\langle O \rangle)$ at all.

The mean of O is obtained by averaging $\overline{O}_k^J$ over all jackknife bins:

$$\overline{O} = \frac{1}{K} \sum_{k=1}^{K} \overline{O}_k^J. \tag{4.31}$$

Both averages satisfy Eq. (4.3), i.e., $\langle \overline{O}_k^J \rangle = \langle \overline{O} \rangle = \langle O \rangle$. Since almost all data are reused in the calculation of the individual jackknife averages, i.e., the jackknife bins have large overlap, the averages will be strongly correlated, but in a trivial way. Most importantly at this point, these strong correlations and the large data sets used let the individual means $\overline{O}_k^J$ hardly deviate from each other, but also from $\overline{O}$. Therefore, we can make the same assumption that led us to Eq. (4.29), namely that $|\overline{O}_k^J - \langle O \rangle| / |\overline{O}_k^J| \ll 1$. We can immediately adopt the result (4.29) to find

$$\langle f(\overline{O}_k^J) \rangle - f(\langle O \rangle) = \frac{1}{2} f''(\langle O \rangle) \sigma_{\overline{O}^J}^2 + \dots, \tag{4.32}$$

where the exact variance of an individual jackknife average is given by

$$\sigma_{\overline{O}^J}^2 = \left\langle \left(\overline{O}_k^J - \langle O \rangle \right)^2 \right\rangle = \frac{M}{M - M^B} \sigma_{\overline{O}}^2 = \frac{1}{M - M^B} \sigma_O^2. \tag{4.33}$$

Equation (4.32) will not change, if we replace $f(\overline{O}_k^J)$ by its jackknife average

$$\bar{f}^J = \frac{1}{K} \sum_{k=1}^{K} f(\overline{O}_k^J), \tag{4.34}$$

because $\langle \bar{f}^J \rangle = \langle f(\overline{O}_k^J) \rangle$. If we compare

$$\langle \bar{f}^J \rangle - f(\langle O \rangle) = \frac{1}{2(M - M^B)} f''(\langle O \rangle) \sigma_O^2 + \dots, \tag{4.35}$$

with Eq. (4.29), we see that the structural similarities of the right-hand sides can be utilized to introduce an estimator

$$\tilde{f}(\overline{O}) = \frac{1}{M^B} \left[M f(\overline{O}) - (M - M^B) \bar{f}^J \right] = K f(\overline{O}) - (K-1) \bar{f}^J, \tag{4.36}$$

which has a reduced bias of order $1/M^2$.

For the estimation of the error of $f(\overline{O})$, we must first calculate the variance

$$\sigma_{f(\overline{O})}^2 = \left\langle \left[f(\overline{O}) - \langle f(\overline{O}) \rangle \right]^2 \right\rangle. \tag{4.37}$$

Taking into account again that $\overline{O}$ and $\langle O \rangle$ differ only slightly, we can Taylor-expand the functions, yielding:

$$\sigma_{f(\overline{O})}^2 = \left\langle \left[f(\langle O \rangle) + f'(\langle O \rangle)(\overline{O} - \langle O \rangle) + \dots - \langle f(\langle O \rangle) \rangle - f'(\langle O \rangle)(\langle \overline{O} \rangle - \langle O \rangle) - \dots \right]^2 \right\rangle$$

$$= [f'(\langle O \rangle)]^2 \sigma_{\overline{O}}^2 + \dots = \frac{1}{M} [f'(\langle O \rangle)]^2 \sigma_O^2 + \dots \tag{4.38}$$

In the evaluation of the expression in the first line, we have once more utilized that $\langle f(\langle O \rangle) \rangle = f(\langle O \rangle)$ and $\langle \overline{O} \rangle = \langle O \rangle$. The result (4.38) is not yet particularly helpful, because $f'(\langle O \rangle))$ is unknown to us. We introduce the estimator for the variance of $f(\overline{O}^J)$ by

$$\sigma_{\bar{f}^J}^2 = \overline{\left[f(\overline{O}^J) - \bar{f}^J \right]^2} = \frac{1}{K} \sum_{k=1}^{K} \left[f(\overline{O}_k^J) - \bar{f}^J \right]^2 \qquad (4.39)$$

and calculate its expected value. To this end, we expand $f(\overline{O}^J)$ about $\langle O \rangle$ by making use of $\overline{O}_k^J \approx \langle O \rangle$ and obtain:

$$\langle \sigma_{\bar{f}^J}^2 \rangle = [f'(\langle O \rangle)]^2 \langle \tilde{\sigma}_{\overline{O}^J}^2 \rangle + \dots, \qquad (4.40)$$

where the expected value $\langle \tilde{\sigma}_{\overline{O}^J}^2 \rangle$ for the measured variance of $\overline{O}^J$ is similar, but not identical to the exact jackknife variance given in Eq. (4.33). We find

$$\langle \tilde{\sigma}_{\overline{O}^J}^2 \rangle = \left\langle \left(\overline{O}_k^J - \overline{O} \right)^2 \right\rangle = \frac{1}{K-1} \sigma_{\overline{O}}^2 = \frac{1}{M(K-1)} \sigma_O^2 = \frac{1}{K} \sigma_{\overline{O}^J}^2. \qquad (4.41)$$

Inserting this result into Eq. (4.40) yields

$$(K-1) \langle \sigma_{\bar{f}^J}^2 \rangle = \frac{1}{M} [f'(\langle O \rangle)]^2 \langle \sigma_O^2 \rangle + \dots. \qquad (4.42)$$

Since the leading term on the right-hand side coincides with that of Eq. (4.38), we conclude that

$$\varepsilon_{f(\overline{O})} = \sqrt{(K-1)\sigma_{\bar{f}^J}^2} = \sqrt{\frac{K-1}{K} \sum_{k=1}^{K} \left[f(\overline{O}_k^J) - \bar{f}^J \right]^2} = \sqrt{\sigma_{f(\overline{O})}^2} + \mathcal{O}(1/M) \qquad (4.43)$$

is an optimal estimator for the error of $f(\overline{O})$, because its bias is only of order $1/M$.[3] For the frequently needed special case $f(\overline{O}) = \overline{O}$, Eq. (4.43) becomes

$$\varepsilon_{\overline{O}} = \sqrt{\frac{K-1}{K} \sum_{k=1}^{K} \left(\overline{O}_k^J - \overline{O} \right)^2}. \qquad (4.44)$$

Our considerations in the last section are founded on basic elements of probability theory and we did not have to make special assumptions about the time series or data sets. Having said that and having taken into account that all data sets we can create are finite, the entire problem of statistical errors and systematic bias affects *all* data obtained in experiments and computer simulations. The error estimation methods discussed in this section are sufficient to judge the quality of these data.[4]

---

[3] Note that the statistical error is typically of order $1/\sqrt{M}$ and thus much larger than the bias, if M is sufficiently large.

[4] There exist also other resampling methods. The most prominent one among these is the bootstrap method, where correlations are destroyed by randomly picking data points from the original set and by creating new sets by repeating this procedure several times. The newly generated data sets are not necessarily identical in their content, because there is a finite probability that the same data point is picked several times and assigned to the same set. The analysis of these reshuffled sets is done in a similar way as we described it for the jackknife method.

4.3 Systematic data smoothing by using Bézier curves

The estimation of errors, as described in the previous section, is the only way to verify the quality of statistical data obtained by experiment or computer simulation. With the methods described, it is simple to obtain a reliable error estimate for a single quantity such as an expectation value $\overline{O}$ of a quantity O. However, it is often also desirable to interpret the curve behavior of a function in its argument space. The microcanonical analysis introduced in Section 2.7, for example, requires precise information about the monotony of energy-dependent quantities, such as the microcanonical entropy $S(E)$. We can, of course, measure the entropy for each energy bin and obtain $S_{\mathrm{meas}}(E)$ and average over many samples or employ the jackknife method to get $\overline{S}(E)$. However, $S_{\mathrm{meas}}(E)$ as well as $\overline{S}(E)$ are discrete sets of stochastic numbers, but no functions. Therefore, it might be useful to consider a method that uses a stochastic series of data to create a smooth, continuous, i.e., *analytic*, estimate for the unknown function $S(E)$. Such is made possible in a very systematic way by using Bézier curves [83, 84] or splines, which are quite popular in computer graphics and product design, because any desired smooth shape can be generated by means of only a few control points.[5]

4.3.1 Construction of a Bézier curve

Figure 4.4 shows an example of a Bézier curve created by using five control points, $\mathbf{P}_0$ to $\mathbf{P}_4$. To start constructing the Bézier curve, the control points are connected consecutively, in the order given by their indices, by straight lines first (note that the order in which control points are connected matters). By this a complete but nonsmooth curve (or better a polygon) is formed. Furthermore, we introduce the path parameter $t \in [0, 1]$, which helps access each point on these legs and on the Bézier curve itself. The point $\mathbf{B}_0^{(1)}$ in the figure, for example, is given by

$$\mathbf{B}_0^{(1)}(t) = \mathbf{P}_0 + t(\mathbf{P}_1 - \mathbf{P}_0) \tag{4.45}$$

and is smoothly parametrized by t. The lower index of any point $\mathbf{B}_i^{(k)}$ refers to the leg it belongs to and the upper index denotes the degree of the construction level. If there are n control points, then the $(n-1)$ "walkers" of the first construction stage are parametrized by

$$\mathbf{B}_i^{(1)}(t) = \mathbf{P}_i + t(\mathbf{P}_{i+1} - \mathbf{P}_i) = (1-t)\mathbf{P}_i + t\mathbf{P}_{i+1}, \quad i = 0, \ldots, n-2. \tag{4.46}$$

In the second step, we repeat this procedure by connecting the successive points $\mathbf{B}_i^{(1)}(t)$, yielding $(n-2)$ new legs. The points located at t are therefore given by

---

[5] Bézier splines are smooth curves composed of segments, which are low-degree Bézier curves. In effect, this means that control points other than the start and endpoint of the entire curve are also elements of the composed curve. This is very useful for shape design, but not for data smoothing, where the smooth function shall not be restrained by meeting intermediate control (i.e., original data) points. Therefore, we here only consider high-degree Bézier curves, in which only the first and the last control point define the endpoints of the curve.

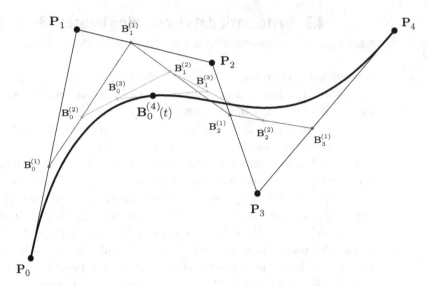

Fig. 4.4   Bézier curve (of degree 4), constructed with five control points $\mathbf{P}_0$, $\mathbf{P}_1$, . . . , $\mathbf{P}_4$. All points $\mathbf{B}_i^{(k)}$ are located at $t = 0.4$ on the respective legs.

$$\mathbf{B}_i^{(2)}(t) = t\mathbf{B}_{i+1}^{(1)}(t) + (1-t)\mathbf{B}_i^{(1)}(t)$$
$$= (1-t)^2\mathbf{P}_i + 2(1-t)t\mathbf{P}_{i+1} + t^2\mathbf{P}_{i+2}, \quad i = 0,\ldots,n-3. \qquad (4.47)$$

For various reasons, this is already a remarkable result. Calculating the derivative

$$\frac{d}{dt}\mathbf{B}_i^{(2)}(t) = 2(1-t)(\mathbf{P}_{i+1} - \mathbf{P}_i) + 2t(\mathbf{P}_{i+2} - \mathbf{P}_{i+1}), \qquad (4.48)$$

we find that the tangent of $\mathbf{B}_i^{(2)}(t)$ at t is a continuous interpolation between the original leg vectors $(\mathbf{P}_{i+1} - \mathbf{P}_i)$ and $(\mathbf{P}_{i+2} - \mathbf{P}_{i+1})$, which are the limiting directions for $t = 0$ and $t = 1$, respectively. Furthermore, we see that these limiting tangents trivially meet in $\mathbf{P}_i$. From these considerations, we may conclude that $\mathbf{B}_i^{(2)}(t)$ is a smooth curve segment. In other words, three control points are already sufficient to create a smooth curve. Because of the form of the expression (4.47), Bézier curves of degree 2 are also called quadratic or parabolic Bézier curves. The property that the Bézier curve employs in the endpoints with a tangent coinciding with the original leg vectors is very useful for the controlled design of curves, because any such Bézier segments can be combined arbitrarily to form a well-defined Bézier spline, which thus is also always a smooth curve, but with nonsmooth second derivative.

The above construction procedure can be continued, yielding in stage three

$$\mathbf{B}_i^{(3)}(t) = (1-t)^3\mathbf{P}_i + 3(1-t)^2t\mathbf{P}_{i+1} + 3(1-t)t^2\mathbf{P}_{i+2} + t^3\mathbf{P}_{i+3}, \quad i = 0,\ldots,n-4. \qquad (4.49)$$

This is the representation of the famous cubic Bézier curves, which are basic elements of cubic Bézier splines.

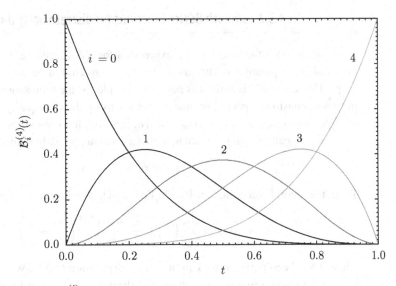

Fig. 4.5 Bernstein polynomials $\mathcal{B}_i^{(4)}(t)$, $i = 0, \ldots, 4$, used for the construction of the Bézier curve shown in Fig. 4.4. Note that $\sum_{i=0}^{4} \mathcal{B}_i^{(4)}(t) = 1$, independently of t.

The example in Fig. 4.4 shows a Bézier curve that was constructed by using five control points. Therefore, it is of degree 4 and parametrized by

$$\mathbf{B}(t) \equiv \mathbf{B}_0^{(4)}(t) = (1 - t)^4\mathbf{P}_0 + 4(1 - t)^3 t\mathbf{P}_1 + 6(1 - t)^2 t^2\mathbf{P}_2 + 4(1 - t)t^3\mathbf{P}_3 + t^4\mathbf{P}_4. \tag{4.50}$$

We observe that a Bézier curve of degree k is always represented by a kth-order polynomial. Even more interesting is that, for t fixed, the expansion coefficients can be interpreted as probabilities with which each of the control points contributes to the location of the point of the Bézier curve that is parametrized by t. In fact, the prefactors remind us of the binomial distribution and the compact representation of any Bézier curve of degree n reads

$$\mathbf{B}(t) = \sum_{i=0}^{n} \mathcal{B}_i^{(n)}(t)\,\mathbf{P}_i, \tag{4.51}$$

where the expansion coefficients are the so-called Bernstein basis polynomials:

$$\mathcal{B}_i^{(n)}(t) = \binom{n}{i}(1 - t)^{n-i}t^i. \tag{4.52}$$

Since $\mathcal{B}_i^{(n)}(0) = \delta_{i0}$ and $\mathcal{B}_i^{(n)}(1) = \delta_{in}$, the endpoints of $\mathbf{B}$ indeed coincide with $\mathbf{P}_0$ and $\mathbf{P}_n$, respectively, but typically no other control point lies on $\mathbf{B}$. The Bernstein polynomials $\mathcal{B}_i^{(4)}(t)$, used for the construction of the exemplified Bézier curve shown in Fig. 4.4, are plotted in Fig. 4.5. For t fixed, the polynomials are normalized,

$$\sum_{i=0}^{n} \mathcal{B}_i^{(n)}(t) = 1 \quad \forall t \in [0, 1], \tag{4.53}$$

and identical with the binomial probability distribution.

4.3.2 Smooth Bézier functions for discrete noisy data sets

Equation (4.51) is the most general expression for a Bézier curve, as it is parametrized by the local curve parameter. This allows to create any kind of curve, even such that include loops. However, in this form it is not very suitable for the representation of functions $y(x)$ in two-dimensional space. Fortunately, for control points y_i equally spaced in x space on the interval $[x_0, x_n]$, $x_i = x_0 + i(x_n - x_0)/n$, it is simple to re-express Eq. (4.51) in a more convenient, scalar form. First, we note that x and t are trivially related with each other:

$$x(t) = x_0 + (x_n - x_0)t. \tag{4.54}$$

Then, the second component of $\mathbf{B}(t)$ in Eq. (4.51), which is now $y(t(x))$, can be written as

$$y_{\text{bez}}(x) = \sum_{i=0}^{n} \binom{n}{i} \left(\frac{x_n - x}{x_n - x_0} \right)^{n-i} \left(\frac{x - x_0}{x_n - x_0} \right)^{i} y_i, \tag{4.55}$$

where t has been replaced by x by inverting expression (4.54). With Eq. (4.55), our goal to obtain a smooth function by means of a discrete set of control points has been accomplished. However, the real power of this expression unveils if noisy data points are used as control points, in which case Eq. (4.55) uniquely maps a stochastic data set into a smoothly interpolating, analytic function. For this reason, it is a particularly useful tool for the analysis of the microcanonical entropy and its derivatives, as described in Section 2.7.

Bézier data manipulation is only beneficial if the expected variance $\sigma_{y_{\text{bez}}(x)}^2$ of the Bézier function at point $x = x_i$ is smaller than the variance $\sigma_{y_i}^2$ of the original data y_i at x_i. Before we discuss this in more detail, we first study an example that allows us to calculate the errors exactly.

For this purpose, let us consider the damped polynomial

$$y(x) = \frac{1}{150} e^{-x^2/7}(x+4)(x+3)(x+2)(x+1)(x-1)(x-2)(x-3) + 2 \tag{4.56}$$

on the interval $x \in [-5, 5)$. This function was chosen because of its several alterations of monotony, which mimics the behavior of the microcanonical temperature $T(E)$ of a small system that experiences a first-order-like transition. This is the most challenging scenario for the Bézier analysis. We now discretize the same function by calculating 5000 data points $y_i = y(x_i)$, where the x_i values are equally spaced in the abovementioned interval. Figure 4.6(a, top) shows the original curve $y(x)$ (dark gray) and its Bézier reconstruction $y_{\text{bez}}(x)$ (black curve), according to Eq. (4.55), on the basis of the discrete data set. The curves are virtually indistinguishable. The exact relative error $\varepsilon_{\text{bez}}(x)$, shown in the bottom plot of Fig. 4.6(a), is almost negligible. The error is largest where the bending of the curves is most pronounced, which is not surprising.

To mimic a noisy data set, similar to what is obtained in experiment or simulation, we now add noise to the data in the following way:

$$y_{\text{rand}}(x_i) = y(x_i)(1 + r_i), \tag{4.57}$$

where $r_i \in [-r_{\max}/2, r_{\max}/2)$ is a random number drawn from a uniform distribution in the given interval. The light gray curves in the top row of Fig. 4.6 are plots of y_{rand} for various

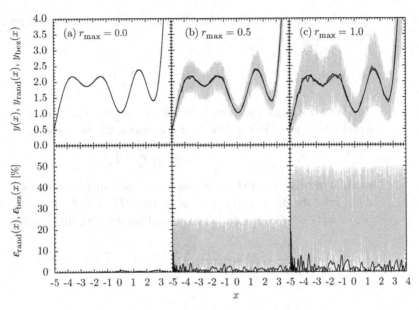

(Top) Function $y(x)$ (dark gray) as defined in Eq. (4.56), with random noise added, $y_{rand}(x)$ (light gray), and Bézier reconstruction $y_{bez}(x)$ (black curve) for fluctuation widths (a) $r_{max} = 0$, (b) $r_{max} = 0.5$, and (c) $r_{max} = 1.0$. (Bottom) Exact errors $\varepsilon_{rand,bez}(x)$ of $y_{rand,bez}(x)$ for the same noise amplitudes.

values of the fluctuation width r_{max}. The black curves in the top row of Figs. 4.6(b) and (c) are the reconstructed functions $y_{bez}(x)$, calculated by using Eq. (4.55). Compared with the original function $y(x)$ (dark gray curves), we can conclude that all qualitative large-scale features of this function are reproduced very well by the Bézier curve. Local fluctuations on short length scales become stronger with increasing noise amplitude, but even in the case shown in Fig. 4.6(c), where the noisy data have up to 50% error compared with the noiseless data, the overall reconstruction works very well. The relative-error curves

$$\varepsilon_{rand,bez}(x) = \left| \frac{y_{rand,bez}(x) - y(x)}{y(x)} \right| \tag{4.58}$$

at the bottom of Figs. 4.6(b) and (c) show that the numerical error of the Bézier curve is typically much smaller than the error of the noisy data. In the following, we will prove that this is generally true for all kinds of data.

We first assume that we have performed M experiments to obtain statistically uncorrelated data sets $y_i^{(k)} = y^{(k)}(x_i)$, $k = 1, \ldots, K$, for each x_i, $i = 0, \ldots, n$. Each data set possesses the unique Bézier curve (4.55)

$$Y_k(x) = \sum_{i=0}^{n} \phi_i^{(n)}(x) y_i^{(k)}, \quad k = 1, \ldots, K, \tag{4.59}$$

where we have introduced the Bernstein polynomial (4.52) in x space in the form

$$\phi_i^{(n)}(x) = \binom{n}{i} \left(\frac{x_n - x}{x_n - x_0} \right)^{n-i} \left(\frac{x - x_0}{x_n - x_0} \right)^{i}. \tag{4.60}$$

Since the exact expectation value $\langle Y \rangle$ is unknown, we introduce the average of the Bézier curves at x by

$$\overline{Y}(x) = \frac{1}{K} \sum_{k=1}^{K} Y_k(x). \tag{4.61}$$

The unknown exact variance of the Bézier curves obtained in $K \to \infty$ experiments is $\sigma_Y^2(x) = \langle Y^2 \rangle(x) - \langle Y \rangle^2(x)$. The finite-$K$ estimator for this variance,

$$\tilde{\sigma}_Y^2 = \overline{Y_k^2} - \overline{Y_k}^2, \tag{4.62}$$

needs to be bias-corrected. According to our discussion in Section 4.2.2, this can easily be done, since the Bézier curves are uncorrelated, $\langle Y_k Y_l \rangle = \delta_{kl} \langle Y^2 \rangle + (1 - \delta_{kl}) \langle Y \rangle^2$, yielding $\langle \tilde{\sigma}_Y^2 \rangle = \sigma_Y^2(1 - 1/K)$. For the bias-corrected estimator we thus obtain

$$\sigma_{Y,c}^2(x) = \frac{K}{K-1} \tilde{\sigma}_Y^2(x) = \frac{K}{K-1} \frac{1}{K} \sum_{k=1}^{K} \left(Y_k(x) - \overline{Y}(x) \right)^2 = \sum_{i=0}^{n} \phi_i^{(n)}(x) \sum_{j=0}^{n} \phi_j^{(n)}(x) A_{ij,c}, \tag{4.63}$$

where we have introduced the bias-corrected estimator for the correlation function of the original data set,

$$A_{ij,c} = \frac{K}{K-1} \left(\overline{y_i y_j} - \overline{y_i}\,\overline{y_j} \right). \tag{4.64}$$

Now we assume that data points at different positions $i \neq j$ in the same data set k are uncorrelated, such that $A_{ij,c} = \delta_{ij} \sigma_{y,c}^2$. With the bias-corrected estimator for the variance of the original data

$$\sigma_{y,c}^2(x_i) = \frac{K}{K-1} [\overline{y^2}(x_i) - \overline{y}^2(x_i)], \tag{4.65}$$

we have found the connection between the variance of the Bézier approximation and the variance of the bare data sets:

$$\sigma_{Y,c}^2(x) = \sum_{i=0}^{n} \left[\phi_i^{(n)}(x) \right]^2 \sigma_{y,c}^2(x_i). \tag{4.66}$$

Remember that x is a continuous variable, whereas x_i is an element of a discrete set of points in x space. Eventually, the statistical error of the average Bézier curve $\overline{Y}(x)$ is given by

$$\varepsilon_Y(x) = \pm \sqrt{\sum_{i=0}^{n} \left[\phi_i^{(n)}(x) \right]^2 \sigma_{y,c}^2(x_i)}. \tag{4.67}$$

It is noteworthy that the statistical error of $\overline{Y}(x)$ depends at any point x on the statistical errors of all "control points," weighed by the square of the Bernstein polynomial associated with the respective position in x space.

What still needs to be done is to prove that the variance of the Bézier approximation is indeed smaller than that of the original data. Only if this can be shown can the Bézier construction really be considered a legitimate tool for systematic smoothing of raw data and reducing local statistical errors.

In Fig. 4.6 (bottom), we see that the error of individual data points can "accidentally" be smaller than the error of the Bézier curve at the same point x. Thus, we cannot prove that the variance of the Bézier curve is always smaller than the variance of the data at any point, but it should be possible to show that the variance $\sigma_{Y,c}^2(x)$ is for all arguments x always smaller than the average variance of the data points, except at the endpoints x_0 and x_n, where $Y(x_{0,n}) = y_{0,n}$ and thus also $\sigma_{Y,c}^2(x_{0,n}) = \sigma_{y,c}^2(x_{0,n})$. We define the average variance of the data in the interval $[x_0, x_n]$ by

$$\{\sigma_{y,c}^2\}_x = \frac{1}{n+1} \sum_{i=0}^{n} \sigma_{y,c}^2(x_i). \tag{4.68}$$

If we are able to prove that

$$\sigma_{Y,c}^2(x) - \{\sigma_{y,c}^2\}_x = \sum_{i=0}^{n} \left(\left[\phi_i^{(n)}(x)\right]^2 - \frac{1}{n+1} \right) \sigma_{y,c}^2(x_i) < 0 \quad \forall x \in (x_0, x_n), \tag{4.69}$$

then the assumption is verified that the variance of the Bézier curve at any point $x \in (x_0, x_n)$ is always smaller than the average variance of the original data points. Defining the minimum and maximum variances of the data by

$$\sigma_{y,c,\min}^2 = \min(\sigma_{y,c}^2(x_i)), \quad \sigma_{y,c,\max}^2 = \max(\sigma_{y,c}^2(x_i)), \quad x_i \in [x_0, x_n] \tag{4.70}$$

and introducing, for brevity,

$$\chi_i^{(n)}(x) = \left[\phi_i^{(n)}(x)\right]^2 - \frac{1}{n+1}, \tag{4.71}$$

it is true that for $x \in (x_0, x_n)$,

$$\sigma_{y,c,\min}^2 \sum_{i=0}^{n} \chi_i^{(n)}(x) \leq \sum_{i=0}^{n} \chi_i^{(n)}(x)\sigma_{y,c}^2(x_i) \leq \sigma_{y,c,\max}^2 \sum_{i=0}^{n} \chi_i^{(n)}(x). \tag{4.72}$$

For x fixed, the Bernstein polynomials are normalized probabilities, i.e., according to Eq. (4.53),

$$\sum_{i=0}^{n} \phi_i^{(n)}(x) = 1 \quad \forall x \in [x_0, x_n], \tag{4.73}$$

and by definition (4.60)

$$\phi_i^{(n)}(x) \begin{cases} < 1 & \text{if } i = 0, 1, \ldots, n \text{ and } x \in (x_0, x_n), \\ = 1 & \text{if } i = 0, x = x_0 \text{ or } i = n, x = x_n. \end{cases} \tag{4.74}$$

Since the endpoints are explicitly excluded in (4.69), we can conclude from the properties of the Bernstein polynomials (4.73) and (4.74) that $\sum_{i=0}^{n} [\phi_i^{(n)}(x)]^2 < 1$ for $x \in (x_0, x_n)$. Consequently,

$$\sum_{i=0}^{n} \chi_i^{(n)}(x) = \sum_{i=0}^{n} \left[\phi_i^{(n)}(x)\right]^2 - 1 < 0, \quad x \in (x_0, x_n). \tag{4.75}$$

That means that the left and the right expressions in the relation (4.72) are negative in this x interval and thus so is the one in the center. With this, our assumption (4.69) is proven.

4.4 Markov processes and stochastic sampling strategies

4.4.1 Master equation

Beside ergodicity, another demand for correct statistical sampling is to ensure that the probability distribution $p(\mathbf{X})$ associated with the desired statistical ensemble is independent of time. This can only be achieved in the simulation, if the *relevant part* of the phase space is sampled sufficiently efficiently to allow for quick convergence toward a stable or, more precisely, stationary estimate for $p(\mathbf{X})$. In most of the Monte Carlo methods, the simulation follows a Markov dynamics, i.e., the update of a given conformation $\mathbf{X}$ to a new one $\mathbf{X}'$ is not influenced by the history that led to $\mathbf{X}$, i.e., the dynamics does not possess an explicit memory. Such a Markov process can be described by the master equation:

$$\frac{\Delta p(\mathbf{X})}{\Delta \tau_0} = \sum_{\mathbf{X}'} [p(\mathbf{X}')t(\mathbf{X}' \to \mathbf{X}; \Delta \tau_0) - p(\mathbf{X})t(\mathbf{X} \to \mathbf{X}'; \Delta \tau_0)], \qquad (4.76)$$

where $t(\mathbf{X} \to \mathbf{X}'; \Delta \tau_0)$ is the transition probability from $\mathbf{X}$ to $\mathbf{X}'$ in a single update (or "time" step $\Delta \tau_0$). Due to particle conservation, it satisfies the normalization condition $\sum_{\mathbf{X}'} t(\mathbf{X} \to \mathbf{X}'; \Delta \tau_0) = 1$, i.e., whatever update we perform, we must end up with a state $\mathbf{X}'$ which is an element of the conformational space. The condition $\Delta p(\mathbf{X})/\Delta \tau_0 = 0$ ensures that the ensemble is in a stationary state if the right-hand side of Eq. (4.76) vanishes. Since the stationarity condition also allows solutions where the distribution function $p(\mathbf{X})$ dynamically changes on cycles, which, however, is not the physical situation in a statistical equilibrium ensemble, we demand more rigorously that the expression in the brackets vanishes. This is called the detailed balance condition. Consequently, the ratio of transition rates is given by

$$\frac{t(\mathbf{X} \to \mathbf{X}'; \Delta \tau_0)}{t(\mathbf{X}' \to \mathbf{X}; \Delta \tau_0)} = \frac{p(\mathbf{X}')}{p(\mathbf{X})} \qquad (4.77)$$

and thus independent of the "time" scale $\Delta \tau_0$, which we therefore omit in the following. From this relation, it follows that it is obviously a good idea to construct an efficient Markov chain Monte Carlo algorithm, i.e., to choose appropriate acceptance probabilities for the Monte Carlo updates to yield the correct transition probability $t(\mathbf{X} \to \mathbf{X}')$, by taking into account the basic microstate probabilities of the statistical ensemble to be simulated. Markov Monte Carlo simulations in the canonical ensemble at fixed temperature T, for example, have to satisfy

$$\frac{t(\mathbf{X} \to \mathbf{X}')}{t(\mathbf{X}' \to \mathbf{X})} = e^{-\beta \Delta E}, \qquad (4.78)$$

where $\Delta E = E(\mathbf{X}') - E(\mathbf{X})$ is the energy difference between the new and the old state. Thus, the transition rate to reach a state $\mathbf{X}'$, supposed to be energetically favored if compared with the initial state $\mathbf{X}$, grows exponentially with $\Delta E < 0$. "Climbing the hill" toward a state with higher energy ($\Delta E > 0$) is, on the other hand, exponentially suppressed. This is in correspondence with the interpretation of the Markov transition state theory. Hence, it is possible to study the kinetic behavior (identification of free-energy barriers, measuring the height of barriers, estimating transition rates, etc.) of a series of processes in equilibrium – for example the folding and unfolding behavior of a protein – by means of Monte Carlo simulations. To quantify the dynamics of a process, i.e., the explicit time dependence, is, however, less meaningful as the conformational change in a single time step depends on the move set and does not follow a physical, e.g., Newtonian, dynamics.[6]

4.4.2 Selection and acceptance probabilities

In order to correctly satisfy the detailed balance condition (4.77) in a Monte Carlo simulation, we have to take into account that each Monte Carlo step consists of two parts. First, a Monte Carlo update of the current state is suggested and second, it has to be decided whether or not to accept it according to the chosen sampling strategy. In fact, both steps are independent of each other in the sense that each possible update can be combined with any sampling method. Therefore, it is useful to factorize the transition probability $t(\mathbf{X} \rightarrow \mathbf{X}')$ into the selection probability $s(\mathbf{X} \rightarrow \mathbf{X}')$ for a desired update from $\mathbf{X}$ to $\mathbf{X}'$ and the acceptance probability $a(\mathbf{X} \rightarrow \mathbf{X}')$ for this update:

$$t(\mathbf{X} \rightarrow \mathbf{X}') = s(\mathbf{X} \rightarrow \mathbf{X}')a(\mathbf{X} \rightarrow \mathbf{X}'). \tag{4.79}$$

The acceptance probability is typically used in the form

$$a(\mathbf{X} \rightarrow \mathbf{X}') = \min\left(1, \sigma(\mathbf{X}, \mathbf{X}')w(\mathbf{X} \rightarrow \mathbf{X}')\right), \tag{4.80}$$

with the ratio of microstate probabilities

$$w(\mathbf{X} \rightarrow \mathbf{X}') = \frac{p(\mathbf{X}')}{p(\mathbf{X})} \tag{4.81}$$

and the ratio of forward and backward selection probabilities

$$\sigma(\mathbf{X}, \mathbf{X}') = \frac{s(\mathbf{X}' \rightarrow \mathbf{X})}{s(\mathbf{X} \rightarrow \mathbf{X}')}. \tag{4.82}$$

The expression (4.80) for the acceptance probability naturally fulfills the detailed-balance condition (4.77). The selection ratio $\sigma(\mathbf{X}, \mathbf{X}')$ is unity, if the forward and backward selection probabilities are identical. This is typically the case for "simple" local Monte Carlo updates. If, for example, the update is a translation of a Cartesian coordinate, $x' = x + \Delta x$, where $\Delta x \in [-x_0, +x_0]$ is chosen from a uniform random distribution, the forward selection for a translation by Δx is equally probable to the backward move, i.e., to translate the

---

[6] The natural way to study the time dependence of Newtonian mechanics is typically based on molecular dynamics methods, which, however, suffer from severe problems to ensure the *correct* statistical sampling at finite temperatures by using thermostats [75, 76]. From a more formal point of view, it is even questionable what "dynamics" shall mean in a thermal system, where even under the same thermodynamic conditions trajectories typically run differently, due to the "random" thermal fluctuations caused by interactions with the huge number [$\mathcal{O}(10^{23})$ per mole] of realistically not traceable heat-bath particles.

particle by $-\Delta x$. This is also valid for rotations about bonds in a molecular system such as torsional rotations about dihedral bonds in a protein. If selection probabilities for forward and backward moves differ, the selection rate is not unity. This occurs frequently in complex, global updates which comprise several steps. Then, the determination of the correct selection probabilities can be difficult and the selection rate typically has to be estimated in test runs first. We will discuss Monte Carlo updates of polymer and protein structures in more detail in Section 4.7.

Note that the overall efficiency of a Monte Carlo simulation depends on both a model-specific choice of a suitable set of moves and an efficient microstate sampling strategy based on $w(\mathbf{X} \rightarrow \mathbf{X}')$.

4.4.3 Simple sampling

The microstate sampling probability $p(\mathbf{X})$ is not necessarily coupled to a certain physical statistical ensemble. The simplest choice is a uniform probability $p(\mathbf{X}) = \text{const}$ independent of ensemble-specific microstate properties. Thus also $w(\mathbf{X} \rightarrow \mathbf{X}') = 1$ and if the Monte Carlo updates satisfy $\sigma(\mathbf{X}, \mathbf{X}') = 1$, the acceptance probability is trivially also unity, $a(\mathbf{X} \rightarrow \mathbf{X}') = 1$, i.e., all generated Monte Carlo updates are accepted, independently of the type of the update. Thus, updates of system degrees of freedom can be performed randomly, where the random numbers are chosen from a uniform distribution. This method is called *simple sampling*. However, its applicability is quite limited. Consider, for example, the estimation of the density of states for a discrete system with this method. After having performed a series of M updates, we will have obtained an energetic histogram $h(E) = M^{-1} \sum_{m=1}^{M} \delta_{E_m, E}$ which represents an estimate for the density of states. The canonical expectation value of the energy can be estimated by $\overline{E} = M^{-1} \sum_{m=1}^{M} E_m e^{-E_m/k_B T} = \sum_E E h(E) e^{-E/k_B T}$. If the microstates are generated randomly from a uniform distribution, it is obvious that we will sample the states $\mathbf{X}$ with an energy $E(\mathbf{X})$ in accordance with their system-specific frequency or degeneracy. High-frequency states thermodynamically dominate in the purely disordered phase. However, near phase transitions toward more ordered phases, the density of states drops rapidly – typically by many orders of magnitude. The degeneracies of the lowest-energy states representing the most ordered states are so small that the thermodynamically most interesting transition region spans even in rather small systems often hundreds to thousands of *orders of magnitude*.[7]

To bridge a region of 100 orders of magnitude between an ordered and a disordered phase by simple sampling would roughly mean to perform about 10^{100} updates in order to find a single ordered state. Assuming that a simple single update would require only a few CPU operations, it will take at least 1 ns on a standard CPU core. Even under this

---

[7] In order to get a sense of the large numbers, consider the 2D Ising model of locally interacting spins on a square lattice which can only be oriented parallel or anti-parallel. For a system with $50 \times 50 = 2500$ spins, the total number of spin configurations is thus $2^{2500} \sim 10^{752}$. The degeneracy of the maximally disordered energetic, paramagnetic macrostate is almost of the same order of magnitude. Since the ferromagnetic ground-state degeneracy is 2 (all spins up or all down), i.e., it is of the order of 10^0, the density of states of this rather small system covers far more than 700 orders of magnitude.

optimistic assumption, it would take more than 10^{83} years to perform 10^{100} updates![8] Thus, for studies of complex systems with sufficiently many degrees of freedom allowing for cooperativity, simple sampling is of very little use.

4.4.4 Metropolis sampling

Because of the dominance of a certain restricted space of microstates in ordered phases, it is obviously a good idea to primarily concentrate in a simulation on a precise sampling of the microstates that form the macrostate under given external parameters such as, for example, the temperature. The canonical probability distribution functions clearly show that within the certain stable phases, only a limited energetic space of microstates is noticeably populated, whereas the probability densities drop off rapidly in the tails. Thus, an efficient sampling of this state space should yield the relevant information within comparatively short Markov chain Monte Carlo runs. This strategy is called *importance sampling*.

The standard importance sampling variant is the Metropolis method [85], where the algorithmic microstate probability $p(\mathbf{X})$ is identified with the canonical microstate probability $p(\mathbf{X}) \sim e^{-\beta E(\mathbf{X})}$ at the given temperature T (or $\beta = 1/k_B T$). Thus, the acceptance probability (4.80) is governed by the ratio of the canonical thermal weights of the microstates:

$$w(\mathbf{X} \to \mathbf{X}') = e^{-\beta[E(\mathbf{X}') - E(\mathbf{X})]}. \tag{4.83}$$

According to Eq. (4.80), a Monte Carlo update from $\mathbf{X}$ to $\mathbf{X}'$ (assuming $\sigma(\mathbf{X}, \mathbf{X}') = 1$) is accepted, if the energy of the new microstate is lower than before, $E(\mathbf{X}') < E(\mathbf{X})$. If the chosen update would provoke an increase of energy, $E(\mathbf{X}') > E(\mathbf{X})$, the conformational change is accepted only with the probability $e^{-\beta \Delta E}$, where $\Delta E = E(\mathbf{X}') - E(\mathbf{X})$. Technically, in the simulation, a random number $r \in [0, 1)$ is drawn from a uniform distribution: if $r \leq e^{-\beta \Delta E}$, the move is still accepted, whereas it is rejected otherwise. Thus, the acceptance probability is exponentially suppressed by ΔE and the Metropolis simulation yields, at least in principle, a time series that is inherently correctly sampled in accordance with the canonical statistics. The arithmetic mean value of a quantity O over the finite Metropolis time series is already an estimate for the canonical expectation value: $\overline{O} = M^{-1} \sum_{m=1}^{M} O_m \approx \langle O \rangle$. In the hypothetical case of an infinitely long simulation ($M \to \infty$), this relation is an exact equality, i.e., the deviation is due to the finiteness of the time series only. However, it is just this restriction to a finite amount of data that limits the quality of Metropolis data. Because of the canonical sampling strategy, reasonable statistics is only obtained in the energetic region that is most dominant for a given temperature, whereas in the tails of the canonical distributions the statistics is rather poor. Thus, there are three physically particularly interesting cases, where Metropolis sampling as a standalone method is not very efficient.

---

[8] Although simple sampling scales perfectly, even highly parallelized simulations on most modern supercomputers with hundreds of thousands of cores are virtually of no help. Given the complexity of problems in statistical physics, it is an illusion to assume that supercomputers alone are sufficient to improve the efficiency of simulations substantially. Much more important is the continued development of efficient simulation strategies and methodologies, as well as algorithms.

First, for low temperatures, where lowest-energy states dominate, the widths of the canonical distributions are extremely small and since $\beta \sim 1/T$ is very large, energetic "uphill" updates are strongly suppressed by the Boltzmann weight $e^{-\beta \Delta E} \to 0$. That means, once caught in a low-energy state, the simulation freezes and it remains trapped in a low-energy state for a long period.

Second, near a second-order phase transition, the standard deviation $\sigma_E = \sqrt{\langle E^2 \rangle - \langle E \rangle^2}$ of the canonical energy distribution function gets very large at the critical temperature T_C, as it corresponds to the maximum (or, in the thermodynamic limit, the divergence) of the heat capacity $C_V = \sigma_E^2/k_B T^2$. Thus, a large energetic space must be precisely sampled ("critical fluctuations"), which requires high statistics. Since in Metropolis dynamics, "uphill moves" with $\Delta E > 0$ are only accepted with a reasonable rate, if at the transition point the ratio $\Delta E/k_B T_C > 0$ is not too large, it can take a long time to reach a high-energy state if starting from the low-energy end. Since near T_C the correlation length diverges like $\xi \sim |\tau|^{-\nu}$ [with $\tau = (T - T_C)/T_C$] and the correlation time in the Monte Carlo dynamics behaves like $t_{corr} \sim |\tau|^{-\nu z}$, the dynamic exponent z allows to compare the efficiencies of different algorithms. The larger the value of z, the less efficient is the method. This drastically reduced simulation dynamics near critical points is called *critical slowing down*. Unfortunately, the standard Metropolis method turns out to be one of the least efficient methods in sampling critical properties of systems exhibiting a second-order phase transition.

The third reason is that the Metropolis method does also perform poorly at first-order phase transitions. In this case, the canonical distribution function is bimodal, i.e., it exhibits two separate peaks with a highly suppressed energetic region in between, since two phases coexist. For the reasons already outlined, it is extremely unlikely to succeed if trying to "jump" from the low- to the high-energy phase by means of Metropolis sampling; it rather would have to explore the valley step by step. Since the energetic region between the phases is entropically suppressed – the number of possible states the system can assume is simply too small – it is thus quite unlikely that this "diffusion process" will lead the system into the high-energy phase, or it will at least take an extremely long time.

However, apart from lowest-energy and phase transition regions, the Metropolis method can be employed successfully, often in combination with reweighting techniques.

4.5 Reweighting methods

4.5.1 Single-histogram reweighting

A standard Metropolis simulation is performed at a given temperature, say T_0. However, it is often desirable to also get quantitative information about the changes of the thermodynamic behavior at nearby temperatures. Since Metropolis sampling is not a priori restricted to a limited phase space, at least in principle, it is indeed theoretically possible to reweight Metropolis data obtained for a given temperature $T_0 = 1/k_B \beta_0$ to a different one, $T = 1/k_B \beta$. The idea is to "divide out" the Boltzmann factor $e^{-\beta_0 E}$ in the estimates for any quantity at the simulation temperature and to multiply it by $e^{-\beta E}$:

$$\langle O \rangle_T = \frac{\left\langle O e^{-(\beta-\beta_0)E} \right\rangle_{T_0}}{\left\langle e^{-(\beta-\beta_0)E} \right\rangle_{T_0}} \approx \overline{O}_T = \frac{\sum_{m=1}^{M} O_m e^{-(\beta-\beta_0)E_m}}{\sum_{m=1}^{M} e^{-(\beta-\beta_0)E_m}}, \qquad (4.84)$$

where we have again considered that the MC time series of length M is finite. In practice, the applicability of this simple reweighting method is rather limited in case the data series was generated in a single Metropolis run, since the error in the tails of the simulated canonical histograms rapidly increases with the distance from the peak. By reweighting, one of the noisy tails will gain the more statistical weight the larger the difference between the temperatures T_0 and T is. In combination with the generalized-ensemble methods to be discussed later in this chapter, however, single-histogram reweighting is the only way of extracting the canonical statistics off the simulated histograms and works perfectly.

4.5.2 Multiple-histogram reweighting

From each Metropolis run, an estimate for the density of states $g(E)$ can easily be calculated. Since the histogram measured in a simulation at temperature T, $h(E; T) = \sum_{m=1}^{M} \delta_{E E_m}$, is an estimate for the canonical distribution function $p_{\text{can}}(E; T) \sim g(E)e^{-\beta E}$, the estimate for the density of states is obtained by reweighting, $\overline{g}(E) = h(E; T)e^{\beta E}$. However, since in a "real" Metropolis run at the single temperature T accurate data can only be obtained in a certain energy interval which depends on T, the estimate $\overline{g}(E)$ is restricted to this typically rather narrow energy interval and by far does not cover the whole energetic region reasonably well.

Thus, the question is whether the combination of Metropolis data obtained in simulations at different temperatures can yield an improved estimate $\overline{g}(E)$. This is indeed possible by means of the multiple-histogram reweighting method [86], sometimes also called "weighted histogram analysis method" (WHAM) [87]. Even though the general idea is simple, the actual implementation is not trivial. The reason is that conventional Monte Carlo simulation techniques such as the Metropolis method cannot yield absolute estimates for the partition sum $Z(T) = \sum_E g(E)e^{-\beta E}$, i.e., estimates for the density of states at different energies $g_i(E)$ and $g_j(E')$ can only be related to each other if obtained in the same run, i.e., $i = j$, but not if performed under different conditions. This is not a problem for the estimation of mean values or normalized distribution functions at fixed temperatures as long as the Metropolis data obtained in the respective temperature threads are used, but interpolation to temperatures where no data were explicitly generated is virtually impossible. Also the multiple-histogram reweighting method does not solve the problem of getting absolute quantities, but at least a "reference partition function" is introduced, which the estimates of the density of states obtained in runs at different simulation temperatures can be related to. Thus, interpolating the data between different temperatures becomes possible.

Basically, the idea is to perform a weighted average of the histograms $h_i(E)$, measured in Monte Carlo simulations for different temperatures, i.e., at β_i (where $i = 1, 2, \ldots, I$ indexes the simulation thread), in order to obtain an estimator for the density of states by combining the histograms in an optimal way:

$$\hat{g}(E) = \frac{\sum_i g_i(E) w_i(E)}{\sum_i w_i(E)}. \qquad (4.85)$$

The exact density of states is given by $g(E) = p_{can}(E; T)Z(T)e^{\beta E}$ and since the normalized histogram $h_i(E)/M_i$ obtained in the ith simulation thread is an estimator for the canonical distribution function $p_{can}(E; T_i)$, the density of states in this thread is estimated by

$$g_i(E) = \frac{h_i(E)}{M_i} Z_i e^{\beta_i E}, \qquad (4.86)$$

where Z_i is the unknown partition function at the ith temperature. Since in Metropolis simulations the best-sampled energy region depends on the simulation temperature, the number of histogram entries for a given energy will differ from thread to thread. Thus, the data of the thread with high statistics at E should in this interpolation scheme get more weight than histograms with fewer entries at E. Therefore, the weight shall be controlled by the errors of the individual histograms. A possibility to determine a set of optimal weights is to reduce the deviation of the estimate $\hat{g}(E)$ for the density of states from the unknown exact distribution $\langle g \rangle(E)$, where the symbol $\langle \ldots \rangle$ is used to refer to this quantity as the true distribution that would have been hypothetically obtained in an infinite number of threads (it should not be confused with a statistical ensemble average here). As usual, the "best" estimate is the one that minimizes the variance $\sigma_{\hat{g}}^2 = \langle (\hat{g} - \langle \hat{g} \rangle)^2 \rangle$. Inserting the relation (4.85) and minimizing with respect to the weights w_i yields the solution

$$w_i = \frac{1}{\sigma_{g_i}^2}, \qquad (4.87)$$

where $\sigma_{g_i}^2 = \langle (g_i - \langle g_i \rangle)^2 \rangle$ is the exact variance of g_i in the ith thread. Because of Eq. (4.86) and the fact that Z_i is an energy-independent constant in the ith thread, we can now concentrate on the discussion of the error of the ith histogram, since $\sigma_{g_i}^2 = \sigma_{h_i}^2 Z_i^2 e^{2\beta_i E}/M_i^2$.

The variance $\sigma_{h_i}^2$ is also an unknown quantity and, in principle, an estimator for this variance would be needed. This would yield an expression that includes the autocorrelation time [86, 87] – similar to the discussion that follows Eq. (4.14). However, to correctly keep track of the correlations in histogram reweighting is difficult and thus also the estimation of error propagation is nontrivial. Therefore, we follow the standard argument based on the assumption of uncorrelated Monte Carlo dynamics (which is typically not perfectly true, of course). The consequence of this idealization will be that the weights (4.87) are not necessarily optimal anymore (the applicability of the method itself is not dependent on the choice of w_i, but the error of the final histogram will depend on the weights).

In order to determine $\sigma_{h_i}^2$ for uncorrelated data, we only need to calculate the probability $P(h_i)$ that in the ith thread a state with energy E (for simplicity we assume that the problem is discrete) is hit h_i times in M_i trials, where each hit occurs with the probability p_{hit}. This leads to the binomial distribution with the hit average $\langle h_i \rangle = M_i p_{hit}$. In the limit of small hit probabilities (a reasonable assumption in general if the number of energy bins is large, and, in particular, for the tails of the histogram), the binomial turns into the Poisson distribution $P(h_i) \to \langle h_i \rangle^{h_i} e^{-\langle h_i \rangle}/h_i!$ with identical variance and expectation value, $\sigma_{h_i}^2 = \langle h_i \rangle$. Insertion into Eq. (4.87) yields the weights

$$w_i(E) = \frac{M_i^2}{\langle h_i \rangle(E) Z_i e^{2\beta_i E}}. \qquad (4.88)$$

Since $\langle h_i \rangle (E)$ is exact, the exact density of states can also be written as

$$g(E) = \frac{\langle h_i \rangle (E)}{M_i} Z_i e^{\beta_i E}, \tag{4.89}$$

which is valid for all threads, i.e., the left-hand side is independent of i. This enables us to replace $\langle h_i \rangle$ everywhere. Inserting expression (4.88) into Eq. (4.85) and utilizing the relation (4.89) to replace $\langle h_i \rangle$, we finally end up with the estimator for the density of states in the form

$$\hat{g}(E) = \frac{\sum_{i=1}^{I} h_i(E)}{\sum_{i=1}^{I} M_i Z_i^{-1} e^{-\beta_i E}}, \tag{4.90}$$

where the unknown partition sum is given by

$$Z_i = \sum_E \hat{g}(E) e^{-\beta_i E}, \tag{4.91}$$

i.e., the set of equations (4.90) and (4.91) must be solved iteratively.[9] One initializes the recursion with guessed values $Z_i^{(0)}$ for all threads, calculates the first estimate $\hat{g}^{(1)}(E)$ using $Z_i^{(0)}$, reinserts this into Eq. (4.91) to obtain $Z_i^{(1)}$, and continues until the recursion process has converged close enough to a fixed point.

There is a technical aspect that should be taken into account in an actual calculation. Since the density of states can even for small systems cover many orders of magnitude and also the Boltzmann factor can become very large, application of the recursion relations (4.90) and (4.91) often results in overflow errors since the common floating-point data types in standard programming languages cannot handle these numbers. At this point, it is helpful to change to a logarithmic representation, which, however, makes it necessary to think about adding up large numbers in logarithmic form. Consider the special but important case of two positive real numbers $a \geq 0$ and $0 \leq b \leq a$, which are too large to be stored such that we wish to use their logarithmic representations $a_{\log} = \log a$ and $b_{\log} = \log b$ instead. Since the result of the addition, $c = a + b$, will also be too large, we introduce $c_{\log} = \log c$ as well. The summation is then performed by writing $c = e^{c_{\log}} = e^{a_{\log}} + e^{b_{\log}}$. Since $a \geq b$ (and thus also $a_{\log} \geq b_{\log}$), it is useful to separate a, and to rewrite the sum as $e^{c_{\log}} = e^{a_{\log}}(1 + e^{b_{\log} - a_{\log}})$. Taking the logarithms yields the desired result, where only the logarithmic representations are needed to perform the summation: $c_{\log} = a_{\log} + \log(1 + x)$, where $x = b/a = e^{b_{\log} - a_{\log}} \in [0, 1]$. The upper limit $x = 1$ is obviously associated to $a = b$, whereas the lower limit $x = 0$ matters if $a \geq 0$, $b = 0$.[10] Since the logarithm of the density of states is proportional to the microcanonical entropy, $S(E) \sim \log g(E)$, the logarithmic representation even has an important physical meaning.

---

[9] Note that for a system with continuous energy space that is partitioned into bins of width ΔE in the simulation, the right-hand side of Eq. (4.91) must still be multiplied by ΔE.

[10] At the lower limit, there is a numerical problem, if $b_{\log} - a_{\log} \ll 0$ (or $x = b/a \ll 1$) is so small that the *minimum* allowed floating-point number is underflown by x. This typically occurs if a and b differ by many tens to thousands of orders of magnitude (depending on the floating-point number precision). In this case, the difference between c and a cannot be resolved, as the error in $c_{\log} = a_{\log} + \mathcal{O}(x)$ is smaller than the numerical resolution; in which case we simply set $c_{\log} = a_{\log}$. If this is not acceptable and a higher resolution is really needed, nonstandard concepts of handling numbers with arbitrary precision could be an alternative.

4.6 Generalized-ensemble Monte Carlo methods

The Metropolis method is the simplest importance sampling Monte Carlo method and for this reason it is a good starting point for the simulation of a complex system. However, it is also one of the least efficient methods and thus one will often have to face the question of how to improve the efficiency of the sampling. One of the most frequently used "tricks" is to employ a modified statistical ensemble within the simulation run and to reweight the obtained statistics after the simulation. The simulation is performed in an artificial *generalized ensemble*.

4.6.1 Replica-exchange Monte Carlo method: Parallel tempering

Although not the most efficient, parallel tempering is the most popular generalized-ensemble method. Advantages are the simple implementation and parallelization on computer systems with many processor cores. The Metropolis method samples conformations of the system in a single canonical ensemble at a fixed temperature, whereas replica-exchange methods like parallel tempering simulate I ensembles at inverse temperatures $\beta_1 < \beta_2 < \ldots < \beta_I$ in parallel (and thus I replicas or instances of the system) [88–91]. In each of the I threads, standard Metropolis simulations are performed. A decrease of the autocorrelation time, i.e., an increase in efficiency, is achieved by exchanging replicas in neighboring threads after a certain number of Metropolis steps are performed independently in the individual threads. The acceptance probability for the exchange of the current conformation $\mathbf{X}$ at β_i and the conformation $\mathbf{X}'$ at β_j is given by

$$a(\mathbf{X} \leftrightarrow \mathbf{X}'; \beta_i, \beta_j) = \min(1, \exp\{-(\beta_i - \beta_j)[E(\mathbf{X}') - E(\mathbf{X})]\}), \qquad (4.92)$$

which satisfies the detailed balance condition. In the generalized ensemble composed of two canonical ensembles at inverse temperatures β_i and β_j, the probability for a state $\mathbf{X}$ at β_i and a state $\mathbf{X}'$ at β_j reads $p(\mathbf{X}, \mathbf{X}'; \beta_i, \beta_j) \sim \exp\{-[\beta_i E(\mathbf{X}) + \beta_j E(\mathbf{X}')]\}$. Thus, parallel tempering samples the product state space of the I canonical ensembles parametrized by $\beta_1, \ldots, \beta_I$. Consequently, the partition function of this generalized ensemble is given by

$$Z_{\text{PT}}(\beta_1, \ldots, \beta_I) = \prod_{i=1}^{I} Z_{\text{can}}(\beta_i) = \prod_{i=1}^{I} \int \mathcal{D}\mathbf{X}_i \, e^{-\beta_i E(\mathbf{X}_i)}. \qquad (4.93)$$

Since the temperature of each thread is fixed, only a small section of the density of states can be sampled in each thread because of the Metropolis limitations. In order to obtain an entire estimate of the density of states, the pieces obtained in the different threads must be combined in an optimal way. This is achieved by subsequent multiple-histogram reweighting as discussed in Section 4.5.2. The main advantage of parallel tempering is its high potential for efficient parallelization. However, the most efficient selection of the temperature set can be a highly sophisticated task. One necessary condition for reasonable exchange probabilities is a sufficiently large overlap of the canonical energy distribution functions in neighboring ensembles. At very low temperatures, the energy distribution is typically a sharp-peaked function. Thus, the density of temperatures must be much higher in the regime of an ordered phase, compared with high-temperature disordered phases. For this reason, the application of the replica-exchange method is often not particularly useful for unraveling the system behavior at very low temperatures or near first-order transitions.

4.6.2 Simulated tempering

Whereas parallel tempering works in the product space of canonical ensembles, simulated tempering virtually dynamically integrates over the temperature space within a single simulation [92, 93]. Each inverse temperature β_i in the discrete set $\beta_1 < \beta_2 < \ldots \beta_2 \ldots < \beta_I$ is assigned a weight $e^{\beta_i F_i}$, which adjusts the partition functions of the canonical ensemble i in a way that an ensemble (or temperature) change within the running simulation can be attempted with reasonable acceptance rate. In fact, the partition function of the expanded generalized ensemble covered by simulated tempering simulations reads:

$$Z_{\mathrm{ST}} = \sum_{i=1}^{I} e^{\beta_i F_i} Z_{\mathrm{can}}(\beta_i) = \sum_{i=1}^{I} e^{\beta_i F_i} \int \mathcal{D}\mathbf{X}\, e^{-\beta_i E(\mathbf{X})}. \qquad (4.94)$$

At constant temperature, a certain number of Metropolis updates are performed, before a temperature update is attempted. The acceptance probability of such an attempt is given by

$$a(\mathbf{X}; \beta_i \rightarrow \beta_j) = \min(1, \exp\{-(\beta_j - \beta_i)E(\mathbf{X}) + \beta_j F_j - \beta_i F_i\}). \qquad (4.95)$$

As in parallel tempering, the spacing between the inverse temperatures affects the efficiency of the dynamics of this method. Additionally, optimal values for the adjustment free energies F_i are not known from the beginning. Therefore, it is necessary to perform adaptive initial Metropolis runs to obtain reasonable estimates for F_i. The acceptance probabilities (4.95) are adjusted well, if the simulation travels frequently back and forth from one end to the other in temperature space. This can be achieved only if the energy histograms of the ensembles at neighboring inverse temperatures have sufficiently large overlap. As a rule of thumb, the overlap $o(\beta_i, \beta_j)$ of two discrete histograms $h_{\mathrm{can}}(E_n; \beta_{i,j})$,

$$o(\beta_i, \beta_j) = 2 \frac{\sum_{n=1}^{N} \min(h_{\mathrm{can}}(E_n; \beta_i), h_{\mathrm{can}}(E_n; \beta_j))}{\sum_{n=1}^{N} [h_{\mathrm{can}}(E_n; \beta_i) + h_{\mathrm{can}}(E_n; \beta_j)]}, \qquad (4.96)$$

should be no less than 0.3 and no greater than 0.7.[11]

4.6.3 Multicanonical sampling

The multicanonical simulation method [94–96] is the most consequent realization of the ideas developed in the context of microcanonical statistical analysis. This method

---

[11] The overlap affects the acceptance probability. Too small overlap values slow down the overall dynamics. If the overlap is too large, the energetically distant regions of the state space are not explored sufficiently. In the former (latter) case, it is more efficient to decrease (increase) the spacing between the respective inverse temperatures.

does not only *aim at* the estimation of the density of states as the previously described generalized-ensemble methods do. Multicanonical sampling dynamics is entirely *based on* sampling governed by the density of states. Consequently driving forward the basic ansatz of simulated tempering to consider the (canonical) temperature as a dynamical variable, in multicanonical simulations the state space is continuously integrated over the (microcanonical) temperature. This close relationship between the fundamental quantities of statistical mechanics and the decoupling of multicanonical sampling from individual canonical ensembles makes it possible to scan the whole phase space within a single simulation with very high accuracy, even if first-order transitions occur.

The basic idea is to deform the Boltzmann energy distribution $p_{\text{can}}(E; T) \propto g(E)e^{-\beta E}$ in such a way that the notoriously difficult sampling of the tails of this distribution is increased and – particularly useful – the sampling rates of the entropically strongly suppressed energetic coexistence regimes in first-order-like transitions and of lowest-energy conformations are improved.

No recursion: Umbrella sampling

The simplest approach is to modify the canonical Boltzmann probability by multiplying it by an artificial weight function $W(\mathbf{X}; T)$,

$$p_{\text{umb}}(\mathbf{X}; T) = \frac{1}{Z_{\text{umb}}} e^{-E(\mathbf{X})/k_B T} W(\mathbf{X}; T). \tag{4.97}$$

The partition function of the new ensemble at temperature T can be written as

$$Z_{\text{umb}} = \int \mathcal{D}\mathbf{X}\, e^{-[E(\mathbf{X})+U(\mathbf{X})]/k_B T}, \tag{4.98}$$

where we have rewritten the weight as $W(\mathbf{X}) = e^{-U(\mathbf{X})/k_B T}$. Therefore, $U(\mathbf{X})$ acts like an additional potential that artificially changes the original model and helps reduce the effect of entropic gaps on the performance of the simulation. This is particularly important in the phase separation regime at first-order-like transitions, where the energy distribution is bimodal (see the schematic illustration in Fig. 4.7). Ideally, the energetic distribution of the new ensemble is very broad, allowing for a substantially increased sampling rate of originally suppressed states, which is why this method is called "umbrella sampling" [97]. Although the simulation is run in a modified ensemble, only expectation values in the original canonical ensemble are of interest. These are obtained by reweighting:

$$\langle O \rangle = \frac{\int \mathcal{D}\mathbf{X}\, O(\mathbf{X}) e^{-[E(\mathbf{X})+U(\mathbf{X})]/k_B T} W^{-1}(\mathbf{X}; T)}{\int \mathcal{D}\mathbf{X}\, e^{-[E(\mathbf{X})+U(\mathbf{X})]/k_B T} W^{-1}(\mathbf{X}; T)} = \frac{\langle OW^{-1} \rangle_{\text{umb}}}{\langle W^{-1} \rangle_{\text{umb}}}. \tag{4.99}$$

The obvious problem is that the canonical distribution function of the original model is unknown in the beginning of the simulation and thus so are the transition barriers. Therefore, a reasonable choice for W requires "educated guessing" and some experience with the system under investigation. It is often suggested to perform a "short" Metropolis simulation of the original system to identify the barriers and then to adjust the weight accordingly. However, one should keep in mind that first-order-like barriers can hardly be identified by

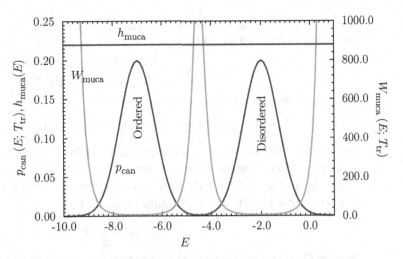

Fig. 4.7 Typical scenario of a first-order transition at transition temperature T_{tr}: Ordered and disordered phases, represented by the peaked sections of the canonical energy distribution $p_{can}(E; T_{tr})$ at low and high energies, are separated by an entropically strongly suppressed energetic region. The multicanonical weight function $W_{muca}(E; T_{tr})$ is chosen in such a way that multicanonical sampling provides a random walk in energy space, independently of (energetic) free-energy barriers. Thus, the energy distribution $h_{muca}(E)$ is ideally constant in the multicanonical ensemble.

means of Metropolis sampling. In such cases, it is desirable to try to determine the weight in a more systematic way, and this leads us to multicanonical sampling.

Recursive multicanonical approach

The general idea is to develop an iterative method for the determination of the weight function, resulting in a modified ensemble which possesses a flat energy distribution (see Fig. 4.7). Hence, the multicanonical weight $W_{muca}(E; T)$ is introduced to satisfy:

$$p_{can}(E; T)W_{muca}(E; T) \sim h_{muca}(E) = \text{const}_{E;T}, \qquad (4.100)$$

where $h_{muca}(E)$ denotes the multicanonical histogram. In the ideal case, it is flat and thus a constant in energy and temperature space. By this construction, the multicanonical simulation performs a random walk in energy space which leads to a rapid decrease of the autocorrelation time in entropically suppressed regions.

Recall that the simulation temperature T does not possess any meaning in the multicanonical ensemble. According to Eq. (4.100), the energy distribution is always constant, independent of temperature. Actually, it is convenient to set it to infinity,[12] in which case

[12] As we will see in the following, the multicanonical recursion starts with an ordinary Metropolis run. Therefore, the performance of the multicanonical simulation in the initial recursions can be improved substantially by setting the simulation temperature to a finite value, such that the most interesting energy regions are sampled right from the beginning of the simulation. Otherwise, it can take several recursions before the walker has dug through and reached the most interesting regions of the energetic state space. At infinite temperature, it starts in the purely random phase, where the density of states is largest, which is energetically typically far away from transition points.

$\lim_{T \to \infty} p_{can}(E; T) \sim g(E)$ and thus $\lim_{T \to \infty} W_{muca}(E; T) \sim g^{-1}(E)$. Then, the acceptance probability (4.80) is governed by

$$w(\mathbf{X} \to \mathbf{X}') = \frac{W_{muca}(E(\mathbf{X}'))}{W_{muca}(E(\mathbf{X}))} = \frac{g(E(\mathbf{X}))}{g(E(\mathbf{X}'))}. \tag{4.101}$$

The weight function can suitably be parametrized as

$$W_{muca}(E) \sim \exp[-S(E)/k_B] = \exp\{-\beta(E)[E - F(E)]\}, \tag{4.102}$$

where $S(E)$ is the microcanonical entropy $S(E) = k_B \ln g(E)$. Since $\beta(E) = k_B^{-1} dS(E)/dE$ is the inverse microcanonical thermal energy (with $\beta(E) = 1/k_B T(E)$, where $T(E)$ is the microcanonical temperature), the microcanonical free-energy scale $f(E) = \beta(E)F(E)$ and $\beta(E)$ are related to each other by the differential equation

$$\frac{df(E)}{dE} = \frac{d\beta(E)}{dE} E. \tag{4.103}$$

Since $\beta(E)$ and $f(E)$ are unknown in the beginning of the simulation, this relation must be solved recursively [95, 96, 98].

Standard recursion

If not already being discrete by the model definition, the energy spectrum must be discretized, i.e., neighboring energy bins have an energetic width ΔE. Thus, for the estimation of $\beta(E)$ and $f(E)$, the following system of difference equations needs to be solved recursively. The starting point is Eq. (4.100) with $p_{can}(E; T) \sim g(E)$, if the simulation temperature is chosen to be infinity. Since the simulated histogram will not be perfectly flat, the estimate for the density of states after the nth recursion will read $\hat{g}^{(n)}(E) \sim h_{muca}^{(n)}(E)/W_{muca}^{(n)}(E)$ such that the dimensionless entropy $s(E) = S(E)/k_B$ can be written as

$$s^{(n)}(E) = \ln \hat{g}^{(n)}(E) = \ln h_{muca}^{(n)}(E) - \ln W_{muca}^{(n)}(E) + c, \tag{4.104}$$

where c is an unimportant constant. Then, the discrete recursive scheme to solve the above set of continuous equations for β, f, s, and finally W_{muca} iteratively looks like this:

$$\beta^{(n+1)}(E) = [s^{(n)}(E) - s^{(n)}(E - \Delta E)]/\Delta E$$
$$= \beta^{(n)}(E) + [\ln h_{muca}^{(n)}(E) - \ln h_{muca}^{(n)}(E - \Delta E)]/\Delta E, \tag{4.105}$$

$$f^{(n+1)}(E) = f^{(n+1)}(E - \Delta E) + [\beta^{(n+1)}(E) - \beta^{(n+1)}(E - \Delta E)](E - \Delta E), \tag{4.106}$$

$$s^{(n+1)}(E) = \beta^{(n+1)}(E)E - f^{(n+1)}(E), \tag{4.107}$$

$$W_{muca}^{(n+1)}(E) = \exp[-s^{(n+1)}(E)]. \tag{4.108}$$

If no better initial guess is available, one typically sets $W_{muca}^{(0)}(E) = 1$ in the beginning. The zeroth iteration thus corresponds to a Metropolis run at infinite temperature, which generates the histogram $h_{muca}^{(0)}(E) = h_{can}(E) =: \hat{g}^{(0)}(E)$. Thus, the histogram is already an estimate for the density of states $\hat{g}^{(0)}(E)$ such that $s^{(0)}(E) = \ln \hat{g}^{(0)}(E)$. Then, the first estimate for the multicanonical weight function $W_{muca}^{(1)}(E)$ can be obtained from the recursion (4.105)–(4.108), which is used to initiate the second recursion, etc. The recursion procedure can be stopped after I recursions, if the weight function has sufficiently converged to a stationary distribution. The number of necessary recursions and also the number of sweeps to be performed at each recursion level is model-dependent. Since the sampled energy space increases from recursion to recursion and the effective statistics of the histogram in each energy bin depends on the number of sweeps, it is a good idea to increase the number of sweeps successively from recursion to recursion. Since the energy histogram should be "flat" after the simulation run at a certain recursion level, an alternative way to control the length of the run is based on a flatness criterion. If, for example, minimum and maximum values of the histogram deviate from the mean histogram value by less than 20%, the run is stopped.

Looking more carefully at the recursive scheme in the form given by the set of equations (4.105)–(4.108) is instructive, as it shows how the multicanonical weights are naturally connected to microcanonical thermodynamic quantities such as temperature, entropy, and free energy as functions of energy. However, by making use of Eq. (4.102), we can reduce this scheme by establishing a relationship between $W_{\mathrm{muca}}(E)$ and β [95]. We simply consider the ratio

$$\frac{W_{\mathrm{muca}}^{(n+1)}(E)}{W_{\mathrm{muca}}^{(n+1)}(E-\Delta E)} = e^{-[s^{(n+1)}(E)-s^{(n+1)}(E-\Delta E)]} = e^{-\beta^{(n+1)}(E)\Delta E} \qquad (4.109)$$

and immediately obtain by using Eq. (4.105):

$$\frac{W_{\mathrm{muca}}^{(n+1)}(E)}{W_{\mathrm{muca}}^{(n+1)}(E-\Delta E)} = e^{-\beta^{(n)}(E)\Delta E}\frac{h_{\mathrm{muca}}^{(n)}(E-\Delta E)}{h_{\mathrm{muca}}^{(n)}(E)} = \frac{W_{\mathrm{muca}}^{(n)}(E)}{W_{\mathrm{muca}}^{(n)}(E-\Delta E)}\frac{h_{\mathrm{muca}}^{(n)}(E-\Delta E)}{h_{\mathrm{muca}}^{(n)}(E)}, \qquad (4.110)$$

which is a direct recursion formula for the multicanonical weights, avoiding the intermediate calculations in the scheme above.

Finally, after the best possible estimate for the multicanonical weight function is obtained, a long multicanonical production run is performed, including all measurements of quantities of interest. From the multicanonical trajectory, the estimate of the canonical expectation value of a quantity O is then obtained at any (canonical) temperature T by:

$$\overline{O}_T = \frac{\sum_t O(\mathbf{X}_t)W_{\mathrm{muca}}^{-1}(E(\mathbf{X}_t))e^{-E(\mathbf{X}_t)/k_{\mathrm{B}}T}}{\sum_t W_{\mathrm{muca}}^{-1}(E(\mathbf{X}_t))e^{-E(\mathbf{X}_t)/k_{\mathrm{B}}T}}. \qquad (4.111)$$

Since the accuracy of multicanonical sampling is independent of the canonical temperature and represents a random walk in the entire energy space, the application of reweighting procedures is lossless. This is a great advantage of the multicanonical method, compared with Metropolis Monte Carlo simulations. A multicanonical simulation virtually samples the system behavior at *all* temperatures simultaneously. In other words, the direct estimation of the density of states is a major advantage, because multiple-histogram reweighting is not needed to obtain this quantity (in contrast to replica-exchange methods).

Error-weighted recursion

Taking a closer look at the right-hand side of the recursion relation (4.105), we realize that the new estimate $\beta^{(n+1)}(E)$ is obtained by adding two contributions of different statistical quality. Whereas $\beta^{(n)}(E)$ contains historic information gathered from the accumulation of histograms of the $n-1$ previous recursions, $\sum_{i=1}^{n-1} h_{\mathrm{muca}}^{(i)}(E)$, the remainder is based on the histogram of the nth recursion only. In other words, for $n \gg 1$, the error of $\beta^{(n)}(E)$ will be much smaller than that of the histogram terms, if the number of sweeps in the nth recursion is not close to the total number of sweeps performed in all previous recursions together, $N^{(n)} \approx \sum_{i=0}^{n-1} N^i$. As we mentioned before, it is generally a good idea to increase the number of sweeps from one iteration level to the next one, but it is hardly possible to satisfy this condition. If $N^{(0)}$ sweeps are performed in the zeroth run, and $N^{(1)} = N^{(0)}$ after the first recursion, then already $N^{(2)} = 2N^{(0)}$ are needed after the second, $N^{(3)} = 4N^{(0)}$ after the third, and so on. This exponential increase amounts to $N^{(n)} = 2N^{(n-1)} = 2^{n-1}N^{(0)}$ sweeps in the nth run. Since hundreds of recursions are not unusual in multicanonical simulations, this approach of balancing statistical errors is certainly not the way to go.

However, if the error is not being kept track of, the consequence can be that the quality of the multicanonical weights will not improve anymore by performing additional recursions, or can even become worse. Therefore, a more sophisticated approach is needed. In the following, we will describe how to weigh the contributions with regard to their actual error. This is known as the error-weighted multicanonical recursion [95, 96] and represents one of the most elegant ways to determine the multicanonical weights (the subsequently described Wang–Landau method is an alternative).

Although the derivation of the recursion is somewhat tricky, the initial idea, the final result, and its implementation are rather simple. The first step is to assume that we already know the optimized estimator $\beta_{\mathrm{opt}}^{(n)}(E)$ for the microcanonical temperature, obtained after the $(n-1)$th recursion. Running the nth multicanonical recursion with it, we can replace Eq. (4.105) by

$$\beta^{(n+1)}(E) = \beta_{\mathrm{opt}}^{(n)}(E) + [\ln h_{\mathrm{muca}}^{(n)}(E) - \ln h_{\mathrm{muca}}^{(n)}(E - \Delta E)]/\Delta E, \qquad (4.112)$$

which yields an estimator for $\beta^{(n+1)}(E)$. This estimator is not yet optimal, because its statistical error is completely governed by the error of the logarithm of the histogram $\ln h_{\mathrm{muca}}^{(n)}(E)$ ($\beta_{\mathrm{opt}}^{(n)}(E)$ does not change in the nth run). Fluctuating and nonfluctuating quantities are combined with the same weight. Since $\beta_{\mathrm{opt}}^{(n)}(E)$ has been obtained using statistics of $(n-1)$ recursions, whereas $\ln h_{\mathrm{muca}}^{(n)}(E)$ contains statistics of a single run only, it is certainly plausible to combine the noisy estimator $\beta^{(n+1)}(E)$ and the previous, already optimized estimator $\beta_{\mathrm{opt}}^{(n)}(E)$ in an error-weighted way. We define

$$\beta_{\mathrm{opt}}^{(n+1)}(E) = \alpha^{(n)}(E)\beta^{(n+1)}(E) + (1 - \alpha^{(n)}(E))\beta_{\mathrm{opt}}^{(n)}(E) \qquad (4.113)$$

$$= \beta_{\mathrm{opt}}^{(n)}(E) + \alpha^{(n)}[\ln h_{\mathrm{muca}}^{(n)}(E) - \ln h_{\mathrm{muca}}^{(n)}(E - \Delta E)]/\Delta E. \qquad (4.114)$$

In order to simplify the representation for the following discussion of histogram errors, we define

$$\gamma^{(i)}(E) = [\ln h_{\mathrm{muca}}^{(i)}(E) - \ln h_{\mathrm{muca}}^{(i)}(E - \Delta E)]/\Delta E. \qquad (4.115)$$

Neither $\beta_{\text{opt}}^{(n)}(E)$ nor $\alpha(E)$ are known at this point, but we want the combination to be optimized with respect to numerical errors and the basic fluctuating quantities needed here are the histograms $h_{\text{muca}}^{(i)}(E)$, obtained at any iterative level $i = 0, 1, \ldots, n$. Based on the same (crude) argument employed in Section 4.5.2, assuming that "hits" of a certain energy bin in the histogram are uncorrelated events, the variance of the histogram corresponds to its average. As a further simplification, we replace the average by the histogram entry itself, such that $\sigma_{h^{(i)}}^2(E) = h_{\text{muca}}^{(i)}(E)$.[13] For the logarithm of the histogram, we thus find
$$\ln\left[h_{\text{muca}}^{(i)} \pm \sqrt{h_{\text{muca}}^{(i)}}\right] = \ln h_{\text{muca}}^{(i)} + \ln\left[1 \pm 1/\sqrt{h_{\text{muca}}^{(i)}}\right] \approx \ln h_{\text{muca}}^{(i)} \pm 1/\sqrt{h_{\text{muca}}^{(i)}}, \text{ such}$$
that the variance reads $\sigma_{\ln h^{(i)}}^2(E) = 1/h_{\text{muca}}^{(i)}(E) = 1/\sigma_{h^{(i)}}^2(E)$. Consequently, utilizing that variances of uncorrelated quantities are additive, we find for the variance of the combination of logarithmic histograms obtained in the ith recursion [Eq. (4.115)]:

$$\sigma_{\gamma^{(i)}}^2(E) = \sigma_{\ln h^{(i)}}^2(E) + \sigma_{\ln h^{(i)}}^2(E - \Delta E) = \frac{h_{\text{muca}}^{(i)}(E) + h_{\text{muca}}^{(i)}(E - \Delta E)}{h_{\text{muca}}^{(i)}(E)h_{\text{muca}}^{(i)}(E - \Delta E)}. \tag{4.116}$$

It is important to note that the definition of the naive estimator (4.112) entails

$$\sigma_{\beta^{(i+1)}}^2(E) = \sigma_{\gamma^{(i)}}^2(E), \tag{4.117}$$

because $\beta_{\text{opt}}^{(i)}$ is constant in the ith run.

Unwinding the original recursion (4.105), we can simply write

$$\beta^{(n)}(E) = \sum_{i=0}^{n-1} \gamma^{(i)}(E). \tag{4.118}$$

In order to optimize $\beta^{(n)}$, we look at the first nontrivial examples. Apparently, $\beta^{(1)}$ is the nonzero starting point of the iteration and the first nontrivial case is $\beta^{(2)}(E) = \gamma^{(0)}(E) + \gamma^{(1)}(E)$. To find the optimized form, we write

$$\beta_{\text{opt}}^{(2)}(E) = \frac{w_0 \gamma^{(0)} + w_1 \gamma^{(1)}}{w_0(E) + w_1(E)}. \tag{4.119}$$

We already know from the optimal arrangement of fragments of the density of states by means of the multi-histogram reweighting method, described in Section 4.5.2, that the optimal error weights are given by the reciprocal variance. Here, we have

$$w_i(E) = \frac{1}{\sigma_{\gamma^{(i)}}^2(E)} = \frac{h_{\text{muca}}^{(i)}(E)h_{\text{muca}}^{(i)}(E - \Delta E)}{h_{\text{muca}}^{(i)}(E) + h_{\text{muca}}^{(i)}(E - \Delta E)}. \tag{4.120}$$

Thus, the weights $w_{0,1}$ determine the optimized estimator $\beta_{\text{opt}}^{(2)}$ completely. Suppose we use this optimized $\beta_{\text{opt}}^{(2)}$ to determine the multicanonical weights according to the multicanonical recursion relations (4.105)–(4.108) and then perform a multicanonical run which yields $h_{\text{muca}}^{(2)}(E)$ or, equivalently, $\gamma^{(2)}(E)$. If we now want to combine $\beta_{\text{opt}}^{(2)}$ and the naive estimator (4.112) by weights (4.120) to make use of the ansatz (4.113) in order to obtain $\beta_{\text{opt}}^{(3)}$,

[13] Note that none of these assumptions is seriously justified. However, any deeper and more correct analysis incorporating autocorrelations would yield only a slightly improved estimator for $\beta_{\text{opt}}^{(n+1)}(E)$. This is not worth the effort, because small "errors" in the multicanonical weight are finally balanced by the multicanonical sampling itself and do not enter into any measured quantity.

we first observe that the overall weight is $w_0 + w_1 + w_2$. We want to let $\beta_{opt}^{(2)}$ contribute its "natural" weights $w_{0,1}$ (in fact, we renormalize $\beta_{opt}^{(2)}$) and couple the naive estimator $\beta^{(3)}$ from Eq. (4.112) to w_2, because from Eq. (4.117) we know that its variance is identical to that of $\gamma^{(2)}$:

$$
\begin{aligned}
\beta_{opt}^{(3)} &= \frac{w_0 + w_1}{w_0 + w_1 + w_2}\beta_{opt}^{(2)} + \frac{w_2}{w_0 + w_1 + w_2}\beta^{(3)} \\
&= \beta_{opt}^{(2)} + \frac{w_2}{w_0 + w_1 + w_2}\gamma^{(2)}.
\end{aligned}
\tag{4.121}
$$

Comparing with the form (4.114) of our ansatz, we find $\alpha^{(2)} = w_2/(w_0 + w_1 + w_2)$. Without further calculation, we can immediately generalize this result for each recursion level, yielding

$$
\alpha^{(n)}(E) = \frac{w_n(E)}{\sum\limits_{i=0}^{n} w_i(E)},
\tag{4.122}
$$

and obtain the desired optimized estimator:

$$
\beta_{opt}^{(n+1)}(E) = \beta_{opt}^{(n)}(E) + \frac{w_n(E)}{w_n(E) + \omega^{(n)}(E)}\ln(h_{muca}^{(n)}(E)/h_{muca}^{(n)}(E - \Delta E))/\Delta E,
\tag{4.123}
$$

where we have introduced the accumulated error weight

$$
\omega^{(n)}(E) = \sum_{i=0}^{n-1} w_i(E) = \omega^{(n-1)}(E) + w_{n-1}(E).
\tag{4.124}
$$

Because of the convenient recursive form, only $\omega^{(n)}(E)$ must be updated after each recursion, but no histograms that have been obtained at lower iteration levels need to be stored. Finally, the optimized multicanonical weights are obtained from the relations (4.105)–(4.108) by replacing $\beta^{(n+1)}$ by $\beta_{opt}^{(n+1)}$.

The optimization of β can be immediately carried over to the reduced recursion scheme (4.110), which simply reads [95]:

$$
\frac{W_{muca}^{(n+1)}(E)}{W_{muca}^{(n+1)}(E - \Delta E)} = \frac{W_{muca}^{(n)}(E)}{W_{muca}^{(n)}(E - \Delta E)}\left(\frac{h_{muca}^{(n)}(E - \Delta E)}{h_{muca}^{(n)}(E)}\right)^{\alpha^{(n)}(E)}.
\tag{4.125}
$$

Since the optimized, error-weighted recursion is much more powerful than the standard recursion in that it provides a smoother and faster convergence in the recursive process of estimating the multicanonical weights, it should generally be favored, even more so as the additional implementation effort is minimal.

Benefit for microcanonical analysis

In Section 2.7, we have discussed the enormous advantage of the microcanonical statistical analysis to identify phase transition points. We particularly emphasized the meaning of the relationship between the density of states, the microcanonical entropy, and the inverse temperature $\beta(E)$ and found that the inflection points of $\beta(E)$ contain all information for the unique identification of the transition points. Multicanonical sampling, as we found earlier, aims at a precise estimation of these microcanonical quantities. Therefore, the multicanonical Monte Carlo method and the microcanonical inflection-point analysis are symbiotically intertwined. In accordance with Eqs. (4.105) and (4.113), the optimal estimate for the microcanonical temperature is obtained by direct measurement of the multicanonical histograms in the production run $n = I$ and it is given by

$$\hat{\beta}(E) = \beta_{\text{opt}}^{(I)}(E) + \ln(h_{\text{muca}}^{(I)}(E)/h_{\text{muca}}^{(I)}(E - \Delta E)])/\Delta E. \qquad (4.126)$$

Changes in the monotony of $\beta(E)$ are best analyzed by considering its derivative with respect to energy,

$$\hat{\gamma}(E) = [\hat{\beta}(E) - \hat{\beta}(E - \Delta E)]/\Delta E. \qquad (4.127)$$

As we have discussed in detail in Section 2.7.3, relevant inflection points of $\beta(E)$ indicating phase transitions correspond with peaks in $\gamma(E)$, and the sign of the peak value determines the class (first or second order) this transition belongs to.

4.6.4 Wang–Landau method

In multicanonical simulations, the weight functions are updated after each iteration, i.e., the weight and thus the current estimate of the density of states are kept constant at a given recursion level. For this reason, the precise estimation of the multicanonical weights in combination with the recursion scheme (4.105)–(4.108) can be a complex and not very efficient procedure. In the method introduced by Wang and Landau [99], the density of states estimate is changed by a so-called modification factor c after each sweep, $g(E) \to c^{(n)}g(E)$, where $c^{(n)} > 1$ is kept constant in the nth recursion, but it is reduced from iteration to iteration. A frequently used ad hoc modification factor is given by $c^{(n)} = \sqrt{c^{(n-1)}} = (c^{(0)})^{1/2^n}$, $n = 1, 2, \ldots, I$, where often $c^{(0)} = e^1 = 2.718\ldots$ is chosen. The acceptance probability and histogram flatness criteria are the same as in multicanonical sampling.

Since the dynamic modification of the density of states in the running simulation violates the detailed balance condition (4.77), the advantage of the high-speed scan of the energy space is paid by a systematic error. However, since the modification factor is reduced with increasing iteration level until it is very small (the iteration process is typically stopped if $c < 1.0 + 10^{-8}$), the simulation dynamics is supposed to sample the phase space according to the stationary solution of the master equation such that detailed balance is (almost) satisfied. Since it is difficult to keep this convergence under control, the optimal method is to use the Wang–Landau method for a very efficient generation of the multicanonical weights, followed by a long multicanonical production run (i.e., at exactly $c = 1$) to obtain the statistical data.

4.7 Elementary Monte Carlo updates

Monte Carlo computer simulations can only be efficient if both parts, the sampling algo-rithm and the set of Monte Carlo updates, are chosen appropriately. We have described some of the most common algorithms in the previous sections in some detail, but we have not yet paid sufficient attention to Monte Carlo updates that are necessary to change the structure of a polymer or protein. Monte Carlo updates of polymer or protein conformations are typically global, which means that any attempt to change the polymer structure locally results in a combination of modifications to degrees of freedom such as atomic coordi-nates (or positions of monomers in coarse-grained models), bond angles, or torsion angles. For typical lattice models with small coordination number, stochastic sampling methods other than combinations of Monte Carlo and sophisticated move sets can be employed. An example for such a method will be described in the next section. Monte Carlo updates of lattice polymer models can be particularly difficult, as violations of detailed balance and/or ergodicity have a stronger impact on simulation results than in the off-lattice case, although one should never rely on that in the latter case, either.

The efficiency of Monte Carlo updates in continuous polymer and protein models strongly depends on the model chosen and can hardly be generalized. For this reason, we will describe only a few of the most basic and popular variants.

The simplest one is the displacement update, in which the coordinates of a monomer at position i in the chain, $\mathbf{x}_i$, are modified by a random shift $\Delta\mathbf{x}_i$. The original conformation of a polymer with chain length N, $\mathbf{X} = \{\mathbf{x}_1, \mathbf{x}_2, \ldots, \mathbf{x}_i, \ldots, \mathbf{x}_N\}$, is changed by the update to $\mathbf{X}' = \{\mathbf{x}_1, \mathbf{x}_2, \ldots, \mathbf{x}_i + \Delta\mathbf{x}_i, \ldots, \mathbf{x}_N\}$. In Cartesian coordinates, the shift is performed by choosing uniformly distributed random numbers in the intervals $\Delta x_i \in [-d_x, +d_x]$, $\Delta y_i \in [-d_y, +d_y]$, and $\Delta z_i \in [-d_z, +d_z]$ (with $d_x, d_y, d_z > 0$). For simplicity, it is common to use a cubic box with boundaries $d_x = d_y = d_z$. Whether the update is accepted or rejected depends on the definition of the transition probability of the Monte Carlo method used. The displacement update is tried for all monomers in the chain (randomly or sequentially). An entire chain update is considered as a Monte Carlo sweep.

The displacement update in Cartesian coordinates is apparently ergodic (each conforma-tion can be reached out of any other) and satisfies the detailed balance condition [the ratio of selection probabilities (4.82) for updates $\mathbf{X} \to \mathbf{X}'$ and $\mathbf{X}' \to \mathbf{X}$ is unity]. The first con-dition is obeyed, because the integral over the volume element $dV = \Pi_{i=1}^N dx_i dy_i dz_i$ yields the entire volume of the conformation space.[14] Detailed balance is satisfied because it is as likely to randomly choose $\Delta\mathbf{x}_i$ for an update $\mathbf{X} \to \mathbf{X}'$ as it is to choose $-\Delta\mathbf{x}_i$ by random selection that, if applied to $\mathbf{X}'$, yields $\mathbf{X}$ again. In other words, the "volume" of the volume element is unaffected by the update: $dV = dV'$.

It is very important to understand that the application of updates based on coordi-nates whose volume elements contain a non-uniform integral measure requires some care to avoid a violation of the detailed balance condition. The most prominent example

---

[14] An update that does not satisfy this condition is not ergodic and cannot be used as a single update. However, a combination of individually non-ergodic updates that as a whole cover the conformation space completely is ergodic.

that causes frequent confusion is the set of spherical coordinates given by the radius r ($0 \leq r < \infty$) of a spherical shell and the angular representation of any point on this shell by polar and azimuthal angles θ ($0 \leq \theta \leq \pi$) and ϕ ($0 \leq \phi < 2\pi$), respectively. The volume element is $dV = \Pi_i \mu(r_i, \theta_i) dr_i d\theta_i d\phi_i$ and the integral measure is obviously not uniform in (r, θ) space: $\mu(r, \theta) = r^2 \sin\theta$ (it is only uniform in ϕ). Effectively, a non-uniform integral measure means that the available number of states does not scale proportionally with the product of shifts of the degrees of freedom. There are various ways to "cure" this problem. The random shifts in the coordinates can be drawn from an appropriately designed non-uniform distribution of random numbers, the non-unity ratio of selection probabilities can be canceled by a weight factor in the transition probability of the Monte Carlo method, or the coordinates are substituted so as to generate a unity integral measure. For spherical coordinates, the necessary substitutions are

$$a = \frac{1}{3}r^3 \quad (0 \leq a < \infty), \qquad b = -\cos\theta \quad (-1 \leq b \leq +1), \qquad (4.128)$$

whereas no substitution is needed for ϕ. Since $da = r^2 dr$ and $db = \sin\theta d\theta$, the spherical volume element is $dV = \mu(a, b, \phi) da \, db \, d\phi$ with uniform integral measure $\mu(a, b, \phi) = 1$. With this substitution done, one can safely use uniformly distributed random numbers drawn from the intervals $\Delta a \in [-d_a, +d_a]$, $\Delta b \in [-d_b, +d_b]$, and $\Delta\phi \in [-d_\phi, +d_\phi]$ to sample the entire phase space correctly.

Another alternative to get rid of non-uniform integral measures is to attach them to quantities that are integrated over. A standard example is the calculation of canonical partition functions or expectation values. Choosing any set of coordinates u, v, w, the partition function can be written as:

$$Z_{\text{can}} \sim \int \Pi_i[\mu(u_i, v_i, w_i) du_i dv_i dw_i] e^{-E/k_B T} = \int \Pi_i[du_i dv_i dw_i] e^{-\tilde{E}/k_B T}, \qquad (4.129)$$

where we have introduced the modified energy function

$$\tilde{E} = E - k_B T \sum_i \ln \mu(u_i, v_i, w_i). \qquad (4.130)$$

This is an instructive example, because it tells us that by ignoring the integral measure in the Monte Carlo update processes we effectively would have simulated a different model with modified entropic properties! The term $S(u_i, v_i, w_i) = k_B \ln \mu(u_i, v_i, w_i)$ is a geometric entropy that accounts for the non-uniformity in the distribution of values of the coordinates of the ith monomer in the chosen coordinate system. It is not a physical entropy, but rather a mathematical counter-term that assigns the physical entropy its correct value.

From the latter discussion, we can also easily identify the previously mentioned ratio of general selection probabilities (4.82) that can be used alternatively in the acceptance criterion (4.80):

$$\sigma(\mathbf{X}, \mathbf{X}') = \frac{\mu(\mathbf{X})}{\mu(\mathbf{X}')}, \qquad (4.131)$$

where $\mu(\mathbf{X})$ and $\mu(\mathbf{X}')$ are the values of the integral measures associated with the original and updated structures, respectively.

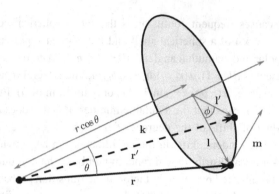

Rotational pivot update of vector **r** about rotation axis **k** by an angle ϕ, yielding the new vector **r**′. The vectors **l, l**′ and **m** are perpendicular to **k; k, l**, and **m** form an orthogonal coordinate system. The angle between **r** and **k** is denoted by θ.

For displacement updates, the use of spherical coordinates is rather overdone, but spherical updates that serve to satisfy geometric constraints in the model are necessary and will be described in the following. It is an obvious limitation of the displacement update that it can only be used in connection with polymer models that allow for changes of the bond length of bonded monomers. The widely used elastic flexible polymer model with FENE bonds (see Sect. 1.6.4) is an example. If the bond is rather stiff, as is quite natural for protein models, the displacement update is not applicable at all. In such cases, the most elementary updates are rotational updates.

One of the most commonly used rotational Monte Carlo moves in simulations of coarse-grained polymer models is the pivot update. This update is the first choice if bond lengths between monomers are fixed (constraint!), but bond angles between adjacent bonds are not. The variant of the pivot update that will be described in the following is based on a randomly selected monomer that serves as the pivot and a rotation axis with randomly chosen direction vector through this pivot point. Then the entire segment of the chain from the monomer acting as the pivot to one of the termini is rotated by a random angle about the rotation axis as shown in Fig. 4.8, but the procedure described in the following for the move of a single monomer in the segment to be rotated must be repeated for each monomer in this segment.

We denote by **r** the vector pointing from the pivot monomer to any monomer in the segment that shall be rotated, and its unit vector by $\mathbf{e}_r = \mathbf{r}/|\mathbf{r}|$. After rotation about the axis **k** (unit vector $\mathbf{e}_k$) by the rotation angle ϕ, the rotated vector connecting the two monomers will be **r**′ (note that the length of the vector remains unchanged: $|\mathbf{r}| = |\mathbf{r}'| =: r$). Introducing the projection angle of $\mathbf{e}_r$ upon $\mathbf{e}_k$ by $\cos\theta = \mathbf{e}_r \cdot \mathbf{e}_k$, it is useful to define the vector **l** (unit vector $\mathbf{e}_l$) that stands perpendicular on **k** (**l** $\perp$ **k**) and points to the same monomer that **r** points to, such that

$$\mathbf{l} = r\,(\mathbf{e}_r - \cos\theta\,\mathbf{e}_k), \tag{4.132}$$

with $|\mathbf{l}| = r\sin\theta$. The orthonormal coordinate system is completed by introducing the unit vector $\mathbf{e}_m$ perpendicular to $\mathbf{e}_k$ and $\mathbf{e}_l$ by $\mathbf{e}_m = \mathbf{e}_k \times \mathbf{e}_l$. The rotation of **r** about **k** to reach $\mathbf{r}'$ causes a rotation of $\mathbf{l}$ by the same angle ϕ *in the plane* perpendicular to $\mathbf{k}$ (spanned by $\mathbf{e}_l$ and $\mathbf{e}_m$), so that the rotated $\mathbf{l}$ vector is given by:

$$\mathbf{l}' = l\,(\cos\phi\,\mathbf{e}_l + \sin\phi\,\mathbf{e}_m), \qquad (4.133)$$

where we have utilized that $l = l'$. Similar to Eq. (4.132) we can write

$$\mathbf{r}' = r\cos\theta\,\mathbf{e}_k + \mathbf{l}' \qquad (4.134)$$

and have hence found the expression for $\mathbf{r}'$:[15]

$$\mathbf{r}' = r\,(\cos\theta\,\mathbf{e}_k + \sin\theta\cos\phi\,\mathbf{e}_l + \sin\theta\sin\phi\,\mathbf{e}_m) \qquad (4.135)$$

with unit vectors:

$$\mathbf{e}_k = \frac{\mathbf{k}}{k}, \quad \mathbf{e}_l = \frac{\mathbf{l}}{l} = \frac{\mathbf{e}_r - \cos\theta\,\mathbf{e}_k}{\sin\theta}, \quad \mathbf{e}_m = \mathbf{e}_k \times \mathbf{e}_l = \frac{\mathbf{e}_k \times \mathbf{e}_r}{\sin\theta}. \qquad (4.136)$$

The singularities for $\theta = 0$ are no problem, since this only occurs in the unlikely case that the randomly chosen rotation axis points exactly in the same direction as $\mathbf{r}$. Since the position of a monomer that lies on the rotation axis is not changed, one simply goes over to the next monomer in the segment that shall be rotated. The special pivot rotation case that the bond vector between the pivot monomer and the adjacent monomer in the chain is considered as the rotation axis represents another common rotational update in its own right: the torsional rotation about a bond axis. In our hierarchy of increasingly complex updates, the torsional rotation is the first choice for models, where bonds *and* bond angles are not thermally excited and, therefore, can be considered as constraints. The only available degrees of freedom are torsional angles. As we have already pointed out, proteins are the most prominent class among those macromolecules that exhibit structural constraints like these.

As depicted in Fig. 4.9(a), the adjacent bond pairs $\mathbf{b}_1$ and $\mathbf{b}_2$, as well as $\mathbf{b}_2$ and $\mathbf{b}_3$, span two planes. Since these planes share one direction ($\mathbf{b}_2$), both planes are simply tilted or rotated about $\mathbf{b}_2$ by an angle ϕ. The rotation axis has the same role as the rotation axis $\mathbf{k}$ in the pivot update. Thus $\mathbf{b}_2 = b_2\mathbf{e}_k$, where b_2 is the length of $\mathbf{b}_2$.

The tilting angle between the two planes is the torsion angle. It can be conveniently expressed as the angle between the surface normals of these planes

$$\mathbf{n}_{12} = \frac{\mathbf{b}_1 \times \mathbf{b}_2}{|\mathbf{b}_1 \times \mathbf{b}_2|}, \qquad \mathbf{n}_{23} = \frac{\mathbf{b}_2 \times \mathbf{b}_3}{|\mathbf{b}_2 \times \mathbf{b}_3|} \qquad (4.137)$$

with $|\mathbf{n}_{12}| = |\mathbf{n}_{23}| = 1$, and it is determined by the scalar product of these normal vectors:

$$\cos\phi = \mathbf{n}_{12} \cdot \mathbf{n}_{23}. \qquad (4.138)$$

It is obvious that the "volume" of the rotation angle space is 2π. However, the cosine function is symmetric in this space and, therefore, the solution of Eq. (4.138) does not

---

[15] This result could have been obtained immediately by appropriate application of the rotation matrix to $\mathbf{r}$, of course.

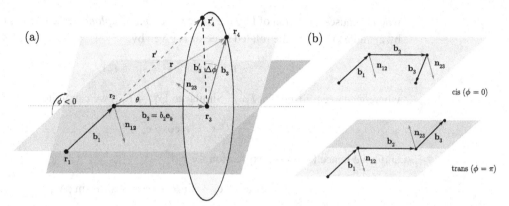

Fig. 4.9 (a) Torsional rotation of bond vector $\mathbf{b}_3$ about the rotation axis $\mathbf{b}_2 = b_2 \mathbf{e}_k$ by an angle $\triangle\phi$. The bond vectors $\mathbf{b}_1$ and $\mathbf{b}_2$ span a plane with normal surface vector $\mathbf{n}_{12}$, whereas $\mathbf{b}_3$ and $\mathbf{b}_4$ lie in a plane with normal $\mathbf{n}_{23}$. (b) The definition of the torsion angle $-\pi < \phi \le \pi$ is such that in the *cis* configuration (normal vectors are parallel, $\mathbf{n}_{12} \parallel \mathbf{n}_{23}$) $\phi = 0$ and in *trans* configuration (anti-parallel normals, $\mathbf{n}_{12} \uparrow\downarrow \mathbf{n}_{23}$) $\phi = \pi$. In these two cases, all bond vectors lie in the same plane. In the configuration shown in (a), $\mathbf{b}_3$ is effectively rotated out of the *cis* configuration by an angle ϕ in negative mathematical orientation, thus $\phi < 0$ in this example.

yield a unique value for ϕ. To resolve this ambiguity, we first define the torsion angle to reside in the interval[16]

$$-\pi < \phi \le \pi. \tag{4.139}$$

There are two special cases, namely, if both planes are identical, as shown in Fig. 4.9(b). If both normals are parallel, $\mathbf{n}_{12} \parallel \mathbf{n}_{23}$ (i.e., $\mathbf{n}_{12} \cdot \mathbf{n}_{23} = 1$), the conformation is said to be in *cis* configuration and the torsion angle is apparently zero, $\phi = 0$. The other case, in which both normals are anti-parallel ($\mathbf{n}_{12} \uparrow\downarrow \mathbf{n}_{23}$ with $\mathbf{n}_{12} \cdot \mathbf{n}_{23} = -1$), is called *trans* configuration with torsion angle $\phi = \pi$.[17] Since the *cis* configuration is the only unique case with respect to Eq. (4.138), we consider any other configuration as if it has been rotated out of *cis*. Therefore, the convention is that ϕ values are positive, if this rotation has a mathematical positive orientation (according to the right-hand grip rule), and ϕ is negative otherwise (the configuration shown in Fig. 4.9 is an example for negative orientation). Thus, the sign of ϕ is identical with the sign of $(\mathbf{n}_{12} \times \mathbf{n}_{23}) \cdot \mathbf{e}_k$, if the configuration is not *cis* or *trans*. We may finally write

$$\phi = \text{sgn}([\mathbf{n}_{12} \times \mathbf{n}_{23}] \cdot \mathbf{e}_k) \arccos(\mathbf{n}_{12} \cdot \mathbf{n}_{23}), \tag{4.140}$$

which is valid and unique for all possible configurations other than *cis* or *trans*.

After this extended but necessary excursion to end up with a proper definition of the torsion angle, we can return to the torsional Monte Carlo update. The goal of this update is to perform a bond rotation by an offset angle $\triangle\phi$ about $\mathbf{b}_2$ that effectively rotates

[16] It is also common to choose alternatively $0 \le \phi < 2\pi$.

[17] The conformation with $\phi = -\pi$ would be associated with the same *trans* configuration as $\phi = +\pi$, which is why we have excluded $\phi = -\pi$ in the interval (4.139).

the monomer, located at position $\mathbf{r}_4$, to its new position $\mathbf{r}'_4$. If we introduce the vector $\mathbf{r} = \mathbf{r}_4 - \mathbf{r}_2$, the angle θ between $\mathbf{r}$ and $\mathbf{b}_2 \sim \mathbf{e}_k$ is given by

$$\cos\theta = \mathbf{e}_r \cdot \mathbf{e}_k, \tag{4.141}$$

where $\mathbf{e}_r = \mathbf{r}/r$ has been used. Drawing a random number $\Delta\phi \in [-d_\phi, +d_\phi]$, where the limit can be chosen (once and then be fixed during a simulation) to be any number $0 < d_\phi < 2\pi$, and adding it to the previous torsion angle yields $\phi' = \phi + \Delta\phi$.[18] Finally, it is necessary to calculate the new coordinates $\mathbf{r}'_4$, which are given by

$$\mathbf{r}'_4 = \mathbf{r}_2 + \mathbf{r}'. \tag{4.142}$$

Collecting all information gathered about the torsional update and considering it a special case of pivot rotation, we can make reference to Eq. (4.135) to calculate $\mathbf{r}'$, where only ϕ needs to be replaced by $\Delta\phi$:

$$\mathbf{r}' = r\left(\cos\theta\, \mathbf{e}_k + \sin\theta \cos\Delta\phi\, \mathbf{e}_l + \sin\theta \sin\Delta\phi\, \mathbf{e}_m\right), \tag{4.143}$$

with the unit vectors (4.136). The torsional update with the same angle $\Delta\phi$ must eventually be applied to all monomers/atoms toward the terminus in the chain leg that shall be rotated.

This completes our short survey of most basic Monte Carlo updates for single polymer or protein chains. For efficiency reasons, it is often necessary to perform different update types in sequence. For example, if different chains interact with each other (e.g., in aggregation processes), or a chain interacts with a static substrate (adsorption), it is useful to perform rigid-body translations and rotations of entire molecules as separate, additional updates. It is also typically most efficient to combine various steps in a single, complex update that remains local, i.e., it affects only a comparatively small part of the chain. One such example is an update based on biased Gaussian steps [100] for proteins, where a series of torsional updates of a few sequential protein backbone dihedral angles is performed in order to ensure that the update does not drastically change the protein conformation (which is likely to be rejected). Such complex updates are particularly sensitive with regard to satisfying the detailed balance condition and require a careful determination of the selection probabilities, which can often only be estimated numerically in test runs.

We now turn to the discussion of alternative sampling strategies, which for discrete models are at least as efficient as sophisticated Monte Carlo approaches.

4.8   Lattice polymers: Monte Carlo sampling vs. Rosenbluth chain growth

Computer simulations of long polymers and peptides are particularly demanding, which is why these simulations are often carried out in finite, discrete spaces such as on underlying lattices. As mentioned earlier, lattice polymers are typically modeled by self-avoiding

---

[18] It is, in principle, no problem if $\phi' = \phi + \Delta\phi$ lies outside the interval (4.139), although for bookkeeping reasons it is recommended to adjust ϕ' by adding or subtracting 2π, respectively, if its value exceeds the interval limits.

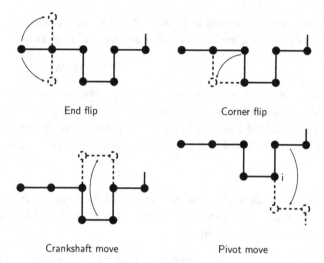

End flip   Corner flip

Crankshaft move   Pivot move

Fig. 4.10   Example for a simple Monte Carlo move set consisting of end and corner flips, crankshaft moves, and pivot updates on a square lattice.

walks. This takes into account the finite volume and the uniqueness of the monomers. A lattice site can hence be occupied by a single monomer only. This has the consequence that the number of very dense conformations of a polymer is by orders of magnitude lower than that of random coil states. In Monte Carlo simulations, particular attention must therefore be dedicated to efficient update procedures which also allow the sampling of dense conformations. So-called move sets consisting of different types of Monte Carlo updates have to be employed to study the structural behavior, e.g., near the Θ point, where polymers subject an attractive interaction collapse from random coils to compact conformations. Monte Carlo updates of lattice polymers are based on semilocal transformations that change the position of a single monomer and a single bond vector (end flips), a single monomer position but two bonds (corner flips), two positions and three bonds (crankshaft), or even more segments [28–31], or nonlocal transformations such as pivot rotations, where the ith monomer serves as pivot point and one of the two partial chains connected with it is rotated about any axis through the pivot [101]. An example of such a move-set is shown in Fig. 4.10.

On regular lattices with small coordination numbers, such as the widely considered simple-cubic lattice, the update of conformations by employing standard moves becomes inefficient, the more dense the conformation is. As it is a general problem of all Monte Carlo simulations of polymer systems, the acceptance rate of changing a dense conformation by semilocal updates decreases drastically and the simulation threatens to get stuck in a specific conformation or to oscillate between two states. The other aspect is that it is very difficult to prove that more complicated structural updates still satisfy detailed balance and ergodicity.

A promising alternative for simulations of lattice polymers and proteins is the completely different approach based on chain growth. The polymer grows by attaching the nth monomer at a randomly chosen nearest-neighbor site of the $(n-1)$th monomer. The

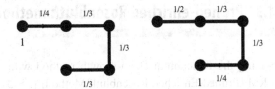

Fig. 4.11   Square-lattice example for the bias implied by Rosenbluth sampling. Both walks shown are grown from the monomer labeled "1." Although the shapes are identical, they are created with different probabilities (left: $p = 1/108$, right: $p = 1/72$).

growth is stopped, if the total length N of the chain is reached or the randomly selected continuation of the chain is already occupied. In both cases, the next chain starts to grow from the first monomer. This simple chain growth is also not yet very efficient, since the number of discarded chains grows exponentially with the chain length.

The performance can be improved with the Rosenbluth chain growth method [33], where first the free nearest neighbors of the $(n-1)$th monomer are determined and then the new monomer is placed to one of the unoccupied sites. Since the probability for the next monomer to be set varies with the number of free neighbors, this implies a bias given by

$$p_n = \left(\prod_{l=2}^{n} m_l\right)^{-1},\qquad(4.144)$$

where m_l is the number of free neighbors to place the lth monomer. The bias is corrected by assigning a Rosenbluth weight factor

$$W_n^{R} = p_n^{-1}\qquad(4.145)$$

to each chain that has been generated by this procedure. An illustrative example for the bias in the Rosenbluth chain-growth method is shown in Fig. 4.11. The two depicted linear chains are grown on a square lattice from both ends (labeled as "1"). According to Rosenbluth sampling, the chain is continued if the number of free neighbor sites is $m \geq 1$. Since the number of free nearest-neighbor places varies, different probabilities for the continuation of the chain occur. Since both conformations are identical, the probability of creation should be the same. This requires the introduction of the correction weights. Although this biased growth is more efficient than simple sampling, this method suffers from attrition too: If all nearest neighbors are occupied, i.e., the chain was running into a "dead end" (attrition point), the complete chain has to be discarded and the growth process has to be started anew. In order to considerably improve the efficiency of Rosenbluth chain growth, a strategy to increase the number of successfully created chains is useful: *"Go with the Winners."*

4.9 Pruned-enriched Rosenbluth method: Go with the winners

Combining the Rosenbluth chain-growth method with population control, as it is done in PERM (Pruned-Enriched Rosenbluth Method) [35–37], leads to a further considerable improvement of the efficiency by increasing the number of successfully generated chains. This method proves particularly useful for studying the Θ point of polymers, since then the Rosenbluth weights of the statistically relevant chains approximately cancel against their Boltzmann probability. The (a-thermal) Rosenbluth weight factor W_n^R is therefore replaced by

$$W_n^{PERM} = \prod_{l=2}^{n} m_l e^{-(E_l - E_{l-1})/k_B T}, \quad 2 \leq n \leq N \quad (E_1 = 0, \quad W_1^{PERM} = 1),$$

where T is the temperature and E_l is the energy of the partial chain $X_l = (x_1, \ldots, x_l)$ created with Rosenbluth chain growth. In PERM, population control works as follows. If a chain has reached length n, its weight W_n^{PERM} is calculated and compared with suitably chosen upper and lower threshold values, $W_n^>$ and $W_n^<$, respectively. For $W_n^{PERM} > W_n^>$, *identical* copies are created which then grow independently. The weight is equally divided among them. If $W_n^{PERM} < W_n^<$, the chain is pruned with some probability, say 1/2, and in case of survival, its weight is doubled. For a value of the weight lying between the thresholds, the chain is simply continued without enriching or pruning the sample. The upper and lower thresholds $W_n^>$ and $W_n^<$ are empirically parametrized. Although their values do not influence the validity of the method, a careful choice can drastically improve the efficiency of the method (the "worst" case is $W_n^> = \infty$ and $W_n^< = 0$, in which case PERM is simply identical with Rosenbluth sampling). An efficient method of parametrization is dynamical adaption of the values [35–40] with respect to the actual number of generated chains c_n with length n and their estimated partition sum

$$Z_n = \frac{1}{c_1} \sum_t W_n^{PERM}(t), \qquad (4.146)$$

where c_1 is the number of growth starts (also called "tours") and t counts the generated conformations with n monomers. Useful choices of the threshold values are

$$W_n^> = C_1 Z_n \frac{c_n^2}{c_1^2}, \quad W_n^< = C_2 W_n^>, \qquad (4.147)$$

where $C_1, C_2 \leq 1$ are constants. For the first tour, $W_n^> = \infty$ and $W_n^< = 0$, i.e., no pruning and enriching.

Results of a simple athermal application of PERM to self-avoiding walks on a simple-cubic lattice are plotted in Fig. 4.12, where the scaling behavior $\langle R_{gyr,ee}^2 \rangle \sim N^{2\nu}$ of the mean square radius of gyration $\langle R_{gyr}^2 \rangle$ and end-to-end distance $\langle R_{ee}^2 \rangle$ with the number of steps N is shown. Data were obtained for chains of $N = 16, 32, \ldots, 32\,768$ steps. For both quantities, the slope of the lines in the logarithmic plot is $\nu = 0.59$, which is close to the precisely known critical exponent $\nu = 0.588 \ldots$ [102].

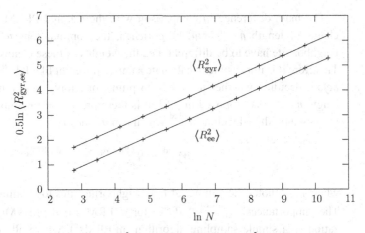

Fig. 4.12 Scaling of mean square radius of gyration $\langle R_{\mathrm{gyr}}^2\rangle$ and end-to-end distance $\langle R_{\mathrm{ee}}^2\rangle$ for self-avoiding walks. Data points refer to results from PERM runs for $N = 16, 32, \ldots, 32\,768$ steps. Lines manifest the respective power-law behaviors.

4.10 Canonical chain growth with PERM

In the so-called "new PERM" variants nPERM$_{\mathrm{is}}^{\mathrm{ss}}$ [new PERM with simple/importance sampling (ss/is)] [38], a considerable improvement is achieved by creating *different* copies, i.e., the chains are identical in $(n-1)$ monomers but have different continuations, instead of completely identical ones, since identical partial chains usually show up a similar evolution. Because of the different continuations, the weights of the copies can differ. Therefore it is not possible to decide about the number of copies on the basis of a joint weight. The suggestion is to calculate first a predicted weight which is then compared with the upper threshold $W_n^>$ in order to determine the number of clones. Another improvement of PERM being followed up since first applications to lattice proteins is that the threshold values $W_n^>$ and $W_n^<$ are no longer constants, but are dynamically adapted with regard to the present estimate for the partition sum and to the number of successfully created chains with length n. The partition sum is proportional to the sum over the weight factors of all conformations of chains with length n, created with a Rosenbluth chain-growth method like, for instance, nPERM$_{\mathrm{is}}^{\mathrm{ss}}$:

$$Z_n = \frac{1}{c_1} \sum_t W_n^{\mathrm{nPERM_{is}^{ss}}}(\mathbf{X}_{n,t}). \qquad (4.148)$$

Here, $\mathbf{X}_{n,t}$ denotes the tth generated conformation of length n. The proportionality constant is the inverse of the number of chain growth starts c_1. Note that due to this normalization it is possible to estimate the degeneracy of the energy states. This is in striking contrast to importance sampling Monte Carlo methods, where the overall constant on the right-hand side of Eq. (4.148) cannot be determined and hence only *relative* degeneracies can be estimated.

The main difference in comparison with the original PERM is that, if the sample of chains of length $n-1$ shall be enriched, the continuations to an unoccupied nearest-neighbor site have to be different, i.e., the weights of these chains with length n can differ. Therefore it is impossible to calculate a uniform weight like W_n^{PERM} as given in Eq. (4.146) *before* deciding whether to enrich, to prune, or simply to continue the current chain of length $n-1$. As proposed in [38], it is therefore useful to control the population on the basis of a predicted weight W_n^{pred} which is introduced as

$$W_n^{\mathrm{pred}} = W_{n-1}^{\mathrm{nPERM}_{\mathrm{is}}^{\mathrm{ss}}} \sum_{\alpha=1}^{m_n} \chi_\alpha^{\mathrm{nPERM}_{\mathrm{is}}^{\mathrm{ss}}}, \qquad (4.149)$$

where m_n denotes the number of free neighboring sites to continue with the nth monomer. The "importances" $\chi_\alpha^{\mathrm{nPERM}_{\mathrm{is}}^{\mathrm{ss}}}$ differ for nPERMss and nPERMis. Due to its characterization as a simple sampling algorithm (nPERMss), where all continuations are equally probable, and as a method with importance sampling (nPERMis), the importances may be defined as

$$\chi_\alpha^{\mathrm{nPERMss}} = 1, \quad \chi_\alpha^{\mathrm{nPERMis}} = \left(m_n^{(\alpha)} + \frac{1}{2}\right) e^{-\beta(E_n^{(\alpha)} - E_{n-1})}. \qquad (4.150)$$

The expression for nPERMis involves the energy $E_n^{(\alpha)}$ of the choice $\alpha \in [1, m_n]$ for placing the nth monomer and the number of free neighbors $m_n^{(\alpha)}$ of this choice which is identical with m_{n+1}, provided the αth continuation was indeed selected for placing the nth monomer. Since $\chi_\alpha^{\mathrm{nPERMis}}$ contains information beyond the nth continuation of the chain, nPERMis controls the further growth better than nPERMss. The predicted weight for the nth monomer is now used to decide how the growth of the chain is continued. If the predicted weight is bigger than the current threshold, $W_n^{\mathrm{pred}} > W_n^>$, and $m_n > 1$, the sample of chains is enriched and the number of copies k is determined according to the empirical rule $k = \min[m_n, \mathrm{int}(W_n^{\mathrm{pred}}/W_n^>)]$. Thus, $2 \le k \le m_n$ different continuations will be followed up. Using nPERMss, the k continuations are chosen randomly with equal probability among the m_n possibilities, while for nPERMis the probability of selecting a certain k-tuple $A = \{\alpha_1, \ldots, \alpha_k\}$ of different continuations is given by

$$p_A = \frac{\sum\limits_{\alpha \in A} \chi_\alpha}{\sum\limits_{A} \sum\limits_{\alpha \in A} \chi_\alpha}. \qquad (4.151)$$

Considering the probabilities p_A as partial intervals of certain length, arranging them successively in the total interval $[0, 1]$ (since $\sum_A p_A = 1$), and drawing a random number $r \in [0, 1)$, one selects the tuple whose interval contains r. This tuple of different sites is then chosen to continue the chain. The corresponding weights are [38]:

$$W_{n,\alpha_j}^{\mathrm{nPERM}_{\mathrm{is}}^{\mathrm{ss}}} = W_{n-1}^{\mathrm{nPERM}_{\mathrm{is}}^{\mathrm{ss}}} \frac{m_n}{k \binom{m_n}{k} p_A} e^{-\beta\left(E_n^{(\alpha_j)} - E_{n-1}\right)}, \qquad (4.152)$$

where $j \in \{1, \ldots, k\}$ is the index of the α_jth continuation within the tuple A. In the special case of simple sampling this expression reduces to $W_{n,\alpha_j}^{\mathrm{nPERMss}} = W_{n-1}^{\mathrm{nPERMss}} m_n$

$\exp[-\beta(E_n^{(\alpha_j)} - E_{n-1})]/k$. If the predicted weight is less than the lower threshold, $W_n^{\text{pred}} < W_n^<$, however, the growth of this chain is stopped with probability 1/2. In this case, one traces the chain back to the last branching point, where the growth can be continued again, or, if there are no branching points, a new tour is started. If the chain survives, the continuation of the chain follows the same procedure as described above, but now with $k = 1$, where Eq. (4.151) simplifies to $p_A = \chi_\alpha / \sum_\alpha \chi_\alpha$ since $A = \{\alpha\}$. In this case the weight of the chain is taken twice. For $W_n^< \le W_n^{\text{pred}} \le W_n^>$, the chain is continued without enriching or pruning (once more with $k = 1$).

The first tour, where the nth monomer is attached for the first time, is started with bounds set to $W_n^> = \infty$ and $W_n^< = 0$, thus avoiding enrichment and pruning. For the following tours, one uses (4.147). The constant $C_1 \le 1$ controls the number of successfully generated chains per tour. For the lower bound a typical choice is $C_2 = 0.2$ [38].

4.11 Multicanonical chain-growth algorithm

The efficiency of PERM depends on the simulation temperature. Therefore, a precise estimation of the density of states requires separate simulations at different temperatures. Then, the density of states can be constructed by means of the multiple-histogram reweighting method [86]. Although this is a powerful method, it is difficult to keep track of the statistical errors involved in the individual histograms obtained in the simulations.

An alternative approach, in which the density of states $g(E)$ is obtained within a *single* simulation without the necessity of a subsequent multi-histogram reweighting, is the combination of PERM with multicanonical sampling, the so-called *multicanonical chain-growth method* [39, 40], which will be described in the following. It should be noted that there is another efficient, related method, called *flat-histogram PERM* [103], which is based on a different ansatz, but it also aims at an improved sampling of "rare" conformations, compared to PERM.

4.11.1 Multicanonical sampling of Rosenbluth-weighted chains

For a given temperature, the $\text{nPERM}_{\text{is}}^{\text{ss}}$ algorithms yield accurate canonical distributions over some orders of magnitude. In order to construct the entire density of states, standard reweighting procedures may be applied, requiring simulations for different temperatures [86]. The low-temperature distributions are, however, very sensitive against fluctuations of weights which inevitably occur because the number of energetic states is low, but the weights are high. Thus, it is difficult to obtain a correct distribution of energetic states, since this requires a reasonable number of hits of low-energy states. Therefore, we assign the chains an additional weight, the multicanonical weight factor W_n^{flat}, chosen such that all possible energetic states of a chain of length n possess almost equal probability of realization. The first advantage is that states having a small Boltzmann probability compared to others are hit more frequently. Second, the multicanonical weights introduced in that manner are proportional to the inverse canonical distribution at temperature T, $W_n^{\text{flat}}(E) \sim 1/p_n^{\text{can},T}(E)$, with respect to the inverse density of states

$$W_n^{\text{flat}}(E) \sim g_n^{-1}(E) \tag{4.153}$$

for $T \to \infty$. Thus, only one simulation is required and a multi-histogram reweighting is not necessary. The multicanonical weight factors are unknown in the beginning and have to be determined iteratively.

Before we go into the technical aspects of this method, let us first discuss it more formally. The energy-dependent multicanonical weights (4.153) are trivially introduced into the partition sum as suitable "decomposition of unity" in the following way:

$$Z_n = \frac{1}{c_1} \sum_t W_n^{\text{nPERM}_{\text{is}}^{\text{ss}}}(\mathbf{X}_{n,t}) W_n^{\text{flat}}(E(\mathbf{X}_{n,t})) \left[W_n^{\text{flat}}(E(\mathbf{X}_{n,t})) \right]^{-1}. \tag{4.154}$$

Since multicanonical sampling effectively works at infinite canonical temperature, we use (4.153) to express the partition sum that then coincides with the total number of all possible conformations as

$$Z_n = \frac{1}{c_1} \sum_t g_n(E(\mathbf{X}_{n,t})) W_n(\mathbf{X}_{n,t}) \tag{4.155}$$

with the combined weight

$$W_n(\mathbf{X}_{n,t}) = W_n^{\text{nPERM}_{\text{is}}^{\text{ss}}}(\mathbf{X}_{n,t}) W_n^{\text{flat}}(E(\mathbf{X}_{n,t})). \tag{4.156}$$

Taking this as the probability for generating chains of length n, $p_n \sim W_n$, leads to the desired flat distribution $H_n(E)$, from which the density of states is obtained by

$$g_n(E) \sim \frac{H_n(E)}{W_n^{\text{flat}}(E)}. \tag{4.157}$$

The canonical distribution at *any* temperature T is calculated by simply reweighting the density of states to this temperature, $p_n^{\text{can},T}(E) \sim g_n(E) \exp(-E/k_B T)$.

4.11.2 Iterative determination of the density of states

In the following, we will describe the recursion method for the multicanonical weights, from which an estimate for the density of states is obtained. Since there is no information about an appropriate choice for the multicanonical weights in the beginning, we set them in the zeroth iteration for all chains $2 \leq n \leq N$ and energies E equal to unity, $W_n^{\text{flat},(0)}(E) = 1$, and the histograms to be flattened are initialized with $H_n^{(0)}(E) = 0$. These assumptions render the zeroth iteration a pure nPERM$_{\text{is}}^{\text{ss}}$ run.

Setting $\beta = 1/k_B T = 0$ (and $k_B = 1$) from now on, the accumulated histogram of all generated chains of length n,

$$H_n^{(0)}(E) = \sum_t W_{n,t}^{\text{nPERM}_{\text{is}}^{\text{ss}}} \delta_{E_t E}, \tag{4.158}$$

is a first estimate of the density of states. In order to obtain a *flat* histogram in the next iteration, the multicanonical weights

$$W_n^{\text{flat},(1)}(E) = \frac{W_n^{\text{flat},(0)}(E)}{H_n^{(0)}(E)} \quad \forall\, n, E \tag{4.159}$$

are updated and the histogram is reset, $H_n^{(1)}(E) = 0$.

The first and all following iterations are multicanonical chain-growth runs and proceed along similar lines as described above, with some modifications. The prediction for the new weight follows again (4.149), but the importances χ_α^{is} (4.150) are in the ith iteration introduced as

$$\chi_\alpha^{\text{is},(i)} = \left(m_n^{(\alpha)} + \frac{1}{2}\right) \frac{W_n^{\text{flat},(i)}(E_n^{(\alpha)})}{W_{n-1}^{\text{flat},(i)}(E_{n-1})}. \tag{4.160}$$

In the simple sampling case, we still have $\chi_\alpha^{\text{ss},(i)} = 1$. If the sample is enriched ($W_n^{\text{pred}} > W_n^{>}$), the weight (4.152) of a chain with length n choosing the α_jth continuation is now replaced by

$$W_{n,\alpha_j}^{\text{ss,is}} = W_{n-1}^{\text{ss,is}} \frac{m_n}{k \binom{m_n}{k} p_A} \frac{W_n^{\text{flat},(i)}(E_n^{(\alpha_j)})}{W_{n-1}^{\text{flat},(i)}(E_{n-1})}, \tag{4.161}$$

where in the simple sampling case (ss) p_A and the binomial factor again cancel each other. If $W_n^{<} \le W_n^{\text{pred}} \le W_n^{>}$, an nth possible continuation is chosen (selected as described for the enrichment case, but with $k = 1$) and the weight is as in Eq. (4.161). Assuming that $W_n^{\text{pred}} < W_n^{<}$ and that the chain has survived pruning (as usual with probability 1/2), we proceed as in the latter case and the chain is assigned twice that weight. The upper threshold value is now determined in analogy to Eq. (4.147) via

$$W_n^{>} = C_1 Z_n^{\text{flat}} \frac{c_n^2}{c_1^2}, \tag{4.162}$$

where

$$Z_n^{\text{flat}} = \frac{1}{c_1} \sum_t W_{n,t}^{\text{ss,is}} \tag{4.163}$$

is the estimated partition sum according to the new distribution provided by the weights (4.161) for chains with n monomers. Whenever a new iteration is started, Z_n^{flat}, c_n, $W_n^{<}$ are reset to zero, and $W_n^{>}$ to infinity (i.e., to the upper limit of the data type used to store this quantity). If a chain of length n with energy E was created, the histogram is increased by its weight:

$$H_n^{(i)}(E) = \sum_t W_{n,t}^{\text{ss,is}} \delta_{E_t E}. \tag{4.164}$$

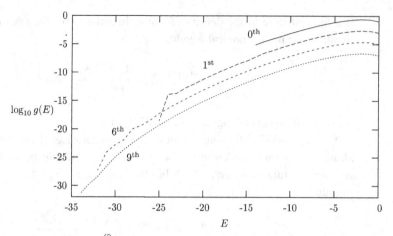

Fig. 4.13 Estimates for the density of states $g_{42}^{(l)}(E)$ for an exemplified heteropolymer with 42 monomers after several recursion levels. Since the curves would fall on top of each other, we have added, for better distinction, a suitable offset to the curves of the 1st, 6th, and 9th runs. The estimate of the 0th run is normalized to unity. From [40].

From iteration to iteration, this histogram approaches the desired flat distribution $H_n(E)$ and after the final iteration $i = I$, the density of states is estimated by

$$g_n^{(I)}(E) = \frac{H_n^{(I)}(E)}{W_n^{\text{flat},(I)}(E)}, \quad 2 \leq n \leq N, \quad (4.165)$$

in analogy to (4.157).

Figure 4.13 shows how the estimate for the density of states of a lattice protein with 42 monomers evolves with increasing number of iterations. The 0th iteration is the initial pure nPERMis run at $\beta = 0$. However, this does not render a proper image of the abilities of nPERMis which works much better at finite temperatures. Iterations 1 to 8 are used to determine the multicanonical weights over the entire energy space $E \in [-34, 0]$. Then, the 9th iteration is the measuring run which gives a very accurate estimate for the density of states covering about 25 orders of magnitude.

The parameter C_1 in Eq. (4.162) controls the pruning/enrichment statistics and thus how many chains of complete length N are generated per tour. For the simulation of the above-mentioned example, $C_1 = 0.01$ was used such that on average 10 complete chains were successfully constructed within each tour [39, 40]. With this choice, the probability for pruning the current chain or enriching the sample is about 20%. In almost all started tours c_1, at least one chain achieves its complete length. Thus the ratio between successfully finished tours and started tours is very close to unity, assuring that the algorithm performs with very good efficiency.

Unlike typical applications of multicanonical or flat histogram algorithms in importance sampling schemes, where all energetic states become equally probable such that the dynamics of the simulation corresponds to a random walk in energy space, the distribution to be flattened in the multicanonical chain-growth method is the histogram that accumulates the weights of the conformations. Hence, if the histogram is flat, a small number of high-weighted conformations with low energy E has the same probability as a large number of appropriate conformations with energy $E' > E$ usually carrying lower weights. Therefore, the number of actual low-energy hits remains lower than the number of hits of states with high energy. In order to accumulate enough statistics in the low-energy region, a comparatively large number of generated conformations in the measuring run is required.

4.12 Random number generators

Stochastic sampling methods such as those described in the previous sections require uniformly distributed random numbers, obtained by random number generators (RNGs). There are many possibilities to create sets of random numbers, which differ in their quality, complexity, speed, and reproducibility. "True" RNGs are typically hardware devices, which, for example, measure fluctuations of environmental parameters such as temperature, pressure, electromagnetic fields, etc. The sequence of random numbers generated by means of such fluctuations is unique and not reproducible. The advantage of true RNGs is that segments of the sequence of random numbers do not repeat; their period is infinity. The generation of these random numbers cannot be controlled, i.e., even if the initial state of the device could be prepared to be the same in all runs, the sequence of the random numbers would always be different. This irreproducibility of random-number sequences is not always desirable, e.g., debugging of implementations of stochastic algorithms is difficult.

Another class of RNGs – typically software-based – generates so-called pseudo-random numbers. These are generated algorithmically according to a well-defined mathematical rule. Since their implementation is simple, almost all stochastic computer simulations use pseudo-random numbers. Generally, pseudo-random numbers possess a finite period. Obviously, the period must be larger than the number of "time steps" in a simulation. The initial state (seed) is well-defined and can be controlled. Restarting a pseudo-RNG with the same seed generates exactly the same sequence of random numbers, i.e., the sequence is reproducible.

Many strategies for generating pseudo-random numbers have been introduced, but it is beyond the scope of this book to discuss details. However, it is important to use pseudo-RNGs with great care and one should be familiar with the basics of their properties before using them in computer simulations [104, 105]. The quality of the random number depends sensitively on the parameters used, and quality control is not trivial. Therefore, it is recommended to use one of the established generators, instead of designing one or "playing" with the parameters of the established ones. Among the most popular are linear congruential and shift register RNGs.

Linear congruential RNGs are based on the recursion for integer random numbers r_i

$$r_{i+1} = (ar_i + c) \bmod m, \quad i \geq 0, \tag{4.166}$$

where a, c, and m are integer constants.[19] A popular example is the 48-bit generator *drand48*, which is included in the standard C library. In this case, $a = 5DEECE66D_{16} \approx 2.5 \times 10^{10}$, $c = B_{16} = 11$, and $m = 2^{48} \approx 2.8 \times 10^{14}$. The generator is initialized by setting the seed $r_0 \in [0, m - 1]$. The random number is then normalized, such that $\hat{r}_i = r_i / m \in [0, 1)$. The period is also determined by $m \sim \mathcal{O}(10^{14})$, which is comparatively short for large-scale Monte Carlo simulations.

The period can be increased, if two or more previously generated random numbers are used to generate a new one. One such class of RNGs is formed by the lagged Fibonacci generators

$$r_i = (r_{i-k} \text{ op } r_{i-l}) \bmod m, \quad i > k > l, \tag{4.167}$$

where op stands for an operation such as $+, -, \times$, or the bitwise XOR.[20] Linear n-bit feedback shift register algorithms are based on the XOR operation,

$$r_i = (r_{i-p} + r_{i-q}) \bmod 2 = r_{i-p} \oplus r_{i-q}, \tag{4.168}$$

and can possess the maximum period $2^n - 1$ for appropriate choices of p and q.

One of the most popular RNGs in scientific computing is the Mersenne Twister [106], which is a twisted generalized feedback shift register RNG and has the amazing period $2^{19937} - 1 \approx 4.3 \times 10^{6001}$. Combining RNGs can also be efficient. An example is RAN-MAR [107], which combines a lagged Fibonacci with an arithmetic sequence generator and achieves a period of $2^{144} \approx 2.2 \times 10^{43}$. Most Monte Carlo results presented in this book were obtained by using RANMAR variants.

4.13 Molecular dynamics

In molecular dynamics simulations of polymers, Newton's equation of motion for the nth monomer with mass $m^{(n)}$ ($n = 1, 2, \ldots, N$) and forces $\mathbf{F}^{(n)}$ acting on this monomer,

$$m^{(n)} \ddot{\mathbf{x}}^{(n)} = \mathbf{F}^{(n)}, \tag{4.169}$$

is numerically integrated. Newton dynamics is energy-conserving and the energy E is fixed by the initial conditions given by the coordinates $\mathbf{x}^{(n)}(t = 0)$ and velocities $\mathbf{v}^{(n)}(t = 0) = \dot{\mathbf{x}}^{(n)}(t = 0)$. The most frequently used integration scheme is the Stoermer–Verlet algorithm [75, 108–112].

From a thermodynamic point of view, molecular dynamics samples the microcanonical ensemble. In order to incorporate canonical thermal fluctuations by coupling the polymer system to a heat bath of canonical temperature T by means of a thermostat, Newton's equations of motion must be modified. It is not the system energy E that has to be constant, but the canonical expectation value $\langle E \rangle$ at the given temperature. Thus, the task of the thermostat is twofold: to keep the temperature constant and to sample the thermal

---

[19] The modulo operation $f \bmod g$ returns the remainder of the division f/g.

[20] The exclusive-or (XOR or $\oplus$) operation, applied to two bits, yields 1, if both bits are unlike; otherwise the result is 0.

fluctuations in line with canonical statistical mechanics, induced by the heat bath, correctly.[21] Among the thermostats that satisfy these conditions, at least theoretically, are the Nosé–Hoover thermostat [113–115], the Andersen thermostat [116], and the Langevin thermostat [109, 111].

The Nosé–Hoover thermostat provides a deterministic dynamics by introducing additional virtual coupling degrees of freedom. The deterministic higher-dimensional hyperdynamics appears stochastic in the projection upon the lower-dimensional position/velocity phase space of the particle. Theoretically, the scattered intersection points of the Nosé–Hoover trajectories with the planes of the particle phase space have the correct canonical statistical distribution. Unfortunately, the parametrization of the Nosé–Hoover thermostat is not straightforward (the coupling degrees of freedom possess effective masses, which are unknown) and autocorrelation and equilibration times depend on the parameter settings. This makes it difficult to control an actual Nosé–Hoover molecular dynamics simulation [76].

The Andersen thermostat is very simple. After each time step δt, each monomer experiences a random collision with a fictitious heat-bath particle with a collision probability $p_{\text{coll}} = \nu \delta t$, where ν is the collision frequency. If the collisions are assumed to be uncorrelated events, the collision probability at any time t is Poissonian, $p_{\text{coll}}(\nu, t) = \nu \exp(-\nu t)$. In the event of a collision, each component of the velocity of the hit particle is changed according to the Maxwell–Boltzmann distribution $p(v_i) = \exp(-mv_i^2/2k_{\text{B}}T)/\sqrt{2\pi mk_{\text{B}}T}$ ($i = 1, 2, 3$). The width of this Gaussian distribution is determined by the canonical temperature. Each monomer behaves like a Brownian particle under the influence of the forces exerted on it by other particles and external fields. In the limit $t \to \infty$, the phase-space trajectory will have covered the complete accessible phase-space, which is sampled in accordance with Boltzmann statistics. Andersen dynamics resembles Markovian dynamics described in the context of Monte Carlo methods and, in fact, from a statistical mechanics point of view, it reminds us of the Metropolis Monte Carlo method.

Another popular thermostat used in molecular dynamics simulations is the Langevin thermostat. It covers the heat-bath coupling part of the Langevin equation by friction and Gaussian random forces $\mathbf{f}$. The Langevin equation basically describes the dynamics of a Brownian particle in solvent under the influence of external forces $\mathbf{F}$. Its simplest form therefore reads:

$$m^{(n)}\ddot{\mathbf{x}}^{(n)} = \mathbf{F}^{(n)} - \gamma \dot{\mathbf{x}}^{(n)} + \mathbf{f}^{(n)}. \tag{4.170}$$

The random force has zero mean at any time, $\langle \mathbf{f}(t) \rangle = \mathbf{0}$, and the force correlation function is $\langle f_i(t)f_j(t') \rangle = 2\gamma k_{\text{B}}T\delta_{ij}\delta(t - t')$. One typically assumes that the solvent is a continuous viscous liquid with low Reynolds number (i.e., it consists of particles smaller than the Brownian particle). In this case Stokes' law can be used to replace the friction coefficient by $\gamma = 6\pi \eta a$, where η is the viscosity of the liquid and a is the radius of a solvent particle.

As we see, the necessary coupling between the system and the heat bath/solvent diminishes the predictability (and time reversibility) of particle trajectories. Consequently, the

---

[21] Analogously, a barostat that is coupled to the system would be necessary to keep the pressure constant, if this is desired.

often claimed advantage of molecular dynamics methods to yield "time-resolved" information about processes needs to be taken with some care. Even if one restarts a simulation with the same initial conditions (but different seed for the random forces – these are not under our control!), the trajectory will look different than in the original run. This is quite a natural result: repeated observation of a Brownian particle under the microscope would tell us the same story. Thus, *any* equilibrium molecular dynamics simulation at finite temperature will only yield the same statistical information that we can also obtain in Monte Carlo simulations. Since in many cases Monte Carlo "jumps" in phase space capture this information much faster than "creeping" molecular dynamics runs (i.e., the autocorrelation times are often notably slower in Monte Carlo simulations [76]), it can be more beneficial to use Monte Carlo instead of molecular dynamics to simulate structural transitions.[22]

---

[22] An important exception are atomic simulations of proteins in explicit solvent, where individual particles (e.g., water molecules) represent the solvent. Monte Carlo updates of the system would too frequently collide with solvent particles which causes unacceptably high rejection rates of trial moves. In this case, the cooperative many-body motion is more efficiently simulated by integrating the equations of motion of each particle step by step. It can even be most efficient to combine Monte Carlo and molecular dynamics. The most prominent hybrid method is replica-exchange molecular dynamics (REMD), where Langevin simulations run in different threads at various temperatures and the replicas are exchanged after some time steps with the exchange transition probability (4.92). Replica-exchange is effectively a Monte Carlo step.

First insights to freezing and collapse of flexible polymers

5.1 Conformational transitions of flexible homopolymers

In this chapter, we will discuss general similarities and differences in crystallization and collapse transitions under the influence of finite-size effects for flexible polymer chains. The analysis of conformational transitions a single polymer in solvent can experience is surprisingly difficult. In good solvent (or high temperatures), solvent molecules occupy binding sites of the polymer and, therefore, the probability of noncovalent bonds between attractive segments of the polymer is small. The dominating structures in this phase are dissolved or random coils. Approaching the critical point at the Θ temperature, the polymer collapses and in a cooperative arrangement of the monomers, globular conformations are favorably formed. At the Θ point, which has already been studied over many decades, the infinitely long polymer behaves like a Gaussian chain, i.e., the effective repulsion due to the volume exclusion constraint is exactly balanced by the attractive monomer–monomer interaction. Below the Θ temperature, the polymer enters the globular phase, where the influence of the solvent is small. Globules are very compact conformations, but there is little internal structure, i.e., the globular phase is still entropy-dominated. For this reason, a further transition toward low-degenerate energetic states is expected to happen: the freezing or crystallization of the polymer. Since this transition can be considered as a liquid-solid phase separation process, it is expected to be of first order, in contrast to the Θ transition, which exhibits characteristics of a second-order phase transition [117, 118].

The complexity of this problem is exposed in the quantitative description of these processes. From the analysis of the corresponding field theory [119] it is known that for the Θ transition the upper critical dimension is $d_c = 3$, i.e., multiplicative and additive logarithmic corrections to the Gaussian scaling are expected and, indeed, predicted by field theory [120–123]. However, until now neither experiments nor computer simulations could convincingly provide evidence for these logarithmic corrections. This not only pertains to analyses of different single-polymer models [35, 124–129], but also the related problem of critical mixing and unmixing in polymer solutions [130–134]. It could even be shown that, depending on the intramolecular interaction range, collapse and freezing transition can fall together in the thermodynamic limit [127, 128, 135, 136].[1]

In this chapter, we will investigate collapse and freezing of a single homopolymer restricted to simple cubic (sc) and face-centered cubic (fcc) lattices [137]. The polymer is modeled by an interacting self-avoiding walk (1.36), because only on this level is a

---

[1] This will be discussed in more detail in Section 6.5.

systematic, comparative analysis possible. The primary discussion will be dedicated to the freezing transition, because this case is particularly instructive for the understanding of the difficulty to systematically classify structural behaviors of small polymers, where finite-size effects are relevant. In this context, the collapse transition is well understood, because finite-size scaling works well [35, 124, 131–133, 138–144]. A precise statistical analysis of the part of the conformational space that is relevant in the low-temperature transition regime is intricate as it is widely dominated by highly compact low-energy conformations which are entropically suppressed.

Most promising for studies of interacting self-avoiding walks are the chain-growth methods based on PERM discussed in Chapter 4, which, in their original formulation [35], are particularly useful for the sampling in the Θ regime. For the analysis of the freezing transition, the generalized multicanonical [39, 40] or flat-histogram variants [103] are particularly efficient. The accuracy of results obtained with these algorithms when applied to lattice polymers is manifested by unraveling even finite-length effects induced by symmetries of the underlying lattice.

5.2 Energetic fluctuations of finite-length polymers

In the following, we employ a detailed canonical statistical analysis of structural transitions of interacting self-avoiding walks on regular sc and fcc lattices and discuss the expected behavior in the thermodynamic limit.

5.2.1 Peak structure of the specific heat

Statistical fluctuations of the energy, as expressed by the specific heat, can signal thermodynamic activity. Peaks of the specific heat as a function of temperature are indicators for transitions or crossovers between physically different macrostates of the system. In the thermodynamic limit, the collective activity, which typically influences most of the system particles, corresponds to thermodynamic phase transitions. For a flexible polymer, three main phases are expected: the random-coil phase for temperatures $T > T_\Theta$, where conformations are unstructured and dissolved; the globular phase in the temperature interval $T_m < T < T_\Theta$ (T_m: melting temperature) with condensed but unstructured ("liquid") conformations dominating; and for $T < T_m$ the "solid" phase characterized by locally crystalline or amorphous metastable structures. In computer simulations, only polymers of finite length are accessible and, therefore, the specific heat typically possesses a less pronounced peak structure, as finite-length effects can induce additional signals of structural activity and shift the transition temperatures. These effects, which are typically connected with surface-reducing monomer rearrangements, are even amplified by steric constraints in lattice models as considered here. Although these pseudo-transitions are undesired in the analysis of the thermodynamic transitions, their importance in realistic systems is currently increasing with the high-resolution equipment available in experiment and technology. The miniaturization of electronic circuits on polymer basis and possible

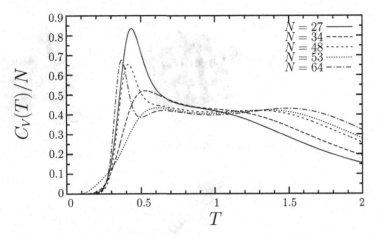

Fig. 5.1 Examples of specific-heat curves (per monomer) for a few exemplified short homopolymers on the sc lattice. From [137].

nanosensory applications in biomedicine therefore require a more in-depth analysis of the finite-length effects in the future.

5.2.2 Simple-cubic lattice polymers

Figure 5.1 shows typical examples of specific heats for very short chains on the sc lattice and documents the difficulty of identifying the phase structure of flexible homopolymers. The 27-mer exhibits only a single dominating peak – which is actually only an sc lattice effect. The reason is that the ground states are cubic ($3{\times}3{\times}3$) and the energy gap toward the first excited states is $\Delta E = 2$.[2] Actually, also the most pronounced peaks for $N = 48$ ($4{\times}4{\times}3$) and $N = 64$ ($4{\times}4{\times}4$) are due to the excitation of perfectly cuboid and cubic ground states, respectively. The first significant onset of the collapse transition is seen for the 48-mer close to $T \approx 1.4$. A clear discrimination between the excitation and the melting transition is virtually impossible. In these examples, solely for $N = 64$ three separate peaks are present. Figures 5.2(a)–5.2(c) show representative conformations in the different pseudophases of the 64-mer. Due to the energy gap, the excitations of the cubic ground state with energy $E = -81$ (not shown) to conformations with $E = -79$ [Fig. 5.2(a)] result in a pseudo-transition which is represented by the first specific-heat peak in Fig. 5.1. The second less-pronounced peak in Fig. 5.1 around $T \approx 0.6 - 0.7$ signals the melting into globular structures, whereas at still higher temperatures $T \approx 1.5$ the well-known collapse peak indicates the dissolution into the random-coil phase. The distribution of the maximum values of the specific heat C_V^{max} with respect to the maximum temperatures $T_{C_V^{\mathrm{max}}}$ is shown in Fig. 5.3. Not surprisingly, the peaks belonging to the

---

[2] This gap is an artifact of the simple-cubic lattice and the resulting excitation transition is a noncooperative effect. Consider, for example, a cuboid conformation with a chain end located in one of the corners. It forms two energetic contacts with the nearest neighbors nonadjacent within the chain. Performing a local off-cube pivot-rotation of the bond the chain end is connected with, these two contacts are lost and no new one is formed.

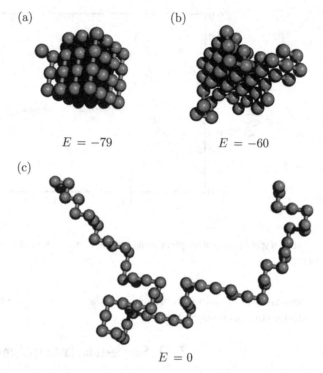

(a)   (b)

$E = -79$ $E = -60$

(c)

$E = 0$

Fig. 5.2   Representative conformations of a 64-mer in the different pseudophases: (a) Excitation from the perfect $4 \times 4 \times 4$ cubic ground state (not shown, $E = -81$) to the first excited crystal state, (b) transition toward globular states, and (c) dissolution into random-coil conformations. From [137].

excitation and freezing transitions ($+$) appear to be irregularly "scattered" in the low-temperature interval $0 < T_{C_V^{\max}} < 0.8$. The height of the peaks indicating the collapse transition of the finite-length polymers ($\odot$) is, on the other hand, monotonously increasing with the collapse-peak temperature.

Figures 5.4(a) and 5.4(b), showing the respective chain-length dependence of the maximum temperatures and maximum specific-heat values, reveal a more systematic picture. At least from the results for the short chains shown, general scaling properties for the freezing transition cannot be read at all. The reason is that the low-temperature behavior of these short chains is widely governed by lattice effects. This is clearly seen by the "sawtooth" segments. Whenever the sc chain possesses a "magic" length N_c such that the ground state is cubic or cuboid (i.e., $N_c \in \mathcal{N}_c = \{8, 12, 18, 27, 36, 48, 64, 80, 100, 125, \dots \}$), the energy gap $\Delta E = 2$ between the ground-state conformation and the first excited state entails a virtual energetic barrier which results in an excitation transition. Since entropy is less relevant in this regime, this energetic effect is not "averaged out" and, therefore, causes a pronounced peak in the specific heat [see Fig. 5.4(b)] at comparatively low temperatures [Fig. 5.4(a)]. This peculiar sc lattice effect vanishes widely by increasing the length by unity, i.e., for chain lengths $N_c + 1$. In this case, the excitation peak either vanishes or remains as a remnant of less thermodynamic significance. The latter appears particularly in those cases where $N = N_c + 1$ with $N_c = L^3$ (with L being any positive integer) is a chain

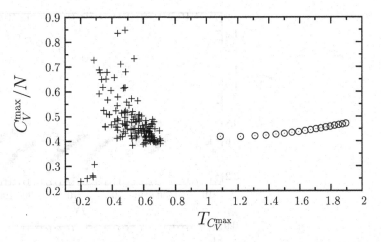

Fig. 5.3 Map of specific-heat maxima for several chain lengths taken from the interval $N \in [8, 125]$. Circles ($\odot$) symbolize the peaks (if any) identified as signals of the collapse ($T_{C_V^{max}} > 1$). The low-temperature peaks ($+$) belong to the excitation/freezing transitions ($T_{C_V^{max}} < 0.8$). The group of points in the lower left corner corresponds to polymers with $N_c + 1$ monomers, where N_c denotes the "magic" lengths allowing for cubic or cuboid ground-state conformations (see Fig. 5.4 and text). From [137].

length allowing for perfectly cubic ground states. Increasing the polymer length further, the freezing peak dominates at low temperatures. Its peak increases with the chain length, whereas the peak temperature decreases. Actually, with increasing chain length, the character of the transition converts from freezing to excitation, i.e., the entropic freedom that still accompanies the melting/freezing process decreases with increasing chain length. In other words, cooperativity is lost: only a small fraction of monomers – residing in the surface hull – is entropically sufficiently flexible to compete with the energetic gain of highly compact conformations. This flexibility is reduced further, the closer the chain length N approaches a number in the "magic" set $\mathcal{N}_c$. If the next length belonging to $\mathcal{N}_c$ is reached, the next discontinuity in the monotonic behavior occurs. Since noticeable "jumps" are only present for chain lengths whose ground states are close to cubes ($N_c = L^3$) or cuboids with $N_c = L^2(L \pm 1)$, the length of the branches in between scales with $\Delta N_c \sim L^2 \sim N_c^{2/3}$. Therefore, a reasonable scaling analysis for $T_{C_V^{max}}(N)$ and $C_V^{max}(N)$ could be performed only for very long chains on the sc lattice, for which, however, a precise analysis of the low-temperature behavior is extremely difficult.

5.2.3 Polymers on the face-centered cubic lattice

The general behavior of polymers on the fcc lattice is comparable to what we observed for the sc polymers. The main difference is that excitations play only a minor role, and the freezing transition dominates the conformational behavior of the fcc polymers at low temperatures. Nonetheless, finite-length effects are still apparent as can be seen in the chain-length dependence of the peak temperatures and peak values of the specific heats plotted in Fig. 5.5(a) and Fig. 5.5(b), respectively. Figure 5.5(a) shows that the locations

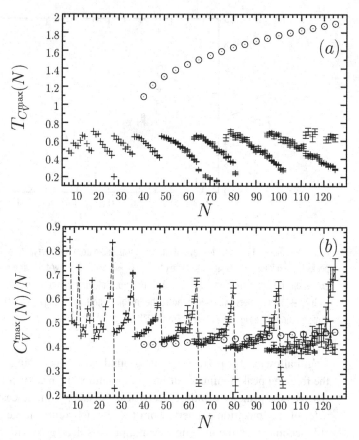

Fig. 5.4 (a) Collapse (⊙) and crystallization/excitation (+) peak temperatures of the specific heat for all chain lengths in the interval $N \in [8, 125]$, (b) values of the specific-heat maxima in the same interval. Θ peaks appear starting from $N = 41$. For the sake of clarity, not all intermediate Θ data points are shown (only for $N = 41, 45, 50, \ldots$). From [137].

of the freezing and collapse transitions clearly deviate with increasing chain lengths and hence we can conclude that for interacting self-avoiding polymers on fcc lattices there is no obvious indication that freezing and collapse could fall together in the thermodynamic limit.

Similar to the sc polymers, the finite-length effects at very low temperatures are apparently caused by the usual compromise between maximum compactness, i.e., maximum number of energetic (nearest-neighbor) contacts, and steric constraints of the underlying rigid lattice. The effects are smaller than in the case of the sc lattice, as there are no obvious "magic" topologies in the fcc case. Ground-state conformations for a few small polymers on the fcc lattice are shown in Fig. 5.6. The general tendency is that the lowest-energy conformations consist of layers of net planes with (111) orientation, i.e., the layers themselves possess a triangular pattern with side lengths equal to the fcc nearest-neighbor distance $\sqrt{2}$ (in units of the lattice constant). This is not surprising, as these conformations are tightly packed, which ensures a maximum number of nearest-neighbor contacts and, therefore,

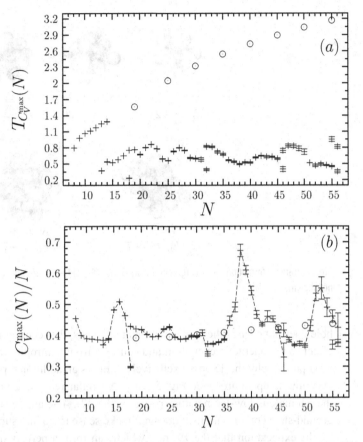

Fig. 5.5 (a) Peak temperatures and (b) peak values of the specific heat for all chain lengths $N = 8, \ldots, 56$ of polymers on the fcc lattice. Circles ($\odot$) symbolize the collapse peaks and low-temperature peaks ($+$) signal the excitation/freezing transitions. The error bars for the collapse transition are typically much smaller than the symbol size. Only for the freezing transition of longer chains, the statistical uncertainties are a little bit larger and visible in the plots. Θ peaks appear starting from $N = 19$. For clarity, Θ data points are only shown for $N = 19, 25, 30, \ldots$. From [137].

lowest conformational energy. An obvious example is the ground-state conformation of the 13-mer as shown in Fig. 5.6(a), which corresponds to the intuitive guess for the most closely packed structure on an fcc lattice: a monomer with its 12 nearest neighbors ("3–7–3" layer structure). A simple contact counting yields 36 nearest-neighbor contacts, which, by subtracting the $N - 1 = 12$ covalent (non-energetic) bonds, is equivalent to an energy $E = -24$. However, this lowest-energy conformation is degenerate. There is another conformation (not shown) consisting of only two "layers," one containing 6 (a triangle) and the other 7 (a hexagon) monomers ("6–7" structure), with the same number of contacts.

A special case is the 18-mer. As Fig. 5.6(b) shows, its ground state is formed by a complete triangle with 6 monomers, a hexagon in the intermediate layer with 7 monomers, and an incomplete triangle (possessing a "hole" at a corner) with 5 monomers ("6–7–5" structure). Although this imperfection seems to destroy all rotational symmetries, it is compensated by an additional symmetry: exchanging any of the triangle corners with the

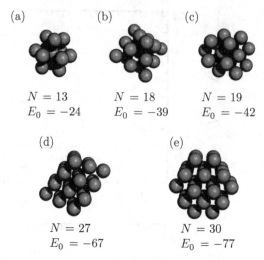

Fig. 5.6   Ground-state conformations and energies of the (a) 13-, (b) 18-, (c) 19-, (d) 27-, and (e) 30-mer on the fcc-lattice (bonds not shown). From [137].

hole does not change the conformation at all! Thus, the seeming imperfection has a similar effect as the energetic excitation and causes a trivial entropic transition. This explains, at least partly, why the 18-mer exclusively exhibits an additional peak in the specific heat at very low temperatures [see Fig. 5.5(a)]. A similar reasoning presumably also applies to the anomalous low-temperature peaks of the 32-, 46-, and 56-mers, but for these larger ground-state conformations it does not make sense to go into such intricate detail.

The expectation that the 19-mer, which can form a perfect shape without any "holes" ("6–7–6" structure), is a prototype of peculiar behavior is, however, wrong. This is due to the existence of degenerate less symmetric ground-state conformations [as the exemplary conformation shown in Fig. 5.6(c)].

The described geometric peculiarities are, however, only properties of very short chains. One of the largest of the "small" chains that still possesses a nonspherical ground state is the 27-mer with the ground-state conformation shown in Fig. 5.6(d). For larger systems, the relative importance of the interior monomers will increase, because of the larger number of possible contacts. This requires the number of surface monomers to be as small as possible, which results in compact, sphere-like shapes. A representative example is the 30-mer shown in Fig. 5.6(e).

5.3 The Θ transition

The scaling behavior of several quantities at and close to the Θ point in three dimensions has been of substantial interest [35, 124, 131–133, 138–144]. Nonetheless, the somewhat annoying result is that the nature of this phase transition is not yet completely understood.

Lattice type	T_Θ	Model	Ref.
sc	3.64 ... 4.13	single chain	[138]
	3.713(7)	single chain	[139]
	3.65(8)	single chain	[141]
	3.716(7)	single chain	[124]
	3.60(5)	single chain	[125]
	3.62(8)	single chain	[126]
	3.717(3)	single chain	[35]
	3.717(2)	polymer solution	[131]
	3.745	lattice theory	[143]
	3.71(1)	polymer solution	[132, 133]
	3.72(1)	single chain	[137]
fcc	8.06 ... 9.43	single chain	[138]
	8.20(2)	single chain	[140, 142]
	8.264	lattice theory	[143]
	8.18(2)	single chain	[137]

Table 5.1 T_Θ values on the sc and fcc lattice from literature.

The associated tricritical $\lim_{n \to 0} O(n)$ field theory has an upper critical dimension $d_c = 3$, but the predicted logarithmic corrections [120–122] could not yet be clearly confirmed. Discussing freezing and collapse on regular lattices, here we mainly focus on the critical temperature T_Θ for polymers on the sc and on the fcc lattice. Values of T_Θ on sc and fcc lattices from various studies are compiled in Table 5.1.

As our main interest is dedicated to the expected difference of the collapse and freezing temperatures, we will focus here on the discussion of the scaling behavior of the finite-size deviation of the maximum specific-heat temperature of a finite-length polymer from the Θ temperature, $T_c(N) - T_\Theta$ [127, 128, 130, 132, 133]. For polymer solutions, Flory–Huggins mean-field theory [145] suggests

$$\frac{1}{T_{\text{crit}}(N)} - \frac{1}{T_\Theta} \sim \frac{1}{\sqrt{N}} + \frac{1}{2N}, \tag{5.1}$$

where $T_{\text{crit}}(N)$ is the critical temperature of a solution of chains of finite length N and $T_\Theta = \lim_{N \to \infty} T_{\text{crit}}(N)$ is the collapse transition temperature. In this case, field theory [120] predicts a multiplicative logarithmic correction of the form $T_{\text{crit}}(N) - T_\Theta \sim N^{-1/2}[\ln N]^{-3/11}$. Logarithmic corrections to the mean-field theory of single chains are known, for example, for the finite-chain Boyle temperature $T_B(N)$, where the second virial coefficient vanishes. The scaling of the deviation of $T_B(N)$ from T_Θ reads [124]:

$$T_B(N) - T_\Theta \sim \frac{1}{\sqrt{N}(\ln N)^{7/11}}. \tag{5.2}$$

However, the mean-field-motivated fit without explicit logarithmic corrections,

$$T_c(N) - T_\Theta = \frac{a_1}{\sqrt{N}} + \frac{a_2}{N}, \qquad (5.3)$$

also has been found to be consistent with numerical data [127, 128]. Up to corrections of order $N^{-3/2}$, Eq. (5.3) is equivalent to

$$\frac{1}{T_c(N)} - \frac{1}{T_\Theta} = \frac{\tilde{a}_1}{\sqrt{N}} + \frac{\tilde{a}_2}{N}, \qquad (5.4)$$

which had been suggested by numerical data obtained in grand canonical analyses of lattice homopolymers and the bond-fluctuation model [130, 132, 133].

The situation remains diffuse as there is still no striking evidence for the predicted logarithmic corrections (i.e., for the field-theoretical tricritical interpretation of the Θ point) from experimental or numerical data. A scaling analysis of the N-dependent collapse transition temperatures can be performed by identifying collapse peak temperatures $T_c(N)$ of individual specific-heat curves, and extrapolating from it the $N \to \infty$ limit T_Θ. Inverse critical temperatures obtained in independent long-chain nPERMss chain-growth simulations (sc: up to $N_{\max} = 32\,000$, fcc: up to $N_{\max} = 4\,000$) [137] are plotted in Fig. 5.7 as functions of $N^{-1/2}$. Fits according to the ansatz (5.4) are also shown and consistent with the data. Optimal fit parameters using the data in the intervals $200 \leq N \leq 32\,000$ (sc) and $100 \leq N \leq 4\,000$ (fcc) are $T_\Theta^{\mathrm{sc}} = 3.72(1)$, $\tilde{a}_1 \approx 2.5$, and $\tilde{a}_2 \approx 8.0$ (sc) and $T_\Theta^{\mathrm{fcc}} = 8.18(2)$, $\tilde{a}_1 \approx 1.0$, and $\tilde{a}_2 \approx 5.5$ (fcc).

For the single-chain system, field theory [121] predicts the specific heat to scale at the Θ point like $C_V(T = T_\Theta)/N \sim (\ln N)^{3/11}$. Short-chain simulations [141] do not reveal a logarithmic behavior at all, whereas for long chains a scaling closer to $\ln N$ is found [35]. The situation is similar for structural quantities such as the end-to-end distance and the radius of gyration.

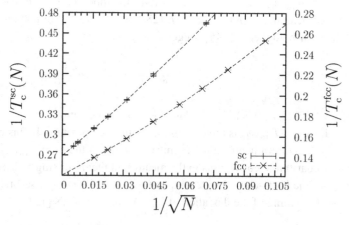

Fig. 5.7   Inverse collapse temperatures for several chain lengths on sc ($N \leq 32\,000$, left scale) and fcc lattices ($N \leq 4\,000$, right scale). Drawn lines are fits according to Eq. (5.4). From [137].

Summarizing this discussion, if logarithmic corrections as predicted by tricritical field theory are present at all, even chain lengths $N = 32\,000$ on an sc lattice and $N = 4\,000$ monomers on the fcc lattice are too small to observe deviations from the mean-field picture. In this case multiplicative and additive logarithmic corrections to scaling are hidden in the fit parameters of the "mean-field-like fits." The subleading additive corrections are expected to be of the form $\ln(\ln N)/\ln^2 N$ [123]. They thus not only disappear very slowly – they are also even of the same size as the leading scaling behavior, which makes it extremely unlikely to observe the logarithmic corrections in computational studies at all [123]. Similar additive logarithmic scaling is also known, for example, from studies of the two-dimensional XY spin model [146].

5.4 Freezing and collapse in the thermodynamic limit

From the results for the freezing and the collapse transition, we conclude that for self-avoiding interacting polymers with stiff bonds on regular lattices both transitions remain well separated in the extrapolation toward the thermodynamic limit. The existence of a liquid phase depends on the range of attraction in the models used. Considering a more general square-well contact potential between nonbonded monomers in our parametrization,

$$v(r) = \begin{cases} \infty & r \le 1, \\ -1 & 1 < r \le \lambda, \\ 0 & \lambda < r, \end{cases} \tag{5.5}$$

the attractive interaction range is simply $R = \lambda - 1$. There is a crossover threshold known for colloids interacting via Lennard-Jones-like and Yukawa potentials, where different solid phases can coexist, $R_c^{(1)} \approx 0.01$ [147–149]. Following Ref. [150], there is also another phase boundary, namely between stable and metastable colloidal vapor–liquid (or coil–globule) transitions, in the range $0.13 < R_c^{(2)} < 0.15$. Other theoretical and experimental approaches yield slightly larger values, $R_c^{(1)} \approx 0.25$ [147, 151–153]. Below $R_c^{(2)}$, the liquid (globule) phases is only metastable. The specific bond-fluctuation model used in Ref. [127] corresponds to $R = 0.225$, i.e., it lies in the crossover regime between the stable and metastable liquid phases [128]. Consequently, the crystallization and collapse transition merge in the infinite-chain limit and a stable liquid phase was only found in a subsequent study of a bond-fluctuation model with larger interaction range [128].

Qualitatively, analogous to the behavior of colloids, the separate stable crystal, globule, and random-coil (pseudo)phases can be explained. Since the range of interactions seems to play a crucial, quantitative role, it is interesting to question to what extent the colloidal picture in the compact crystalline and globular phases is systematically modified for polymers with different nonbonded interaction ranges, where steric constraints (through covalent bonds) are a priori not negligible. We return to the discussion of the influence of the nonbonded interaction range upon the formation of structural phases more systematically in Section 6.5.

6

Crystallization of elastic polymers

A central result of the discussion in the last chapter was the strong influence of finite-size effects on the freezing behavior of flexible polymers constrained to regular lattices. Thus, (unphysical) lattice effects interfere with (physical) finite-size effects and the question remains what polymer crystals of small size could look like. Since all effects in the freezing regime are sensitive to system or model details, this question cannot be answered in general. Nonetheless, it is obvious that the surface exposed to a different environment, e.g., a solvent, is relevant for the formation of the whole crystalline or amorphous structure. This is true for any physical system. If a system tries to avoid contact with the environment (a polymer in bad solvent or a set of mutually attracting particles in vacuum), it will form a shape with a minimal surface. A system that can be considered as a continuum object in an isotropic environment, like a water droplet in the air, will preferably form a spherical shape.

But what if the system is "small" and discrete? Small crystals consisting of a few hundred cold atoms, e.g., argon [154], but also as different systems as spherical virus hulls enclosing the coaxially wound genetic material [155, 156] exhibit an icosahedral or icosahedral-like shape. But why is just the icosahedral assembly naturally favored?

The capsid of spherical viruses is formed by protein assemblies, the protomers, and the highly symmetric morphological arrangement of the protomers in icosahedral capsids reduces the number of genes that are necessary to encode the capsid proteins. Furthermore, the formation of crystalline facets decreases the surface energy, which is particularly relevant for small atomic clusters. Another point is that the arrangement of a finite number of constituents (atoms, proteins, monomers) on the facets of an icosahedron optimizes the interior space filling and thus reduces the total energy of the system. This is also the reason why it can be expected that elastic polymers favor icosahedral shapes as well [52,53]. This kind of crystallization and finite-size dependent deviations from it is what we will discuss in more detail in this chapter.

6.1 Lennard-Jones clusters

In linear elastic polymers, bonds that connect monomers are not stiff. Within a certain range, bonded monomers can adapt their distance to external perturbations without much energetic effort. One can imagine the bond as a rather floppy spring. For this reason, crystalline structures of elastic polymers with nonbonded monomers interacting with each other will exhibit similarities compared to atomic or Lennard-Jones clusters, provided the

Fig. 6.1 (a) Anti-Mackay and (b) Mackay growth overlayer [158] on the facet of an icosahedron. From [53].

energy and length scales of bonded and nonbonded interactions enable this. It is therefore instructive to review the size-dependent features of small Lennard-Jones clusters.

Many-particle systems governed by van der Waals forces are typically described by the Lennard-Jones (LJ) pair potential (1.34),

$$V_{LJ}(r_{ij}) = 4\epsilon \left[\left(\frac{\sigma}{r_{ij}} \right)^{12} - \left(\frac{\sigma}{r_{ij}} \right)^{6} \right], \tag{6.1}$$

where r_{ij} is the distance between two atoms located at $\mathbf{r}_i$ and $\mathbf{r}_j$ $(i,j = 1,\ldots,N)$, respectively. An important example are atomic clusters, whose structural properties have been the subject of numerous studies, mainly focusing on the identification of ground-states and their classification. It has been estimated [157] that icosahedral-like LJ clusters are favored for systems with $N < 1\,690$ atoms. Larger systems prefer decahedral structures until for $N > 213\,000$ face-centered cubic (fcc) crystals dominate. In the small-cluster regime, energetically optimal icosahedral-like structures form by atomic assembly in overlayers on facets of an icosahedral core. There are two generic scenarios (see Fig. 6.1): either hexagonal closest packing (hcp) is energetically preferred (anti-Mackay growth), or the atoms in the surface layer are arranged as in the fcc-shaped tetrahedral segment of the interior icosahedron (Mackay growth) [158]. The surprisingly strong dependence of structural liquid–solid transitions on the system size has its origin in the different structure optimization strategies. "Magic" system sizes allow for the formation of most stable complete icosahedra $(N = 13, 55, 147, 309, 561, 923, \ldots)$ [159]. Except for a few exceptional cases – for $13 \leq N \leq 147$ these are $N = 38$ (fcc truncated octahedron), 75–77 (Marks decahedra), 98 (Leary tetrahedron), and 102–104 (Marks decahedra) [157] – LJ clusters typically possess an icosahedral core and overlayers are of Mackay $(N = 31$–$54, 82$–$84, 86$–$97,$ 99–$101, 105$–$146, \ldots)$ or anti-Mackay $(N = 14$–$30, 56$–$81, 85, \ldots)$ type [160].

It thus seems obvious that there are strong analogies in liquid–solid and solid–solid transitions of LJ clusters and classes of flexible polymers [52, 53, 161–165], but also between colloid and polymer systems (see, e.g., [166]). For this reason, we will now discuss some basic properties of generic icosahedral structures.

6.2 Perfect icosahedra

Since we will consider polymer chains with lengths in the icosahedral regime, it is instructive to review properties of perfect icosahedra. The surface of the icosahedron consists of

Table 6.1 Exemplified numbers of particles in the interior bulk (N_n^{bulk}) and in the surface layer (N^{surf}) for icosahedral shapes with $n \leq 6$ Mackay layers surrounding the central particle; $N_n = N_n^{\text{bulk}} + N^{\text{surf}}$ is the total "magic" number of particles. Also given is the surface-to-volume ratio $r_{\text{sv}} = N_n^{\text{surf}}/N_n^{\text{bulk}}$.

n	N_n^{bulk}	N_n^{surf}	N_n	r_{sv}
1	1	12	13	12.00
2	13	42	55	3.23
3	55	92	147	1.67
4	147	162	309	1.10
5	309	252	561	0.82
6	561	362	923	0.65

20 triangular faces, where two of each share an edge and five a common vertex, i.e., in total there are 30 edges and 12 vertices. Thus, the smallest number of LJ particles necessary to form a stable, perfectly icosahedral shape is $N_1 = 13$ (12 on the surface, one in the center). The next larger icosahedral structure possesses two overlayers which are of Mackay type [see Fig. 6.1(b)]. In this case, the top layer is composed of 60 identical triangles sharing 42 vertices. Therefore, $N_2 = 55$ particles are needed to construct an icosahedron with two layers surrounding the central particle. The numbers 13 and 55 are "magic numbers" because only systems with precisely these sizes can form perfectly icosahedral shapes. Putting the third complete layer on top requires $N_3 = 147$ particles, which is the next magic number in this hierarchy. This can be continued *ad infinitum*. Exemplified magic numbers of particles N_n, the number of particles in the interior bulk (i.e., without the top layer) N_n^{bulk}, and in the top layer N_n^{surf} are listed in Table 6.1 for icosahedra with up to six overlayers.

Interpreting the process of structure formation of a small system of LJ particles as a conformational transition, it is particularly interesting to consider the structural behavior at the surface as it is for these system sizes more relevant than the bulk effects. In the thermodynamic limit, of course, the phase transition will be driven mainly by the crystal formation in the bulk (fcc structure in the case of LJ particles). However, the systems we are going to study in this chapter are so small that the general aspects of nucleation transitions for very large systems are no longer valid. The crystallization of small systems depends extremely on the precise system size – a lesson that has already been taught in Section 5.2 – and this is due to surface effects.

For this reason it is instructive to study the surface-to-volume ratio of these icosahedral structures. As can be concluded from the numbers listed in Table 6.1, the surface-to-volume ratio $r_{\text{sv}} = N_n^{\text{surf}}/N_n^{\text{bulk}}$ not surprisingly decreases with the number of particles N_n, but this happens slowly. The magic numbers listed in Table 6.1 follow a clear hierarchy. The number of particles in the surface layer can be derived from the Mackay overlayer growth principle as depicted in Fig. 6.1(b). Since the order of the layer n corresponds to $(n + 1)$ particles lining up at each edge, the new top layer contains $(n - 1)(n - 2)/2$ particles in the center of each facet, surrounded by $(n-1)$ non-vertex particles in the edges and on the vertices. The total number of particles in the surface layer is:

$$N_n^{\text{surf}} = 20\frac{1}{2}(n-1)(n-2) + 30(n-1) + 12 = 10n^2 + 2. \tag{6.2}$$

The number of particles in the bulk is easier to calculate by the recursive relation

$$N_n^{\text{bulk}} = N_{n-1}^{\text{bulk}} + N_{n-1}^{\text{surf}} = N_{n-1}^{\text{bulk}} + 10n^2 - 20n + 12, \tag{6.3}$$

as the growth of the bulk volume with n is associated with adding the number of surface particles of the former surface layer of order $(n-1)$ to the previous volume. The ansatz $N_n^{\text{bulk}} = an^3 + bn^2 + cn^2 + d$ is suitable to solve the difference equation originating from Eq. (6.3). This yields

$$N_n^{\text{bulk}} = \frac{10}{3}n^3 - 5n^2 + \frac{11}{3}n - 1, \tag{6.4}$$

where we have made use of the numbers listed in Table 6.1 to determine the parameters. Finally, the magic numbers are obtained from $N_n = N_n^{\text{surf}} + N_n^{\text{bulk}}$ and can be expressed as

$$N_n = \frac{10}{3}n^3 + 5n^2 + \frac{11}{3}n + 1. \tag{6.5}$$

In the limit of large systems ($n \to \infty$), $N_n^{\text{bulk}} \sim N_n$ and $N_n^{\text{surf}} \sim N_n^{2/3}$. Thus, in this limit, the surface-to-volume ratio vanishes like

$$r_{\text{sv}} \sim n^{-1}, \tag{6.6}$$

or, with Eq. (6.5),

$$r_{\text{sv}} \sim N_n^{-1/3}. \tag{6.7}$$

This rather slow decrease gives a strong indication for the expectation that even the crystallization process of comparatively large systems with hundreds to thousands of particles will be significantly influenced by strong surface effects. This is not only true for magic LJ clusters; it is a general result, which is also valid for systems that do not form perfectly icosahedral shapes. As we will show in the following systematic analysis for elastic and flexible LJ polymers, size-dependent peculiarities are relevant for the individual crystallization processes of these systems, too.

6.3 Liquid–solid transitions of elastic flexible polymers

6.3.1 Finitely extensible nonlinear elastic Lennard-Jones polymers

The crystallization of polymers can be nicely studied by means of a model that enables the formation of icosahedral or icosahedral-like global energy minimum shapes that are virtually identical to LJ clusters of the same size. This means that bonds between monomers must be highly elastic such that the energetic excitation barriers for changes of the bond length within a certain range are extremely small. Therefore, elastic polymers can be

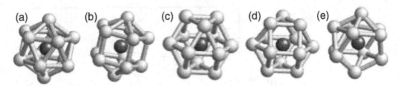

Fig. 6.2 Possible shell conformations ($r_c = 2.5\sigma$) of a monomer possessing 12 (a)–(d) or 11 (e) neighbors: (a) icosahedron, (b) elongated pentagonal pyramid [168], (c) cuboctahedron (fcc), (d) triangular orthobicupola (hcp) [168], (e) incomplete icosahedron. Sticks illustrate shell contacts, not bonds. From [53].

modeled by a combination of the FENE potential (1.38) with (1.39) and a slightly modified version of the Lennard-Jones potential (1.34). The energy of a conformation therefore reads

$$E = \frac{1}{2} \sum_{\substack{i,j=1 \\ i \neq j}}^{N} V_{\mathrm{LJ}}^{\mathrm{mod}}(r_{ij}) + \sum_{i=1}^{N-1} V_{\mathrm{bond}}^{\mathrm{sym}}(r_{ii+1}). \tag{6.8}$$

The LJ potential is truncated and shifted and reads

$$V_{\mathrm{LJ}}^{\mathrm{mod}}(r_{ij}) = V_{\mathrm{LJ}}(\min(r_{ij}, r_c)) - V_{\mathrm{LJ}}(r_c), \tag{6.9}$$

with $V_{\mathrm{LJ}}(r_{ij})$ from Eq. (1.34). As usual, r_{ij} denotes the distance between the ith and jth monomer and r_c is the cutoff distance.[1]

6.3.2 Classification of geometries

Structural changes between different geometries appear to be a characteristic feature of the behavior of LJ clusters at low temperatures. Since we expect a similar behavior also for elastic polymers, it is interesting to classify the geometrical structures in order to specify conformational transitions. One possibility to do so is, e.g., to introduce certain bond orientation parameters [167]. Here, we restrict ourselves to information provided by the contact map, where two monomers are considered as being in contact if their distance is smaller than a threshold value r_{contact}.

The total number of monomer contacts is not an appropriate measure for classification since in the interior of a frozen polymer every monomer usually has exactly 12 neighbors. Instead, the contacts between those 12 neighbors reflect their arrangement, which corresponds to the local geometry. In Fig. 6.2(a)–(d), different conformations of basic cells composed of a monomer and its 12 neighbors are shown. Counting the contacts between the neighbors is a simple but efficient way to characterize different types: only the icosahedral cell [Fig. 6.2(a)] reveals 30 shell contacts corresponding to the 30 edges of an icosahedron. It is always found in the center of any icosahedral conformation. Consequently, if there is no such basic element, the global geometry cannot be icosahedral. On the other hand,

---

[1] In the actual generalized Monte Carlo simulations [52, 53] that yielded the results discussed in the following, the parameter settings were as follows. The LJ energy scale was set to $\epsilon = 1$ and the length scale was given by $\sigma = 2^{-1/6} r_0$ with the minimum-potential distance $r_0 = 0.7$. The LJ cutoff distance was chosen to be $r_c = 2.5\sigma$. FENE potential parameters were $R = 0.3$ and $K = 40$.

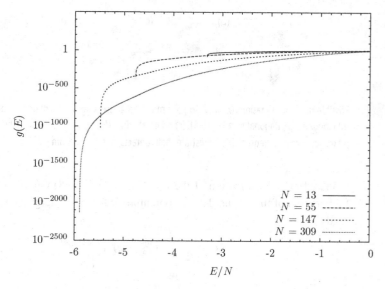

Fig. 6.3　Density of states for polymers forming complete icosahedra. From [53].

icosahedral cells are also formed by a sufficiently large anti-Mackay overlayer at the corners of the icosahedral core. If the number of outer monomers is too small, defects in the icosahedral cell occur [Fig. 6.2(e)], where the central monomer and its 11 neighbors form 25 shell contacts. The total number of both structures n_{ic} is a suitable "order" parameter [52, 53] which allows a classification of the global geometry at low temperatures, given roughly by

$$n_{ic} \begin{cases} = 0, & \text{non-icosahedral} \\ = 1, & \text{icosahedral + Mackay,} \\ \geq 2, & \text{icosahedral + anti-Mackay.} \end{cases} \tag{6.10}$$

More precisely, if $n_{ic} = 0$, the polymer forms a non-icosahedral structure, e.g., it is decahedral or fcc-like; $n_{ic} = 1$ indicates icosahedral geometry with Mackay overlayer or a complete icosahedron which might possess a few monomers bound in anti-Mackay type. Finally, for $n_{ic} \geq 2$, the monomers form an icosahedral core with a considerably extended anti-Mackay overlayer. The probabilities $p_{n_{ic}}(T)$ for the different values of n_{ic} as a function of temperature provide the necessary information to reveal structural transitions.

Figure 6.2(b) shows an elongated pentagonal pyramid with 25 shell contacts which is the basic module of five-fold symmetry axes in icosahedra and decahedra. It also occurs along the edges of an icosahedral core, which is covered by an anti-Mackay overlayer. Hence, it appears in icosahedral conformations of polymers with $N \geq 31$ monomers and in decahedral structures. Besides, it is formed at the edges of the central tetrahedron in conformations with a tetrahedral symmetry. An example is the ground state of the LJ cluster with $N = 98$ or the low-energy minima for $N = 159$ and $N = 234$. In consequence, the total number of elongated pentagonal pyramids n_{epp} can be used to distinguish decahedral and tetrahedral structures. Figures 6.2(c) and (d) show conformations that hardly differ because both of them possess 24 neighbor–neighbor contacts and occur in almost all geometries considered here. Only the truncated fcc-octahedron (i.e., the ground state of the 38-mer) does

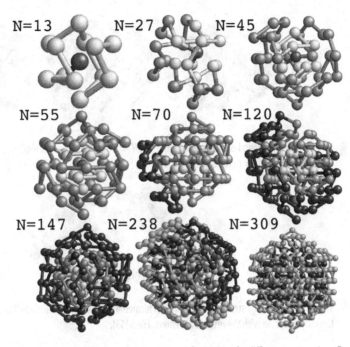

Fig. 6.4 Putative icosahedral or icosahedral-like ground-state state conformations for different system sizes. From [53].

not exhibit triangular orthobicupolae [Fig. 6.2(d)]. Cuboctahedra [Fig. 6.2(c)] are related to fcc- and triangular orthobicupolae [Fig. 6.2(d)] to hcp-packing. For the classification of the formation of crystalline structures, we will in the following mainly focus on the analysis of the number of complete and defected icosahedral cells, n_{ic}. Sophisticated, e.g., multi-canonical, simulations with a suitable set of Monte Carlo updates are needed to obtain the expectation values and distributions of these indicators with high accuracy [52, 53]. For the discussion of the thermodynamics of crystallization by means of energetic fluctuations, a precise estimate of the density of states $g(E)$ is necessary as the specific heat is derived from it. In conventional canonical analysis, peaks in the specific heat signalize transitions between structural phases accompanied by strong energetic fluctuations. The density of states [53] is plotted in Fig. 6.3 for several magic chain lengths and covers hundreds to thousands of *orders of magnitude* in the thermodynamically interesting transition regimes.

6.3.3 Ground states

Global energy minimum conformations of the elastic polymers are virtually identical to ground-state configurations of LJ clusters, i.e., for almost all system sizes, the ground state is icosahedral or icosahedral-like (Fig. 6.4). At low temperatures, the covalent, elastic bonds cause only small deviations from the corresponding atomic cluster, since the minimum of the FENE potential is close to the equilibrium distance. The bonds between adjacent monomers arrange themselves in a way that minimizes the bond potential. As a consequence, bonds between different shells of an icosahedral core are rare since the

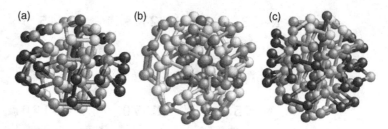

Fig. 6.5 Non-icosahedral ground-state conformations: (a) $N = 75, r_c = 2.5\sigma$ decahedral, (b) $N = 98, r_c = 5\sigma$ tetrahedral, (c) $N = 102, r_c = 5\sigma$ decahedral. From [53].

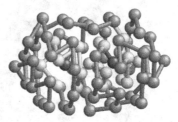

Fig. 6.6 The minimum-energy structure for $N = 87$ is formed by two entangled icosahedral cores ($r_c = 2.5\sigma$). The ends of the chain coincide with the respective centers. From [53].

corresponding monomer distances are smaller than the equilibrium distance and would entail higher bond energies. For longer chains, the central icosahedral cell is strongly compressed. A similar effect occurs in decahedral conformations. Bonds between monomers on the central axis are favorable since their length is close to the optimum. This forces the polymer chain to adapt this axis at low temperatures [Figs. 6.5(a),(c)].

Perfect icosahedra are formed as expected by polymers with "magic" numbers of $N = 13, 55, 147, 309, \ldots$ monomers. If N slightly exceeds a "magic" polymer length, the polymer builds an icosahedral core with an anti-Mackay overlayer, which grows with the chain length. At some point (e.g., $N > 30$ and $N > 80$), the overlayer adopts the structure of the core by changing to Mackay type. Further increase of the number of monomers completes the outer shell and leads to the next perfect icosahedral shape. A few polymers of certain sizes possess ground-state conformations that correspond to different non-icosahedral geometries. One finds a truncated octahedron for $N = 38$ and a decahedral conformation for $N = 75 - 77$ [Fig. 6.5(a)]. Some deviations are caused by the cutoff of the LJ potential, so the chains with $N = 81, 85, 87, 98, 102$ monomers possess lowest-energy structures that do not correspond to the respective clusters unless an untruncated LJ potential is considered. Using the cutoff $r_c = 2.5\sigma$, icosahedral ground states with Mackay overlayer for $N = 81, 85, 98, 102$ are found, and for $N = 87$ a conformation with two merged icosahedral cores (Fig. 6.6).

6.3.4 Thermodynamics of liquid–solid transitions toward complete icosahedra

Chains that form complete icosahedra exhibit a very clear and uniform thermodynamic behavior. Two separate conformational transitions occur and are indicated by peaks in the

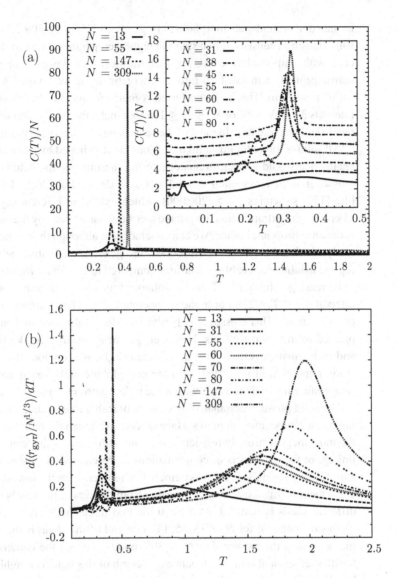

Fig. 6.7 (a) Specific heats (in the inset shifted by a constant value for clarity) and (b) fluctuations of the mean radius of gyration $d\langle r_{gyr}\rangle/dT$ as functions of the temperature for chains of different length. From [52].

specific C_V/N heat and in the fluctuations of the radius of gyration $d\langle r_{gyr}\rangle/dT$. As shown in Figs. 6.7(a) and (b), the icosahedra melt in the interval $0.3 < T < 0.5$ and a liquid-like regime is reached where the polymer still arranges in a globular shape but exhibits no distinct structure. Hence, the icosahedral order parameter $\langle n_{ic}\rangle \approx 1$ changes to $\langle n_{ic}\rangle \approx 0$. The corresponding peak in the normalized specific heat increases rapidly with system size and allows in principle a precise determination of the melting temperature.

There are some characteristic features that can be read from Figs. 6.7(a) and (b): first, the transitions are particularly strong for chains with "magic" length, which possess perfectly icosahedral ground-state morphology (e.g., $N = 13, 55, 147, 309$). A second type of liquid–solid transition consists of two steps, at higher temperatures the formation of an icosahedral core with anti-Mackay overlayer that transforms at lower temperatures by monomer rearrangement at the surface into an energetically more favored Mackay layer ("solid–solid" transition). This is the preferred scenario for most of the chains with lengths in the intervals $31 \leq N \leq 54$ or $81 \leq N \leq 146$ that make the occupation of edge positions in the outer shell unavoidable. In most of the remaining cases, typically anti-Mackay layers form.

All solid–solid or liquid–solid transitions are dominated by finite-size effects. For longer chains one would expect, in analogy to the thermodynamic behavior of LJ clusters, the crossover from icosahedral ground states to decahedral ($N \gtrsim 1\,500$) and later to fcc-like [157] structures ($N \gtrsim 200\,000$), which exhibit a different crystallization behavior. Therefore the extrapolation toward the thermodynamic limit by means of finite-size scaling is not an appropriate choice. We can conclude that although the elastic polymers are entropically restricted by the covalent bonds, the general, qualitative behavior in the freezing regime exhibits noticeable similarities compared to LJ cluster formation.

Increasing T further leads to the collapse transition with transition temperatures in the interval $1.0 \leq T \leq 2.0$. For higher temperatures, the chains arrange randomly in extended conformations. This crossover is hardly signaled in the specific heat curves, as it is suppressed by finite-size effects. However, geometric quantities like the radius of gyration and its fluctuation [Fig. 6.7(b)] give some insight. It is obvious that the liquid–solid transition remains well separated from the coil-globule collapse; moreover the intermediate temperature interval increases within the investigated chain length interval.

The solid phase is dominated by extremely stable icosahedra, which, however, depending on the temperature, exhibit surface defects. Although the mobility of a great majority of the monomers is strongly restricted, there are still changes that can be observed. First, the linkage of the monomers of conformations representative for the ensemble in this regime can still vary strongly. Only at extremely low temperatures the lowest-energy conformation gains thermodynamic relevance. Starting from the center, the number of bonds connecting different shells is reduced. At $T = 0$ the first three shells are connected only by a single bond, such that for $N = 13, 55, 147$ one end of the chain is the central monomer and the other is at the surface. For $N = 309$, bonds between the corners of the third and the fourth shell exist also at $T = 0$, since the length of this bond is roughly equal to alternative bonds within the third shell, which include the monomers at the corners. Second the entire icosahedron undergoes a compactification with decreasing temperature.

6.3.5 Liquid–solid transitions of elastic polymers

In the following, we more systematically analyze the behavior of small elastic polymers in the liquid–solid transition regime. Particular emphasis will be dedicated to the chain-length dependence of geometric changes in the conformations of the polymers while passing the transition line.

The specific-heat curves for elastic polymers with different chain lengths have already been plotted in Fig. 6.7(a). The ground state of the short 13-mer is the energetically stable icosahedron whose almost perfect symmetry is only slightly disturbed by the FENE bonds.

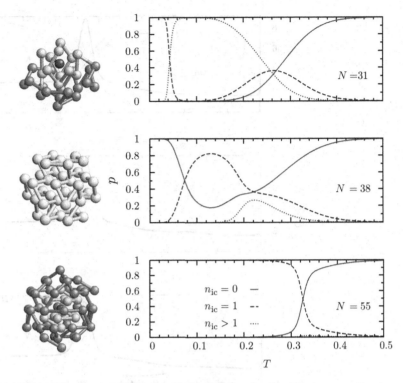

Fig. 6.8   Temperature dependence of the probability of icosahedral (Mackay: $n_{ic} = 1$, anti-Mackay: $n_{ic} > 1$) and non-icosahedral ($n_{ic} = 0$) structures for exemplified polymers with lengths $N = 31, 38, 55$. In the left panel, the corresponding lowest-energy morphologies are shown. From [52].

The melting transition is indicated by the peak at $T = 0.33$. With increasing system size the low-temperature conformations are less symmetric and the peak becomes broader. For $N > 30$, the crossover from Mackay-like ground states to anti-Mackay conformations takes place. The corresponding peak increases with growing N and shifts from $T \approx 0.04$ to higher temperatures. For the next "magic" polymer with $N = 55$, there is only one melting transition left and anti-Mackay-like structures are strongly suppressed. As an effect of the bond elasticity, the melting transitions occur at slightly higher temperatures than in the case of pure LJ clusters [160].

From the structural point of view, it is instructive to study the temperature dependence of the probability $p_{n_{ic}}$ that the "order" parameter, as defined in Eq. (6.10), is $n_{ic} \leq 1$ or $n_{ic} \geq 2$. Figure 6.8 shows the respective populations $p_{n_{ic}}$ for the structural morphologies of the chains with $N = 31, 38, 55$, parametrized by n_{ic}, as a function of temperature. For $N = 31$, one finds that liquid structures with $n_{ic} = 0$ dominate above $T = 0.5$, i.e., no icosahedral cells are present. Decreasing the temperature and passing $T \approx 0.4$, nucleation begins and the populations of structures with icosahedral cells ($n_{ic} = 1$ and $n_{ic} > 1$) increase. As has already been mentioned, the associated energetic fluctuations are also signaled in the specific heat [see inset of Fig. 6.7(a)]. Cooling further, always $n_{ic} > 1$, showing that the remaining monomers build up an anti-Mackay overlayer and create additional icosahedra. Very close to the temperature of the "solid–solid" transition near $T \approx 0.04$, where also

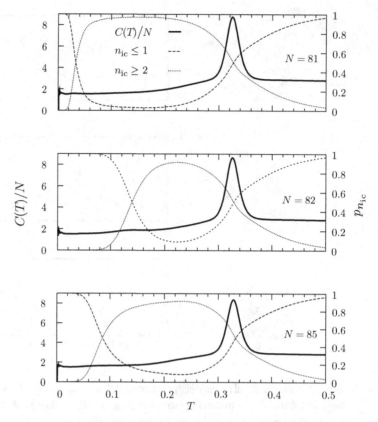

Fig. 6.9   Whereas the Mackay–anti-Mackay crossover is not recognizable in the specific heat curves, the probability of occurrence of specified numbers of icosahedral cells, $p_{n_{ic}}$, reveals the transition temperature ($r_c = 2.5\sigma$). From [53].

the specific heat exhibits a significant peak, the transition from anti-Mackay to Mackay overlayer occurs and the ensemble is dominated by frozen structures containing a single icosahedral cell surrounded by an incomplete Mackay overlayer.

The exceptional case of the 38-mer shows a significantly different behavior. A single icosahedral core forms and in the interval $0.08 < T < 0.19$ icosahedra with Mackay overlayer are dominant. Although the energetic fluctuations are weak, near $T \approx 0.08$, a surprisingly strong structural crossover to non-icosahedral structures occurs: the formation of a maximally compact fcc truncated octahedron.

The "magic" 55-mer exhibits a very pronounced transition from unstructured globules to icosahedral conformations with complete Mackay overlayer at a comparatively high temperature ($T \approx 0.33$). Below this temperature, the ground-state structure has already formed and is sufficiently robust to resist the thermal fluctuations.

The behavior of chains containing between 55 and 147 monomers again generally corresponds to that of LJ clusters [160] with the differences that the melting transition occurs at higher temperatures and that the Mackay–anti-Mackay transition temperature increases much faster with system size. Besides there are a few chains with ground states of types different from LJ clusters which are induced by the truncation of the LJ potential, as

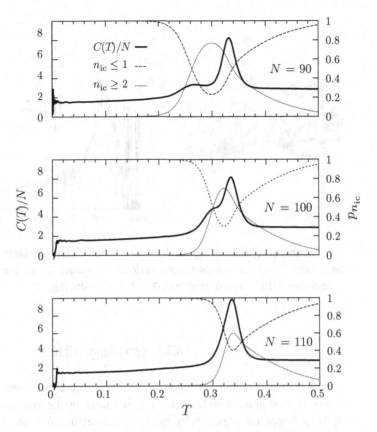

Fig. 6.10 Already for $N \geq 110$ the entire solid phase is dominated by Mackay ($n_{ic} = 1$) conformations ($r_c = 2.5\sigma$). From [53].

mentioned above. Anti-Mackay ground states are found up to system sizes of $N = 80$. In contrast to the clusters, $N = 81$ and $N = 85$ possess a global-energy minimum of Mackay type. Hence a solid–solid transition is encountered also in these cases. However, no peaks indicating the solid–solid transition are seen in the specific heat for $N = 81, 82, 85$, and the crossover to anti-Mackay conformations can only be identified in changes of n_{ic}. In Fig. 6.9, the probability $p_{n_{ic}}$ is plotted for these cases.

While one would expect higher transition temperatures for growing system size, this prediction fails in the case of $N = 85$, where the anti-Mackay energy minimum is almost as deep as the ground state due to an optimal arrangement of the outer monomers. Nevertheless, for $N \geq 90$, the transition shifts rapidly to higher temperatures (Fig. 6.9) and manifests in the specific heat as well. Whereas in the case of pure LJ clusters the two peaks remain separated up to sizes of 130 atoms [160], we observe both transitions merging already for $N \approx 100$. In contrast to the polymers, the Mackay–anti-Mackay transition temperature of clusters even decreases near $N = 120$, presumably because the anti-Mackay overlayer is almost complete, leading to a spherical and therefore stable conformation. This indicates that in this polymer model anti-Mackay structures lose weight compared to atomic clusters. It is also worth noting that for $N = 75$ one finds a crossover between decahedral and icosahedral conformations, which is indicated by a small peak at $T \approx 0.8$.

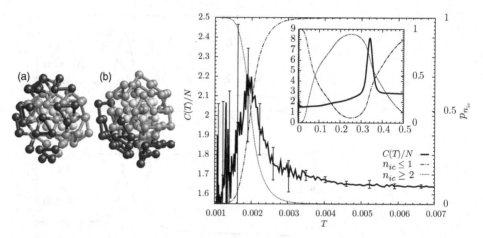

Fig. 6.11 (Left) Conformations of minimal energy for $N = 85$ with (a) anti-Mackay and (b) Mackay overlayer. (Right) The polymer with $N = 85$ and untruncated LJ potential ($r_c = \infty$) changes from anti-Mackay ($T < 0.002$) to Mackay conformations ($0.002 < T < 0.1$) and back ($0.1 < T < 0.34$). After [53].

6.3.6 Long-range effects

At this point we cannot yet judge whether the differences compared to the behavior of atomic clusters at medium temperatures are caused by the truncation of the LJ potential or by the polymer topology, i.e., by the bonds connecting to beads. To answer this question, we will discuss the change in long-range ordering when modifying the cutoff r_c. First, let us consider the polymer with $N = 85$ monomers employing the original LJ potential, where $r_c = \infty$. The corresponding LJ cluster is the largest with an anti-Mackay ground state in the interval $55 < N < 147$ [158]. Without cutoff, any deviation of polymer crystallization compared to the cluster behavior is caused by the bond potential only. In contrast to the previous discussion with truncated potential ($r_c = 2.5\sigma$), the anti-Mackay ground state is retained. The 30 outer monomers completely cover 10 faces of the icosahedral core and build an energetically favored structure [Fig. 6.11(a)]. One might notice that there are no bonds connecting monomers on different faces directly since the bond length would be too far from the potential minimum. This means that for very low energies only a few bond configurations are allowed and that the anti-Mackay state is much less metastable than the Mackay state [Fig. 6.11(b)] for which many more bond configurations with low energies are possible. This leads to an entropic dominance of the latter in the temperature interval $0.002 < T < 0.08$ as is visible in Fig. 6.11; a solid–solid transition that has not been reported for atomic clusters. This transition is also signaled by the specific heat at $T \approx 0.002$, since at this temperature the energetically favored anti-Mackay state prevails. The transition back to anti-Mackay conformations extends over a wider temperature interval around $T \approx 0.08$ and cannot be localized in the specific heat. Again, the change from a LJ cluster to a polymer leads to a greater prominence of Mackay conformations in this regime.

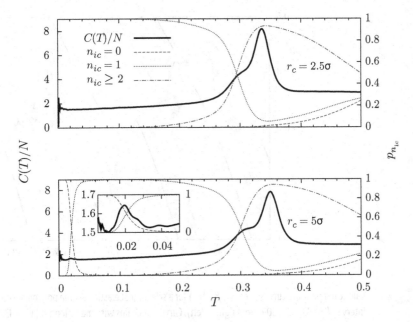

Fig. 6.12   The solid–solid transition to the decahedral ground state for the 102-mer is induced by long-range effects. From [53].

In order to study the influence of the cutoff on the thermodynamics at higher temperatures we now consider the doubled cutoff, $r_c = 5\sigma \approx 3.12$, and compare the exemplified system with $N = 102$. For the corresponding cluster, two separate transitions were observed at medium temperatures, which stands in contrast to the polymer with original cutoff ($r_c = 2.5\sigma$), where both transitions merge. Applying the enlarged cutoff we observe only a slight shift in the peak positions whereas the major differences compared to the behavior of the LJ cluster, as demonstrated in Ref. [160], persist (Fig. 6.12). It is still impossible to distinguish both conformational transitions clearly by means of the specific heat, i.e., the temperature domain where anti-Mackay-like conformations prevail is very small. One may conclude that this difference compared to the cluster behavior (i.e., the suppression of anti-Mackay states) is mainly an effect of the bonds.

The second interesting result is the different ground states. As in the case of unbounded clusters with untruncated LJ potential, the ground-state conformation is decahedral for $N = 102$ [Fig. 6.5(c)]. We can also identify the solid–solid transition toward Mackay structures: the decahedral-icosahedral crossover occurs at the temperature $T \approx 0.02$ (see Fig. 6.12), with a relatively prominent signal in the specific heat.

To conclude, there is some evidence that during the change from atomic LJ clusters to polymers, Mackay-like structures are more favored independently of a truncation of the LJ potential. This results in a shift of the Mackay–anti-Mackay (or Mackay overlayer melting) transition to higher temperatures and the formation of a Mackay-dominated temperature interval, as was discussed for the 85-mer with untruncated LJ potential (see also the discussion in Refs. [52, 53]).

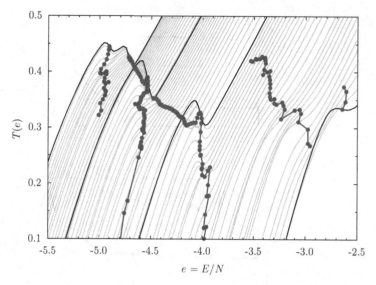

Fig. 6.13   Caloric temperature curves $T(e) = \beta^{-1}(e)$ for a selection of elastic, flexible polymers with chain lengths in the interval $N = 13, \ldots, 309$ (from right to left). Curves for chains with magic length ($N = 13, 55, 147, 309$) are bold. The relevant inflection points, indicating the conformational transitions on the basis of our analysis method, are marked by ● symbols. From [61].

6.4 Systematic analysis of compact phases

It has been demonstrated in Section 2.7.3 that microcanonical inflection-point analysis offers the possibility to study all structural transitions in a finite system uniquely and systematically. We here employ this method to estimate the transition points for the entire set of elastic Lennard-Jones homopolymers with $N = 13, \ldots, 309$ monomers [61].

Figure 6.13 shows the caloric temperature curves for elastic polymers with various chain lengths in the liquid and solid regimes, calculated from highly accurate density of states estimates obtained in sophisticated multicanonical Monte Carlo simulations [53]. The identified inflection points associated with conformational transitions are indicated by ● symbols. As expected, there is no general and obvious relation of the behavior of chains with slightly different lengths. This is due to the still dominant finite-size effects of the polymer trying to reduce their individual surface-to-volume ratio, which therefore strongly depends on optimal monomer packings in the interior and on the surface of the conformations. In the previous sections, we have found that for chains of moderate lengths ($N \leq 147$ [53, 165]), the different behavior can be traced back to the monomer arrangements on the facets of icosahedral structures, known as Mackay and anti-Mackay overlayers. Solid–solid transitions between Mackay and anti-Mackay structures are also possible under certain conditions in these systems [53, 165]. This can be seen in Fig. 6.14, where all transition temperatures $T_{\mathrm{tr}}(N) = \beta_{\mathrm{tr}}^{-1}(N)$ for liquid–solid and solid–solid transitions are plotted in dependence of the chain length N.[2] Symbols ● indicate first-order

---

[2]   The inflection-point analysis for $\beta(E)$ can also easily be applied to the coil–globule transition, which occurs at much higher temperatures (for this reason, transition points are not inserted in Fig. 6.14).

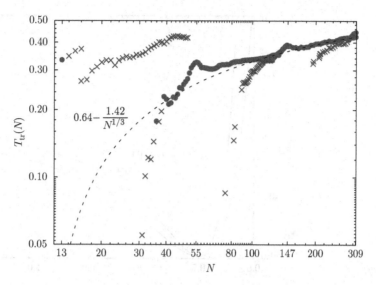

Fig. 6.14 Transition temperatures $T_{\mathrm{tr}}(N)$ of conformational transitions for small elastic polymers with chain lengths $N = 13, \ldots, 309$ in the liquid–solid and solid–solid transition regimes, obtained from inflection-point analysis. First-order transition points are marked by symbols •, second-order transition points by symbols ×. Also shown is a fit for the liquid–solid transition temperature toward the thermodynamic limit $N \to \infty$ (dashed line). From [61].

transitions, which for $N > 38$ can be associated to the respective liquid–solid transitions, whereas symbols × mark second-order transitions.

If the associated transition temperatures are smaller than the liquid–solid transition temperatures, the symbols indicating second-order behavior belong to solid–solid transitions, e.g., transitions between geometrical shapes with Mackay or anti-Mackay overlayers. Note the different behavior for "magic" chain lengths $N_{\mathrm{magic}} = 13, 55, 147, 309, \ldots$, in which icosahedral Mackay ground states form. Figure 6.14 also gives evidence for the convergence of the solid–solid and liquid–solid transition temperatures when N approaches a magic length. This behavior repeats for each N interval that finally ends at a certain magic length N_{magic}, where both transitions merge into a single first-order liquid–solid transition. The influence of the solid–solid effects weakens with increasing system size, while the liquid–solid transition remains a true phase transition in the thermodynamic limit. Inserted into the plot is a fit function $T_{\mathrm{tr}}(N) = T^{\mathrm{ls}} - aN^{-1/3}$, which suggests an estimate for the thermodynamic phase transition temperature $T_{\mathrm{tr}}^{\mathrm{ls}} \approx 0.64$.

6.5 Dependence of structural phases on the range of nonbonded interactions

We will now investigate how the occurrence and interplay of the structural phase of elastic, flexible polymers depends on the mutual interaction range between nonbonded monomers [127, 128, 135, 136]. For this purpose, we consider the single-polymer model (6.8), consisting of the FENE potential (1.39) for the bonded interaction and a truncated Lennard-Jones

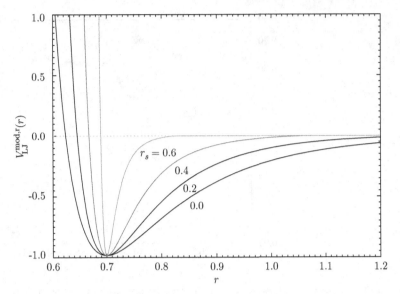

Fig. 6.15   Modified Lennard-Jones potential $V_{\mathrm{LJ}}^{\mathrm{mod,r}}(r)$ for different values of r_s.

potential for the interaction of nonbonded monomers. Similar to Eq. (6.9), we introduce it as

$$V_{\mathrm{LJ}}^{\mathrm{mod,r}}(r_{ij}) = V_{\mathrm{LJ}}^{\mathrm{r}}(\min(r_{ij}, r_c)) - V_{\mathrm{LJ}}^{\mathrm{r}}(r_c), \tag{6.11}$$

with the interaction-range-dependent LJ potential of the form [136]

$$V_{\mathrm{LJ}}^{\mathrm{r}}(r_{ij}) = 4\epsilon \left[\left(\frac{\sigma}{r_{ij} - r_s} \right)^{12} - \left(\frac{\sigma}{r_{ij} - r_s} \right)^{6} \right]. \tag{6.12}$$

In the following, we use units in which $\epsilon = 1$ and $\sigma(r_s) = (r_0 - r_s)/2^{1/6}$ (with $r_0 = 0.7$), for simplicity. The cut off radius is set to $r_c(r_s) = 2.5\sigma(r_s) + r_s$, such that $V_{\mathrm{LJ}}^{\mathrm{mod,r}}(r_{ij}) \equiv 0$ for $r_{ij} > r_c$ and $V_{\mathrm{LJ}}^{\mathrm{r}}(r_c) = (-3983616/244140625)\epsilon \approx -0.016317\epsilon$. The parameter r_s controls the width of the potential. The qualitative influence of r_s on the shape of the potential is shown in Fig. 6.15 for various values of r_s.

The parameter r_s is somewhat counterintuitive, as the width of the potential well decreases with increasing value of r_s. We consider the potential (6.9) ($r_s = 0$) as the limiting case with the maximum width, i.e., we concentrate on short-range effects ($r_0 > r_s > 0$) only, which are most interesting in this context. For the subsequent analysis it is, therefore, more convenient to introduce the potential width δ as a new parameter. It is defined through the square-well potential

$$V_{\mathrm{sq}}(r) = \begin{cases} \infty & \text{if } r \leq r_1, \\ -\epsilon_{\mathrm{sq}} & \text{if } r_1 < r < r_2, \\ 0 & \text{if } r \geq r_2, \end{cases} \tag{6.13}$$

with the constant $\epsilon_{\mathrm{sq}} = \epsilon/2 + V_{\mathrm{LJ}}^{\mathrm{r}}(r_c)$ such that $\epsilon_{\mathrm{sq}} = (236173393/488281250)\epsilon \approx 0.483683\epsilon$. The limiting distances of the well, r_1 and r_2, are given by $V_{\mathrm{LJ}}^{\mathrm{mod,c}}(r_1) =$

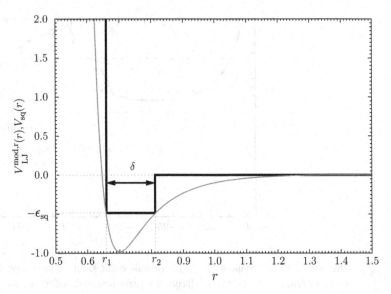

Fig. 6.16 The potential width $\delta = r_2 - r_1$ of $V_{LJ}^{mod,r}(r)$ is defined by the width of the associated square-well potential $V_{sq}(r)$ with depth $-\epsilon_{sq}$, such that $V_{LJ}^{mod,r}(r_{1,2}) = -\epsilon_{sq}$.

$V_{LJ}^{mod,c}(r_2) = -\epsilon_{sq}$, independently of r_s (see Fig. 6.16). The relationship between r_s and the potential width δ is linear:

$$\delta = r_2 - r_1 = \lambda_0(r_0 - r_s). \tag{6.14}$$

The proportionality constant is given by

$$\lambda_0 = 2^{1/6}\left[\left(1 + \frac{1}{\sqrt{2}}\right)^{1/6} - \left(1 - \frac{1}{\sqrt{2}}\right)^{1/6}\right] \approx 0.312382. \tag{6.15}$$

The maximum value of δ corresponds to the width of the unmodified Lennard-Jones potential ($r_s = 0$) and reads $\delta_{max} = \lambda r_0$ (which is ≈ 0.218667 in the units used here).

Figure 6.17 shows the inverse microcanonical temperature curves $\beta(E)$ for a polymer with 90 monomers for three different δ values. These curves are interesting in many regards and also tell us how important it is to perform systematic analyses of physical effects by varying a single parameter only, here the interaction range δ. By making use of the inflection-point analysis of $\beta(E)$, as described in Section 2.7.3, we identify two transitions in the plotted energy region. These are the second-order collapse or Θ transition at higher energies (circles), and the first-order inflection points (triangles), indicating the liquid–solid transition. This behavior is very similar to our discussion in previous sections.

However, what is different here is that, by decreasing the interaction range, the second-order transition approaches the first-order transition point. It intrudes into the energetic regime, bridged by the latent heat, that accommodates the first-order transition. In an infinitely large system, i.e., in the thermodynamic limit, we would simply speak of a "merger": the onset of a triple point, at which the liquid phase ceases to exist. In the small system considered here, the feature looks similar, but allows for different interpretations.

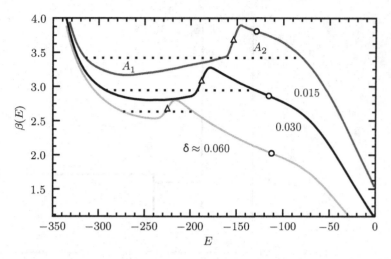

Fig. 6.17 Inverse microcanonical temperatures $\beta(E)$ of a flexible, elastic 90-mer for three values of δ. Triangles (positive peaks in the first derivative $\gamma(E)$)) and circles (negative γ peaks) indicate the inflection points that correspond to first-order and second-order transitions, respectively. Dotted lines are Maxwell constructions; A_1 and A_2 are coinciding areas enclosed by $\beta(E)$ and the Maxwell line. From [136].

First, the "merger" requires the first-order and the second-order transition to be identical, or, more precisely, to be indistinguishable. This is obviously not the case here. The transitions get closer, but are still separate. The second observation regards the Maxwell construction, which is exact in the thermodynamic limit. For the finite system, we can still perform this construction, but we are not actually forced to do so. There is no unphysical behavior in the $\beta(E)$ curves near the nucleation transition here (compare this with our discussions of Maxwell constructions in Section 2.7.2) that would need to be corrected by Maxwell construction.

We even observe that the location of the first-order inflection point deviates stronger from the intersection point of the Maxwell line with the $\beta(E)$ curve the smaller δ becomes. We can calculate the surface entropy, which corresponds to the areas $A_1 = A_2$, and see that these barriers become larger by decreasing the interaction range. The latent heat (the width of the Maxwell line) also increases. The strong antisymmetry of the microcanonical intruder, caused by the close second-order transition, is also an indication that the Maxwell-like interpretation is no longer particularly useful for systems like this. Another feature that resists conventional interpretation is that there is a crossover between the second-order collapse transition and the first-order nucleation transition in microcanonical temperature space: for δ sufficiently small, the "collapse" occurs at a lower temperature than the "nucleation"! This sounds weird, but the temperature should not be interpreted in the traditional way. Regarding the energy, everything is still as expected. The collapse transition occurs at higher energy than the liquid–solid transition and this does not change when approaching the limit $\delta \rightarrow 0$. The basic quantity is the energy, not the temperature, which is only a function of energy related to entropic changes. So, what the crossover really means is that,

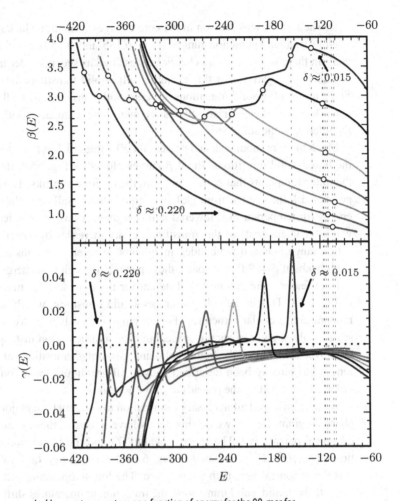

Fig. 6.18 (Top) Microcanonical inverse temperature as a function of energy for the 90-mer for $\delta \approx 0.015, 0.030, 0.060, 0.095, 0.125, 0.155, 0.185, 0.220$ (from top to bottom). Circles mark the inflection points. (Bottom) Derivative of the inverse temperature, $\gamma(E) = d\beta(E)/dE$, for the same values of the parameter δ. From [136].

below a certain threshold value of the interaction range δ, the entropic change has become larger in the case of the collapse transition, compared with the nucleation transition.

We will now investigate more systematically how the interaction range δ affects the structural transitions of this polymer. Figure 6.18(top) shows $\beta(E)$ for various values of δ, and its derivative, $\gamma(E) = d\beta(E)/dE$, is plotted in Fig. 6.18(bottom). The inflection points of $\beta(E)$, which are relevant for the identification of structural transitions, are marked by circles. The leftmost β curve belongs to $\delta = \delta_{\max} \approx 0.22$, which is the width of the standard LJ potential. In this case, we find a signal at very low energy, which is so weak that it is only reflected by a shoulder (or, more precisely, by an inflection point) in γ.[3] It is a solid–solid transition, where the monomer arrangement on the surface of the incomplete

---

[3] In inflection-point analysis, we refer to it as a third-order transition.

outer shell of the icosahedron in the core changes from anti-Mackay to Mackay type. For slightly smaller interaction ranges ($\delta \approx 0.185$ and 0.155), we find these transitions as well. In these cases they can clearly be classified as second-order transitions, because the γ peaks are negative. For this example, solid–solid transitions occur only for $\delta > 0.12$. Since these solid–solid transitions are essentially surface effects, they disappear for large chain lengths, where surface effects are negligible, compared with the volume effects in the interior of the solid body.

Much more pronounced, and present for all values of δ in the range considered here, are the liquid–solid transitions. In all β curves plotted in Fig. 6.18, these are represented by the inflection points that correspond with the positive γ peaks. The liquid–solid transition points mark the change from globular (liquid) to crystalline (solid) conformations, if the energy (or temperature) is lowered in this regime. As we have noted earlier, the energetic width (or latent heat) of this transition rapidly increases by lowering the effective inter-action range between nonbonded monomers. Whereas the transition points change only slightly about $\beta = 3.0$ in β space, they span a large region in energy space.

Both features are completely different for the third kind of transition that we identify in Fig. 6.18. This is the collapse or gas–liquid transition, which separates the phases of random and globular structures. For all parameter values δ, we see that they vary and scatter only slightly in energy space, but span a large region in β space. The merging vs. crossover behavior of the liquid–solid and gas–liquid transitions at very short interaction ranges has already been described above. All gas–liquid transition points are of second order, as indicated by the (broad) negative γ peaks.

We have itemized all necessary information to construct the major parts of the structural phase diagram. Before we do this, let us have a look at the transition behavior from the canonical perspective. The specific heat $c_V(T)$ and the fluctuation of the radius of gyra-tion, $d\langle r_{\mathrm{gyr}}\rangle(T)/dT$, are plotted in Fig. 6.19 for a selection of δ values. The temperature T is the canonical heat-bath temperature. The low-temperature peaks in the specific heat indicate the liquid–solid transitions; the transition temperatures shift slightly to lower tem-peratures with increasing interaction range. What is more interesting is the needle-like shape of the peak. As we already know from our discussion of the microcanonical results, the liquid–solid transitions are first-order transitions with coexisting liquid and solid phase. At these small values of δ the latent heat is particularly high, which explains the large fluctuation width that corresponds to the specific heat. The fluctuations of the radius of gyration in the same temperature region are also noteworthy and tell us that there is a further compactification of globular structures toward crystalline conformations.

The Θ collapse transition cannot be clearly identified in the specific heat curves. It is represented by the very broad plateau-like shoulders seen in Fig. 6.19(a) at tempera-tures higher than the liquid–solid transition point. A much more pronounced indicator for this transition is the radius of gyration. The fluctuations of this quantity possess a clear maximum signaling the transition. Nevertheless, the width of these peaks is also very large and increases with the interaction range. We discussed earlier that the peak position of high-temperature transitions in small systems sensitively depends on the quantity consid-ered. Therefore, a precise location of transition points by means of canonical quantities is virtually impossible.

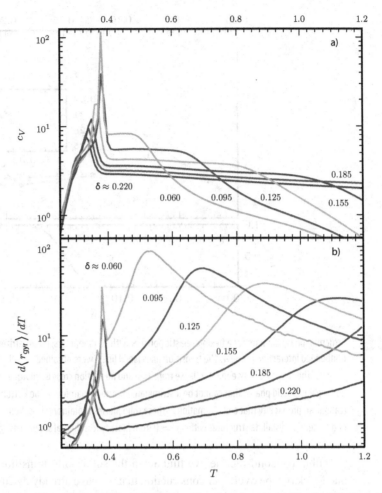

Fig. 6.19 (a) Specific heat $c_V(T)$ and (b) thermal fluctuation of the radius of gyration, $d\langle r_{gyr}\rangle(T)/dT$, of the 90-mer for various δ values as functions of the canonical temperature. After [136].

For the discussion of the structural phase diagram, then, we return to the microcanonical analysis. The complete phase diagram, parametrized by temperature T and interaction range δ, is shown in Fig. 6.20. The transition lines are based on consequent microcanonical inflection-point analysis, as described above. We find three major phases: the "gas phase" G, where random conformations dominate; the "liquid phase" L of preferred globular, compact states; and the "solid phase" S of crystalline structures. The solid phase is subdivided into subphases, in which conformations differ by the type of monomer arrangement in the "crystal." The Θ collapse transition line separates the G and the L phase and meets at low temperature ($T \approx 0.3$) and small interaction range $\delta \approx 0.02$ the liquid–solid transition line. We have already discussed the crossover of these transition lines and its interpretation in the context of microcanonical analysis under the influence of strong finite-size effects. Effectively, the liquid phase disappears for monomer–monomer interaction ranges at very short length scales. The inset shows the crossover in more detail.

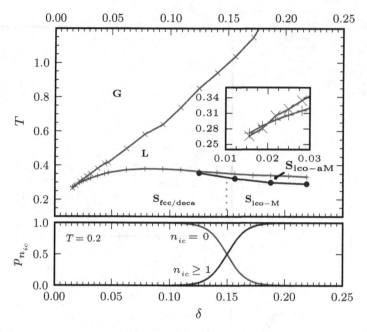

Fig. 6.20 Structural phase diagram for a flexible, elastic polymer with 90 monomers, parametrized by temperature T and nonbonded interaction range δ. The transition lines (solid lines) were obtained by inflection-point microcanonical analysis. The crossover between Θ collapse transition and nucleation cross is enlarged in the inset. The dashed vertical line separates solid phases that cannot be discriminated thermodynamically. The bottom figure shows for $T = 0.2$ the (canonical) probability that a conformation in these solid phases contains n_{ic} icosahedral cores, thereby separating fcc or decahedral crystalline structures with $n_{ic} = 0$ from Mackay icosahedral shapes ($n_{ic} \geq 1$). From [136].

Within the solid regime, we find again the solid–solid transition between anti-Mackay and Mackay type overlayer construction that we have already discussed in Section 6.3. For the 90-mer, this transition only occurs for interaction ranges $\delta > 0.12$. The solid phase S_{ico-aM} is dominated by structures with at least one icosahedral core and an incomplete outer shell of anti-Mackay type (hcp) [cf. Fig. 6.1(a)]. By reducing the temperature further and passing the solid–solid transition line, the packing is optimized and a Mackay-type fcc layer [cf. Fig. 6.1(b)] forms in subphase S_{ico-M}. Note that this solid–solid transition can clearly be found by microcanonical analysis, because the transition is accompanied by a reduction of the rate of entropic change, i.e., $\beta(E)$ varies less in this energetic regime and possesses an inflection point indicating this transition. For this reason, we consider this a structural phase transition, which, as a finite-system surface effect, disappears in the thermodynamic limit.

The vertical dashed line in the phase diagram signals a different kind of solid–solid transition. It turns out that for $\delta < 0.15$ the icosahedral interior becomes energetically less optimal, and it is replaced by a decahedral arrangement of monomers. These structures can also possess extended fcc-packed fractions ($S_{fcc/deca}$).[4]

---

[4] One might even expect a separate fcc phase to be present at extremely small δ-values and temperatures, but if it exists at all, it will be difficult to detect it [169, 170].

However, this transition is not a structural phase transition in the context of the microcanonical thermodynamic interpretation. The reason is that a spontaneous collective re-arrangement of the entire crystal, under the condition of slightly varying or even constant temperature, is extremely unlikely. Although at this transition point icosahedral and decahedral conformations coexist, a transition between them is not directly possible. The solid structure of one of the crystalline types has to be melted first, before it can be crystallized again, in which case it might follow a different transition path toward the other crystalline shape.

Below the phase diagram in Fig. 6.20, the canonical probability $p_{n_{ic}}$ that a conformation possesses a single or more icosahedral cores ($n_{ic} \geq 1$), or no icosahedral core ($n_{ic} = 0$), is plotted. The former case indicates that the conformation has icosahedral geometry, whereas in the latter, the crystalline conformation is of decahedral or fcc type. For the 90-mer considered here, the crossover between decahedral/fcc structures and icosahedral conformations occurs at $\delta \approx 0.15$.

The major results obtained for the 90-mer are qualitatively largely independent of system size. This pertains to the general structure of the phase diagram, but not the solid subphases. The crossover between the collapse and the nucleation transition is a pure finite-size effect and disappears for long polymers.

7

Structural phases of semiflexible polymers

In our discussion of flexible polymers of finite length, we have seen that finite-size effects essentially influence the formation of "crystalline" structural phases. We will now investigate how the arrangement of monomers in this crystalline regime changes, if the flexibility is successively reduced by bond–bond correlations. The class of linear macromolecules that effectively fills the gap between purely flexible and stiff chains is called *semiflexible polymers*. The most prominent representatives in nature are DNA and RNA, but also many proteins possess an effective stiffness. Examples are myosin fibers and actin filaments, which are sufficiently stiff to prevent self-interaction. In these cases, the continuum wormlike-chain model (1.22) is reasonable to investigate thermal fluctuations. However, if the persistence length is small enough, self-interaction occurs, which can cause conformational transitions, depending on environmental conditions. For the basic introduction to semiflexible polymers and their modeling, the reader may consult Section 1.6.

Meanwhile, many examples of single- and double-stranded nucleic acids are known, where the structural behavior deviates from the one predicted by the wormlike-chain model. In these cases, the persistence length is not constant, but depends on chain length and external conditions such as temperature and salt concentrations [171–173]. Consequently, the persistence length is no longer an appropriate length scale for the description of structural behavior. Finite-size effects induced by small chain lengths and the actual discrete nature of polymer chains (allowing for "kinks" and cyclization in DNA [174–177]), as well as excluded-volume effects causing an effective thickness due to extended side chains (e.g., bottle-brush polymers [178] and proteins [171]) are also potential sources for deviations from the standard wormlike-chain behavior. Interactions influence orientational order [179] and non-equilibrium properties [180, 181], and entanglement [182] in melts and liquid crystals of semiflexible polymers.

Therefore, it is beneficial to systematically investigate how the chain stiffness influences the structural behavior and how the simple wormlike-chain behavior separates from the complex behavior of highly flexible polymers. In the following we will discuss the hyperphase diagram for all classes of polymers between the flexible and the stiff limits [183, 184].

7.1 Structural hyperphase diagram

In the most generic model to describe discrete semiflexible polymers with self-interaction the energy of a conformation with N monomers is written as

$$E = \sum_{i=1}^{N} \sum_{j>i+1}^{N} V_{\mathrm{nb}}(r_{ij}) + \sum_{i=1}^{N} V_{\mathrm{b}}(r_{i\,i+1}) + \sum_{k=1}^{N-2} V_{\mathrm{bend}}(\theta_k), \qquad (7.1)$$

where $V_{\mathrm{nb}}(r_{ij})$ is the Lennard-Jones potential (1.34) for the interaction between nonbonded pairs of monomers and $V_{\mathrm{b}}(r_{i\,i+1})$ is the potential of adjacent bonded monomers in the linear chain. In the simulations yielding the results that we will discuss in this chapter, the variant (1.41) was used [184]. Finally, according to the discussion in Section 1.6.2, the potential due to bending is given by $V_{\mathrm{bend}}(\theta_k) = \kappa(1 - \cos\theta_k)$, where θ_k is the bending angle formed by the two bonds that connect the monomers k, $k+1$, and $k+2$. By varying the bending stiffness κ, all classes of linear polymers, ranging from the entirely flexible ($\kappa = 0$) to completely stiff chains ($\kappa = \infty$), can be mimicked.[1]

In the following, we will investigate the generic influence of the bending rigidity κ on the ability of classes of single elastic polymers to form stable compact structural phases. In the example we discuss in detail, $N = 30$. Thus, the system exhibits sufficiently high cooperativity enabling the formation of stable structural phases, although finite-size effects are not negligible. It therefore possesses essential features that are common to a large class of biomolecules. The general phase structure remains intact for larger systems, however, one has to keep in mind that the finite-size effects essentially affect the formation of structured (pseudo)phases of such small systems and that geometric assemblies of monomers depend on the chain length.

For the identification of the structural phases, we investigate thermal fluctuations of energy and radius of gyration as functions of T and κ by canonical statistical analysis. The quantities we consider here are defined by temperature derivatives of the mean energy, in the form of the specific heat

$$c_V(T,\kappa) = \frac{1}{N} \frac{\partial \langle E \rangle(T,\kappa)}{\partial T} = \frac{1}{Nk_{\mathrm{B}}T^2} [\langle E^2 \rangle(T,\kappa) - \langle E \rangle^2(T,\kappa)], \qquad (7.2)$$

and of the square radius of gyration,

$$\frac{\partial \langle R_{\mathrm{g}}^2 \rangle(T,\kappa)}{\partial T} = \frac{1}{k_{\mathrm{B}}T^2} [\langle ER_{\mathrm{g}}^2 \rangle(T,\kappa) - \langle E \rangle(T,\kappa)\langle R_{\mathrm{g}}^2 \rangle(T,\kappa)]. \qquad (7.3)$$

Both quantities can be considered as thermodynamic landscapes in parameter space (T,κ), where cooperative behavior (resulting in structural transitions) is signaled by regions of high thermal activity ("ridges"). It should be noted that this kind of comparative canonical analysis leaves a systematic uncertainty in the estimation of transition lines, yielding a complete but rather qualitative picture of the transition behavior. This is due to the finite size of the system and, for this reason, structural phases identified here should not be confused with phase transitions in the thermodynamic sense.

The energetic and structural representations of the conformational phase diagram are shown in Figures 7.1 (specific heat) and 7.2 (structural fluctuations). Both landscapes

[1] The simulations were performed by employing a variant of the Wang–Landau algorithm, in which the energy E space and the κ space were sampled simultaneously [184]. The Lennard-Jones parameters for the nonbonded pairs of monomers were $\epsilon = 1$ and $\sigma = 2^{-1/6}$. The FENE parameters were set to $R_0 = 1.2$ and $K = 2$, whereas the LJ parameters of the bonded monomers were estimated in such a way that the equilibrium bond length was unity, i.e., $V_{\mathrm{b}}(r_{i\,i+1} = 1) = \min$.

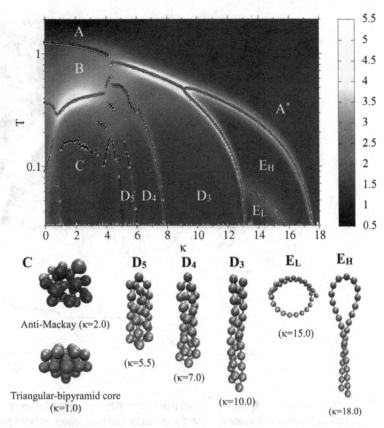

Fig. 7.1 Surface plot of the specific heat (7.2) for classes of polymers with $N = 30$ monomers as a function of temperature T and stiffness κ (k_B set to unity). Brighter shades correspond to higher thermal activity, signaling structural transitions. For a large number of κ values, locations of peaks and "shoulders" are emphasized by circles and squares, respectively, for easier identification of transition points. Conformational phases are labeled: A, random-coil; A*, random-rod-like; B, liquid-globular; C, solid-globular; D_m, rod-like bundles with m segments; E, toroidal. Below the hyperphase diagram, some exemplary conformations, representing the most interesting structural phases, are also shown. From [184].

feature essentially the same information, implying the robustness of the statistical analysis and the interpretation of the results. Below Fig. 7.1, minimum-energy conformations, representative for the structured phases (C, D, E) in different κ intervals, are depicted. Peak and shoulder positions have been identified in the specific heat and $\partial \langle R_g^2 \rangle / \partial T$ landscapes. Their locations are marked by symbols in the surface maps in Figures 7.1 and 7.2, revealing transition lines between a number of unique states as a function of T and κ. The obtained structural hyperphase diagram can then be clearly separated into three major regions: random coils and rods (A), liquid globules (B), and a variety of structured phases (C, D, E).

The most intricate conformational macrostates belong to the compact solid, globular phase C. In this regime, the polymer maximizes the number of pairwise monomer–monomer contacts and is therefore highly energy-dominated. Since monomers in the center

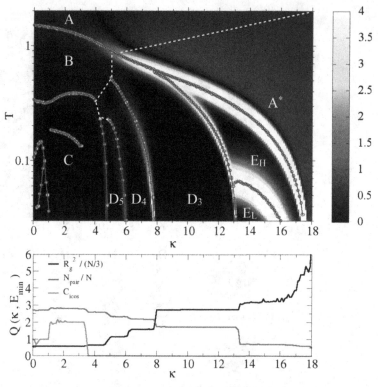

Fig. 7.2 Top: Surface plot of $d\langle R_g^2\rangle(T,\kappa)/dT$, with peaks indicated by circles, and subsequent lines highlighting conformational phase boundaries. Dashed lines indicate transitions suggested by the Khalatur parameters and the specific heat. Bottom: Square radius of gyration R_g^2, relative number of pairwise monomer–monomer contacts N_{pair}/N, and the number of icosahedral cores C_{icos} for the low-energy conformations in dependence of κ, providing insight into differences of structural properties in the low-T phases. From [184].

of the globule can only possess a maximum number of 12 neighbors, structure formation is guided by a competition of surface and volume effects. Small changes in T or κ result in restructuring. Thus, C is actually a composition of several rather glass-like subphases, dominated by energetically metastable states. Close to the flexible limit ($\kappa = 0$), finite polymers tend to form structures with local symmetries. This is the regime of the perfectly icosahedral structures known for flexible polymers with "magic" chain lengths $N = 13, 55, 147, \ldots$, as discussed in Chapter 6. The minimum-energy conformations of the 30-mer contain one ($\kappa < 1.0$) or two icosahedral cores ($1.0 < \kappa < 3.5$); see Fig. 7.2 (bottom).

The characteristic structures in the D_m phases are bundles of rod-like fibers, where m is the number of fibers in the bundle (the number of turns is $m-1$). The number of turns/fibers is governed by the competition of the energetic gain in forming monomer–monomer contacts and the energetic "penalties" caused by the turns. Therefore, the precise substructure of this phase strongly depends on N. Since turns become energetically more costly for larger values of κ, the number of fibers decreases correspondingly with increasing κ, along with the number of monomer–monomer contacts. This can be seen in Fig. 7.2 (bottom), where the relative number of pairwise contacts N_{pair}/N for the lowest-energy conformations is plotted in dependence of κ. This quantity exhibits a noteworthy, but rather slight, step-like decrease in the D phase (note the pronounced step-like increases in the radii of gyration in this regime shown in the same figure). Therefore, the length of the bundle $N_{bundle} \approx N/m$ can easily be estimated by the ratio of the relevant competing energy scales: $N_{bundle} \approx \kappa/\epsilon$ (where $\epsilon = 1$ in our units), such that the degree m of the D phases is empirically given by $N\epsilon/\kappa$ [see examples of D_m conformations in Fig. 7.1 (bottom)]. Note that as N increases, higher-order geometries with larger contact numbers become present, e.g., cylindrical hcp-based conformations appear for $N \geq 45$, in which case the simple relationship between m and κ no longer holds.

Eventually, toroidal loops are present in phase E_L and hairpins in E_H, very characteristic for semiflexible polymers near the wormlike-chain regime under the influence of nonbonded interactions. These structures resemble spontaneously cyclized double-stranded DNA structures that are deemed essential for gene regulation processes in cells [174, 175]. Toroids are dominating the E_L region. These structures undergo subtle conformational changes, which occur typically in the inner radius of the conformation. However, larger chains exhibit significant changes in stacking of loops within a toroid. A clear distinction between these cases is somewhat intricate, even more so as structures from the boundary region of D_3 and E_H mix. The expected phase D_2, i.e., a hairpin with sharp bend, is not particularly dominant in this region.

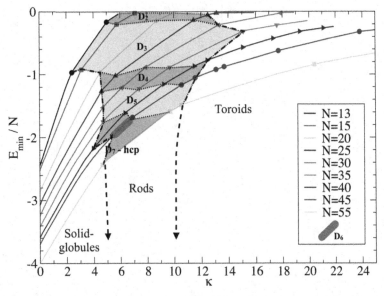

Fig. 7.3 Energies of lowest-energy conformations for chains with $13 \leq N \leq 55$ in dependence of κ. Symbols indicate points of qualitative structural changes (e.g., from solid to a rod, or within rods, D_3 to D_4, etc.). Dashed lines with arrowheads are only visual indicators of possible behavior for $N > 55$, and should not be considered any formal extrapolation. From [184].

7.2 Variation of chain length

Supporting our conclusion that the phase structure obtained for $N = 30$ is qualitatively similar for other system sizes, Fig. 7.3 shows the κ dependence of energies of conformations in the ordered phases for chain lengths in the interval $13 \leq N \leq 55$. This is particularly relevant for the structural phases dominated by finite-size effects such as the D phases. D_2 through D_5 are consistent with previous discussions, but this pattern is upset for $N \geq 35$, where a jump to D_7 (hexagonal close packing) occurs, with only a single instance of D_6 occurring for $N = 45$ (within the resolution of exemplary chain lengths N).

The conclusion that we draw from the discussion in this chapter is that the structural behavior of self-interacting semiflexible polymers cannot be adequately described by the wormlike-chain model that does not allow for the description of structural transitions. As we have seen, structural transitions also occur for semiflexible polymers and have to be taken into account. As in the case of purely flexible polymers, finite-size effects are responsible for a variety of "solid," i.e., ordered, crystalline phases. This implies that the detailed structural behavior needs to be considered in the understanding of biomolecular processes on short length scales. It is therefore also relevant for the nanofabrication of molecular devices.

8

Generic tertiary folding properties of proteins on mesoscopic scales

After the general discussion of the folding behavior of homopolymers, we will now investigate how the additional disorder, imposed by the sequence of different types of monomers in protein-like heteropolymers, influences the formation of stable structural phases. Assuming again that cooperative behavior is based on the interplay of attractive interaction and stochastic behavior, we will study structure formation on global scales by means of simple non-specific models. This will yield some interesting insights into thermodynamic properties of tertiary folding processes and their interpretation.

8.1 A simple model for a parallel β helix lattice protein

As a first example we consider the conformational transitions accompanying the tertiary folding behavior of the sequence of hydrophobic (H) and polar residues (P)

$$PH_2PHPH_2PHPHP_2H_3PHPH_2PHPH_3P_2HPHPH_2PHPH_2P \qquad (8.1)$$

that in the HP model (1.11) forms a parallel helix in the ground state. It was originally designed to serve as a lattice model of the parallel β helix of *pectate lyase C* [185]. The ground-state energy is known to be $E_{\min} = -34$ and the ground-state degeneracy is $g_0 = 4$ (except translational, rotational, and reflection symmetries) [22, 39, 40].

It is a nice example because, despite the simplicity and limitations of the model, it shows transition features that are very characteristic for heteropolymer folding and apply, in similar form, to realistic proteins as well. In the following, we will discuss the transition behavior in detail by means of statistical canonical analysis, which will give us some general insight into the interpretation of conformational transitions of proteins as finite-size variants of structural phase transitions.

The average structural properties at finite temperatures can be characterized well by the mean end-to-end distance $\langle R_{ee}\rangle(T)$ and the mean radius of gyration $\langle R_{gyr}\rangle(T)$, as shown in Fig 8.1. The pronounced minimum in the end-to-end distance can be interpreted as an indication of the transition between ordered lowest-energy states and disordered but compact globular conformations: the few ground states have similar and highly symmetric shapes (due to the reflection symmetry of the sequence), but the ends of the chain are polar and therefore they are not required to reside close to each other. Increasing the temperature allows the protein to fold into conformations different from the ground states, and contacts between the ends become more likely. The mean end-to-end distance decreases and the protein enters the globule "phase." Further increasing the temperature then leads

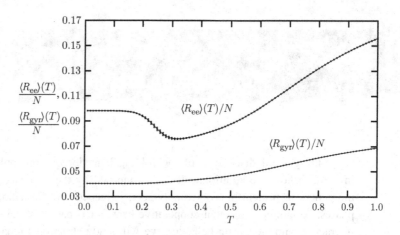

Fig. 8.1   Mean end-to-end distance $\langle R_{ee}\rangle$ and mean radius of gyration $\langle R_{gyr}\rangle$ of the 42-mer. From [40].

to a disentangling of the globules. Random-coil conformations with larger end-to-end distances dominate.

Figure 8.2 shows the specific heat, as well as the derivatives of the mean end-to-end distance and of the mean radius of gyration with respect to the temperature,

$$\frac{d}{dT}\langle R_{ee}\rangle(T) = \frac{1}{T^2}\left[\langle ER_{ee}\rangle(T) - \langle E\rangle(T)\langle R_{ee}\rangle(T)\right], \tag{8.2}$$

$$\frac{d}{dT}\langle R_{gyr}\rangle(T) = \frac{1}{T^2}\left[\langle ER_{gyr}\rangle(T) - \langle E\rangle(T)\langle R_{gyr}\rangle(T)\right]. \tag{8.3}$$

Two temperature regions of conformational activity (shaded in gray), where the curves of the fluctuating quantities exhibit extremal points, can qualitatively be identified. The temperature region of the ground-state–globule transition is bounded by $T_0^{(1)} \approx 0.24$ and $T_0^{(2)} \approx 0.28$. The globule–random-coil transition occurs between $T_1^{(1)} \approx 0.53$ and $T_1^{(2)} \approx 0.70$.

At high temperatures, random conformations are favored. In consequence, in the corresponding entropy-dominated ensemble, the high-degenerate high-energy structures govern the thermodynamic behavior of the macrostates. A typical representative is shown as an inset in the high-temperature pseudophase in Fig. 8.2. Annealing the system (or, equivalently, decreasing the solvent quality), the heteropolymer experiences a conformational transition toward globular macrostates. A characteristic feature of these intermediary "molten" globules is the compactness of the dominating conformations as expressed by a small gyration radius. Nonetheless, the conformations do not exhibit a noticeable internal long-range symmetry and behave rather like a fluid. Local conformational changes are not hindered by strong free-energy barriers. The situation changes by entering the low-temperature (or poor-solvent) conformational phase. In this region, energy dominates over entropy and the effectively attractive hydrophobic force favors the formation of a maximally compact core of hydrophobic monomers. Polar residues are expelled to the surface of the globule and form a shell that screens the core from the (fictitious) aqueous environment.

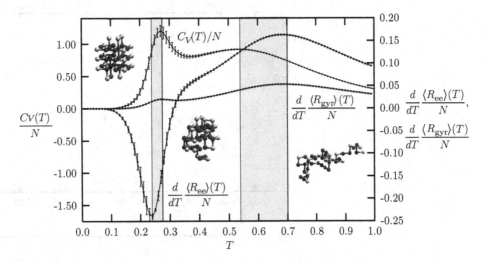

Fig. 8.2   Specific heat C_V/N and derivatives with respect to temperature of mean end-to-end distance $\langle R_{ee}\rangle$ and radius of gyration $\langle R_{gyr}\rangle$ as functions of temperature for the 42-mer. The ground-state–globule transition occurs between $T_0^{(1)} \approx 0.24$ and $T_0^{(2)} \approx 0.28$, while the globule–random coil transition occurs between $T_1^{(1)} \approx 0.53$ and $T_1^{(2)} \approx 0.70$ (shaded areas). From [40].

The existence of the hydrophobic-core collapse, which is accompanied by the creation of a polar shell that screens the hydrophobic core from the solvent, renders the folding behavior of a heteropolymer different from crystallization or amorphous transitions of homopolymers. The reason is the disorder induced by the sequence of different monomer types. The hydrophobic-core formation is the main cooperative conformational transition which accompanies the tertiary folding process of a single-domain protein.

In Fig. 8.3 the canonical distributions $p_{42}^{can,T}(E)$ are plotted for different temperatures in the vicinity of the two transitions. From Fig. 8.3(a) we see that the distributions possess two peaks at temperatures within that region where the ground-state–globule transition takes place. This is interpreted as indication of a "first-order-like" transition, i.e., both types of macrostates coexist in this temperature region [32]. The behavior in the vicinity of the globule–random-coil transition is less spectacular, as can be seen in Fig. 8.3(b). Since the energy distribution shows up one peak only, this transition could be denoted as being "second-order-like" (no coexistence between "phases"). The width of the distributions grows with increasing temperature until it has reached its maximum value which is located near $T \approx 0.7$. For higher temperatures, the distributions become narrower again [40].

Since finite-size scaling is not applicable because of the non-extendable sequences of different types of monomers, "transitions" between protein macrostates have to be distinguished from phase transitions in the strict thermodynamic sense. Conformational transitions for polymers of finite size, such as proteins, are usually weak and therefore difficult to identify. One of the major problems with understanding protein folding, however, is the fact that the amino acid sequence strongly influences the folding characteristics of an individual protein. On the other hand, this is obviously the key to the variety of functions bioproteins can fulfill.

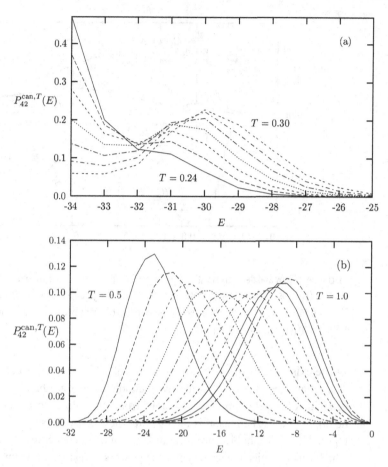

Canonical distributions for the 42-mer at temperatures (a) $T = 0.24, 0.25, \ldots, 0.30$ close to the ground-state–globule transition region between $T_0^{(1)} \approx 0.24$ and $T_0^{(2)} \approx 0.28$, (b) $T = 0.50, 0.55, \ldots, 1.0$. The high-temperature peak of the specific heat in Fig. 8.2 is near $T_1^{(1)} \approx 0.53$, but at $T_1^{(2)} \approx 0.73$ the distribution has the largest width [40]. Near this temperature, the mean radius of gyration and the mean end-to-end distance (see Figs. 8.1 and 8.2) have their biggest slope. From [40].

8.2 Protein folding as a finite-size effect

Understanding protein folding by means of equilibrium statistical mechanics and thermodynamics is a difficult task. A *single* folding event of a protein cannot occur "in equilibrium" with its environment. But protein folding is often considered as a folding/unfolding process with folding and unfolding rates which are constant in a stationary state that defines the "chemical equilibrium." Thus, the statistical properties of an infinitely long sequence of folding/unfolding cycles under constant external conditions (which are mediated by the surrounding solvent) can then also be understood – at least in parts – from a thermodynamic point of view. In particular, folding and unfolding of a protein are conformational transitions and one is tempted to simply take over the conceptual philosophy behind thermodynamic phase transitions, in particular known from "freezing/melting" and "condensation/evaporation" transitions of gases. But this approach is *necessarily* wrong. Thermodynamic phase transitions occur only in the thermodynamic limit, i.e., in infinitely large systems. A protein is, however, a heteropolymer uniquely defined by its *finite* amino acid sequence, which is actually comparatively short and cannot be made longer without changing its specific properties. This is different for polymerized molecules ("homopolymers"), where the infinite-length chain limit can be defined, in principle. The intensely studied collapse or Θ transition between the random-coil and the globular phase is such a phase transition in the truest sense, where a finite-size scaling toward the infinitely long chain is feasible [137]. In this case, a classification of the phase transitions into continuous transitions (where the latent heat vanishes and fluctuations exhibit power-law behavior close to the critical point) and discontinuous transitions (with nonvanishing latent heat) is also possible.

For proteins (or heteropolymers with a "disordered" sequence), finite-size scaling is not available and thus attempts to introduce a classification scheme for transitions accompanying protein folding events are also questionable. Nonetheless, cooperative conformational changes are often referred to as "folding," "hydrophobic-collapse," "hydrophobic-core formation," or "glassy" transitions. All these transitions are defined on the basis of certain parameters, also called "order parameters" or "reaction coordinates," but should not be confused with thermodynamic phase transitions. The onset of finite-system transitions is also less spectacular: their identification on the basis of peaks and "shoulders" in fluctuations of energetic and structural quantities and interpretation in terms of "order parameters" is a rather intricate procedure. Since the different fluctuations do not "collapse" for finite systems, a unique transition temperature often can not be defined. Despite a surprisingly high cooperativity, collective changes of protein conformations are often not happening in a single step. As found in studies of lattice models like the examples presented in this chapter, transition *regions* separate the "pseudophases," where random coils, maximally compact globules, or states with compact hydrophobic core dominate.

Although lattice models such as the HP model are very useful in unraveling generic folding characteristics, they suffer from lattice artifacts, which are less relevant for long chains. In order to obtain a more precise and thus finer resolved image of folding characteristics, it is necessary to "get rid of the lattice" and to allow the coarse-grained protein to fold into the three-dimensional continuum.

8.3 Hydrophobic–polar off-lattice heteropolymers

A manifest off-lattice generalization of the HP model is the AB model (1.13), where the hydrophobic monomers are labeled A and the polar or hydrophilic ones B [14]. The contact interaction is replaced by a distance-dependent Lennard-Jones type of potential accounting for short-range excluded volume repulsion and long-range interaction, the latter being attractive for AA and BB pairs and repulsive for AB pairs of monomers. An additional

Table 8.1 Exemplary sequences [187] used in the study of thermodynamic properties of heteropolymers and the purely hydrophobic homopolymer A_{20}. The number of hydrophobic monomers is denoted by #A.

No.	Sequence	#A
20.1	$BA_6BA_4BA_2BA_2B_2$	14
20.2	$BA_2BA_4BABA_2BA_5B$	14
20.3	$A_4B_2A_4BA_2BA_3B_2A$	14
20.4	$A_4BA_2BABA_2B_2A_3BA_2$	14
20.5	$BA_2B_2A_3B_3ABABA_2BAB$	10
20.6	$A_3B_2AB_2ABAB_2ABABABA$	10
20.7	A_{20}	20

interaction accounts for the bending energy of any pair of successive bonds. This model was first applied in two dimensions [14] and generalized to three-dimensional AB proteins [186], partially with modifications implicitly taking into account additional torsional energy contributions of each bond [187, 188].

We will now investigate thermodynamic properties of AB heteropolymers and compare their folding properties for different sequences. For this purpose, we consider the six exemplified heteropolymers with 20 monomers [187] listed in Table 8.1. The hydrophobicity ($= $#A monomers in the sequence) is identical ($=14$) for the first four sequences 20.1–20.4, while sequences 20.5 and 20.6 possess only 10 hydrophobic residues. Sequence 20.7 is the homopolymer A_{20} consisting of hydrophobic residues only.

In order to identify conformational transitions, let us discuss the heat capacity $C_V(T) = (\langle E^2 \rangle - \langle E \rangle^2)/k_BT^2$ with $\langle E^k \rangle = \sum_E g(E)E^k \exp(-E/k_BT)/ \sum_E g(E) \exp(-E/k_BT)$, where $g(E)$ is the density of states, which can be estimated precisely in multicanonical Monte Carlo simulations. For sequence 20.1, the density of states, covering 70 orders of magnitude, is shown in Fig. 8.4, along with the characteristic multicanonical ("flat") histogram. This strong decrease of the density-of-states curve near the ground-state energy is common to all short heteropolymer sequences. It reflects the isolated character of the ground state within the energy landscape. The specific heat curves for the six 20-mers are shown in Fig. 8.5. A first observation is that the specific heats show up two distinct peaks with the low-temperature peak located at $T_C^{(1)}$ and the high-temperature peak at $T_C^{(2)}$. The values are compiled in Table 8.2.

The sequences considered here are very short and the native fold contains a single hydrophobic core. Interpreting the curves for the specific heats in Fig. 8.5 in terms of conformational transitions, we conclude that the heteropolymers tend to form, within the temperature region $T_C^{(1)} < T < T_C^{(2)}$, intermediate states (often also called traps) comparable with globules in the collapsed phase of polymers. For sequences 20.5 and, in particular, 20.6 the smaller number of hydrophobic monomers causes a much sharper

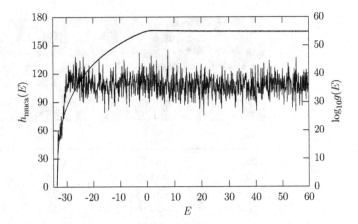

Fig. 8.4   Multicanonical histogram $h_{\mathrm{muca}}(E)$ and logarithm of the density of states $g(E)$ for sequence 20.1. From [47].

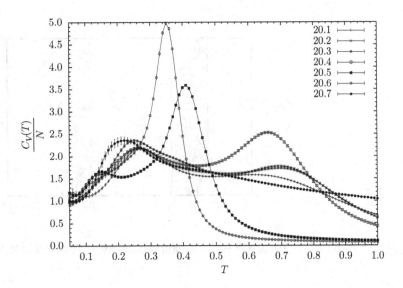

Fig. 8.5   Specific heats of the 20-mers listed in Table 8.1.

transition at $T_C^{(2)}$ than at $T_C^{(1)}$ (where, in fact, the specific heat of sequence 20.6 possesses only a turning point). The pronounced transition near $T_C^{(2)}$ is connected with a dramatic change of the radius of gyration, as can be seen in Fig. 8.7, indicating the collapse from stretched to highly compact conformations with decreasing temperature. The conformations dominant for high temperatures $T > T_C^{(2)}$ are random coils, while for temperatures $T < T_C^{(1)}$ primarily conformations with compact hydrophobic core are favored. In between, there is the intermediary globular "phase." As has already been discussed, these "phases" are not phases in the strict thermodynamic sense, since for heteropolymers the definition of the thermodynamic limit is in principle nonsensical. As a consequence, fluctuating quantities, for example the derivatives with respect to the temperature of the mean

Table 8.2 Minimal energies and temperatures of the maximum specific heats as obtained by multicanonical sampling for the six heteropolymers 20.1–20.6 listed in Table 8.1 [189]. The global maximum of the respective specific heats is indicated by an asterisk ($\star$). The specific heat of sequence 20.6 possesses only one maximum at $T_C^{(2)} \approx 0.35$. The value given for $T_C^{(1)}$ belongs to the pronounced turning point.

No.	$E_{\min}^{\mathrm{MUCA}}$	$T_C^{(1)}$	$T_C^{(2)}$
20.1	−33.77	0.27$\star$	0.61
20.2	−33.92	0.26$\star$	0.69
20.3	−33.58	0.25$\star$	0.69
20.4	−34.50	0.26	0.66$\star$
20.5	−19.65	0.15	0.41$\star$
20.6	−19.32	0.15	0.35$\star$

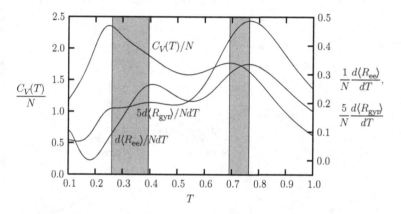

Fig. 8.6   Fluctuations of energy (specific heat), radius of gyration, and end-to-end distance for sequence 20.3. From [189].

radius of gyration, $d\langle R_{\mathrm{gyr}}\rangle/dT$, and the mean end-to-end distance, $d\langle R_{\mathrm{ee}}\rangle/dT$, do not indicate conformational activity at the same temperatures as well as if compared with the specific heat.

Similar to the lattice analyses of heteropolymer folding, conformational transitions of heteropolymers occur within a certain interval of temperatures, not at a fixed critical temperature. The peak temperatures $T_C^{(1)}$ and $T_C^{(2)}$ defined previously for the specific heat are only representatives for the entire intervals. In order to make this more explicit, we consider sequence 20.3 in more detail. In Fig. 8.6, we compare the energetic fluctuations (in form of the specific heat C_V/N) with the respective fluctuations of radius of gyration and end-to-end distance, $d\langle R_{\mathrm{ee,gyr}}\rangle/dT = (\langle R_{\mathrm{ee,gyr}}E\rangle - \langle R_{\mathrm{ee,gyr}}\rangle\langle E\rangle)/k_BT^2$. Obviously, the temperatures with maximal fluctuations are not identical and the shaded areas are spanned over the temperature intervals, where strongest activity is expected. We observe for this example that two such centers of activity can be separated, linked by an intermediary interval

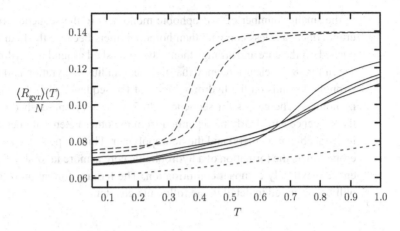

Fig. 8.7 Mean radius of gyration $\langle R_{gyr} \rangle$ as a function of the temperature T for the sequences 20.1–20.4 (solid curves), 20.5, 20.6 (long dashes), and for the homopolymer A_{20} (short-dashed curve). From [189].

of globular traps. In fact, there is a minimum of the specific heat at T_C^{min} between $T_C^{(1)}$ and $T_C^{(2)}$, but the height of the barrier at $T_C^{(2)}$ is rather small and the globules are not very stable.

It is also interesting to compare the thermodynamic behavior of the heteropolymers with the homopolymer consisting of 20 A-type monomers, A_{20}. In contrast to heteropolymers, homopolymers show up a characteristic second-order phase transition between random coil conformations ("good solvent") and compact globules ("poor solvent"), the Θ transition [119] (see also the analysis of the collapse of lattice homopolymers in Section 5.3 and Ref. [137]).

The specific heat of this short homopolymer, also plotted in Fig. 8.5, shows that the collapse from random coils (high temperatures) to globular conformations (low temperatures) happens, roughly, in one step. There is only one energetic barrier as indicated by the single peak of the specific heats. The onset of the separate collapse transition can only be guessed as the specific heat exhibits only a smart shoulder near $T \approx 0.9$.

In Fig. 8.7, the mean radii of gyration as a function of the temperature for the sequences from Table 8.1 are shown in comparison with the homopolymer. For all temperatures in the interval plotted, the homopolymer obviously takes more compact conformations than the heteropolymers, since its mean radius of gyration is always smaller. This different behavior is an indication for a rearrangement of the monomers that is particular for heteropolymers: the formation of the hydrophobic core surrounded by the hydrophilic monomers. Since the homopolymer trivially also takes in the ground state a hydrophobic core conformation (since it only consists of hydrophobic monomers), which is obviously more compact than the complete conformations of the heteropolymers, we conclude that hydrophobic monomers weaken the compactness of low-temperature conformations. Thus, homopolymers and heteropolymers show a different "phase" behavior in the dense phase. Homopolymers fold into globular conformations which are hydrophobic cores with maximum number of hydrophobic contacts. Heteropolymers also form very compact hydrophobic cores which are, of course, smaller than that of the homopolymer due to the smaller number of hydrophobic monomers in the sequence. In total, however, heteropolymers are less compact than homopolymers because the hydrophilic monomers are pushed off the core and arrange themselves in a shell around the hydrophobic core. We also see in Fig. 8.7 a clear tendency that the mean radius of gyration and thus the compactness strongly depends on the hydrophobicity of the sequence, i.e., the number of hydrophobic monomers. The curves for sequences 20.5 and 20.6 (long-dashed curves) with 10 A's in the sequence can clearly be separated from the other heteropolymers in the study (with 14 hydrophobic monomers) and the homopolymer. This supports the assumption that for heteropolymers the formation of a hydrophobic core is more favorable than the folding into an entire maximally compact conformation. The tertiary formation of the hydrophobic core will be discussed in detail in the following chapter.

9

Protein folding channels and kinetics of two-state folding

Folding of linear chains of amino acids, i.e., bioproteins and synthetic peptides, is, for single-domain macromolecules, accompanied by the formation of secondary structures (helices, sheets, turns) and the tertiary hydrophobic-core collapse. While secondary structures are typically localized and thus limited to segments of the peptide, the effective hydrophobic interaction between nonbonded, nonpolar amino acid side chains results in a global, cooperative arrangement favoring folds with compact hydrophobic core and a surrounding polar shell that screens the core from the polar solvent. Systematic analyses for unraveling general folding principles are extremely difficult in microscopic all-atom approaches, since the folding process is strongly dependent on the "disordered" sequence of amino acids and the native-fold formation is inevitably connected with, at least, significant parts of the sequence. Moreover, for most proteins, the folding process is relatively slow (microseconds to seconds), which is due to a complex, rugged shape of the free-energy landscape [190–192] with "hidden" barriers, depending on sequence properties. Although there is no obvious system parameter that allows for a general description of the accompanying conformational transitions in folding processes (as, for example, the reaction coordinate in chemical reactions), it is known that there are only a few characteristic folding behaviors known, such as two-state folding, folding through intermediates, and glass-like folding into metastable conformations [193–199].

Thus, if a classification of folding characteristics is useful at all, strongly simplified coarse-grained models should allow to describe generic statistical [46,47] and kinetic [200] pseudo-universal properties. The reason it appears useful to use a simplified, mesoscopic model such as the AB model is twofold: first, it is believed that tertiary folding is mainly based on effective hydrophobic interactions such that atomic details play a minor role. Second, systematic comparative folding studies for mutated or permuted sequences are computationally extremely demanding at the atomic level and are to date virtually impossible for realistic proteins. We will investigate in the following how by monitoring a simple angular "order" parameter it is indeed possible to identify different complex folding characteristics, even in a coarse-grained heteropolymer model as simple as the AB model. This kind of analysis is comparable to studies of phase transitions based on effective order parameters in other disordered systems such as, e.g., spin glasses, where simplified models are successfully employed [201]. The individual folding trajectories will be characterized by a similarity parameter that is related to the replica overlap parameter used in spin-glass analyses. This is useful as the amino acid sequence induces intrinsic disorder and frustration into the system and therefore a peptide behaves thermodynamically similar to a spin system with a quenched disorder configuration of couplings.

9.1 Similarity measure and order parameter

In order to verify the structural similarities of two conformations $\mathbf{X} = (\mathbf{x}_1, \mathbf{x}_2, \ldots, \mathbf{x}_N)$ and $\mathbf{X}' = (\mathbf{x}_1', \mathbf{x}_2', \ldots, \mathbf{x}_N')$, the most frequently used standard measure is the root mean square deviation (rmsd) of the respective structures,

$$\text{rmsd} = \min \sqrt{\frac{1}{N} \sum_{i=1}^{N} |\tilde{\mathbf{x}}_i - \tilde{\mathbf{x}}_i'|^2}. \tag{9.1}$$

Here, $\tilde{\mathbf{x}}_i = \mathbf{x}_i - \mathbf{x}_0$ and $\tilde{\mathbf{x}}_i' = \mathbf{x}_i' - \mathbf{x}_0'$ denote the positions with respect to the centers of masses $\mathbf{x}_0 = \sum_{j=1}^{N} \mathbf{x}_j/N$ and $\mathbf{x}_0' = \sum_{j=1}^{N} \mathbf{x}_j'/N$ of the ith monomer in the two conformations. Obviously, the rmsd is zero for exactly coinciding conformations, and the larger the value, the worse the coincidence. The minimization of the sum in Eq. (9.1) is performed with respect to a global relative rotation of the two conformations in order to find the best match. The explicit calculation in computer simulations is costly and an efficient implementation is required. One example is the exact quaternion-based optimization procedure [202].

Another, less costly, unique alternative is the introduction of an angular overlap order parameter. Before we define it, let us first discuss the characteristic features of free-energy landscape parametrized by angular degrees of freedom.

The folding process of proteins is necessarily accompanied by cooperative conformational changes. Therefore, it should be expected that one or a few parameters can be defined that enable the description of the structural ordering process. The number of degrees of freedom in most all-atom models is given by the dihedral torsional backbone and side-chain angles. In coarse-grained C^{α} models such as the AB model, the original dihedral angles are replaced by a set of virtual torsional and bond angles. In fact, the number of degrees of freedom is not necessarily reduced in simplified off-lattice models. Therefore, the complexity of the space of degrees of freedom is comparable with more realistic models, and it is also a challenge to identify a suitable order parameter for the folding in such minimalistic heteropolymer models. On the other hand, the computational simplicity of these models allows for a more systematic and efficient analysis of the heteropolymer folding process.

Typical degrees of freedom that uniquely parametrize a polymer conformation are the torsion and the bond angles. In the context of coarse-grained models, these angles are usually designated "virtual," because the coarse-grained monomers accommodate large chemical units. These groups are composed of atoms that also form angles which partially are degrees of freedom on a microscopic scale. The Ramachandran torsion angles in a microscopic representation of a protein (see Fig. 1.6) should not be confused with the virtual torsion angles as defined in Fig. 9.1, where four monomers ("amino acids") form the two planes tilted by the torsion angle Φ. Another consequence is that bond angles become degrees of freedom in a coarse-grained model, whereas they are almost inflexible in atomistic models. The virtual bond angles are given by $\Theta = \pi - \theta$, where θ is the bending angle (see Section 1.6.2).

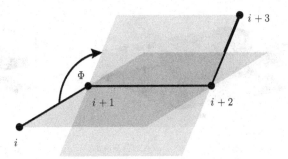

Fig. 9.1   Definition of the torsion angle Φ, which is defined as the angle between the two planes spanned by the monomers i, $i+1, i+2$ and $i+1, i+2, i+3$, respectively. By definition, a *cis* conformation is formed, if $\Phi = 0$, whereas in a *trans* arrangement $\Phi = \pm\pi$. Thus, $\Phi \in [-\pi, \pi]$.

In Fig. 9.2, we show the probability distributions $p_{\text{ang}}(\Theta, \Phi)$ of all successive pairs of Φ and Θ for the exemplified AB sequence 20.3 ($A_4B_2A_4BA_2BA_3B_2A$) that we have already discussed in the previous chapter, at several temperatures.[1] This plot can be considered as the AB analog of the Ramachandran map for real proteins. Although this representation is not appropriate to describe the folding process, which will be rather complicated for this example as described later on, a few interesting features can already be seen from this figure. At the temperature $T = 0.3$, we observe two domains in this landscape, i.e., a structural pre-ordering has already taken place. The distribution is noticeably peaked for bond angles around 90° and torsion angles close to 0°, i.e., almost perfectly planar *cis* conformations are favored in the ensemble as well as segments with bond angles between 60° and 70° for a broad distribution of torsion angles mainly between 40° and 100°. The reason for the large width of the torsion-angle distribution in this region is that the temperature is still too high for fine-structuring within the conformations. Explicit torsional barriers might stabilize these segments even at this temperature but are disregarded in the model.

Decreasing the temperature down to $T = 0.1$, we see that the landscape of this accumulated distribution of the degrees of freedom becomes very complex, and the peaks are much sharper. In fact, close to $T \approx 0.1$, we observe a conformational transition toward the formation of lowest-energy states resembling the ground state. Actually, the complexity of this landscape can be better understood when considering the folding channels in the following, where we will see that this heteropolymer exhibits metastability and therefore rather glassy behavior. A remarkable aspect is the formation of the peaks in the bond-angle distribution at low temperatures close to 60°, 90°, and 120°, as these angles are typical base angles in face-centered cubic crystals. As this concerns only segments of the conformations, the conformational transition is actually not a crystallization. To summarize, distributions of degrees of freedom are not quite useful to describe the folding process. For this reason it is necessary to define a suitable effective system parameter [193, 194].

---

[1] For a polymer with N monomers, there are in this model $(N-2)$ bond and $(N-3)$ torsion angles. In order to form successive pairs of these angles, the last bond angle (counted from the first monomer in the sequence) has been left out in Fig. 9.2.

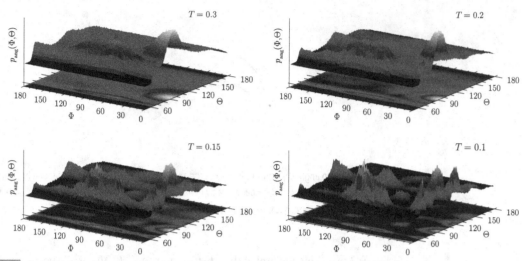

Fig. 9.2   Exemplified bond and torsion angle distributions of sequence $A_4B_2A_4BA_2BA_3B_2A$ at different temperatures. The distributions of the torsion angles are reflection-symmetric and therefore only the positive intervals are shown. From [47].

Such a variant is the angular overlap order parameter [46, 47, 189, 195, 196]. It is based on the idea of defining a simple and computationally low-cost measure for the similarity of two conformations, where the differences of the angular degrees of freedom are calculated. We define the overlap parameter as follows:

$$Q(\mathbf{X}, \mathbf{X}') = 1 - d(\mathbf{X}, \mathbf{X}'). \tag{9.2}$$

With $N_b = N - 2$ and $N_t = N - 3$ being the respective numbers of bond angles Θ_i and torsional angles Φ_i, the angular deviation between the conformations is calculated according to

$$d(\mathbf{X}, \mathbf{X}') = \frac{1}{\pi (N_b + N_t)} \left[\sum_{i=1}^{N_b} d_b \left(\Theta_i, \Theta_i' \right) + \min \left(\sum_{i=1}^{N_t} d_t^+ \left(\Phi_i, \Phi_i' \right), \sum_{i=1}^{N_t} d_t^- \left(\Phi_i, \Phi_i' \right) \right) \right], \tag{9.3}$$

where

$$d_b(\Theta_i, \Theta_i') = |\Theta_i - \Theta_i'|,$$
$$d_t^{\pm}(\Phi_i, \Phi_i') = \min \left(|\Phi_i \pm \Phi_i'|, 2\pi - |\Phi_i \pm \Phi_i'| \right). \tag{9.4}$$

Here we have taken into account that the AB model is invariant under the reflection symmetry $\Phi_i \rightarrow -\Phi_i$. Thus, it is not useful to distinguish between reflection-symmetric conformations and therefore only the larger Q value is considered (which corresponds to the smaller value of the "distance" measure d). Since $-\pi \leq \Phi_i \leq \pi$ and $0 \leq \Theta_i \leq \pi$, the overlap is unity, if all angles of the conformations $\mathbf{X}$ and $\mathbf{X}'$ coincide, else $0 \leq Q < 1$. It should be noted that the average overlap of a random conformation with the corresponding reference state is for the sequences considered close to $\langle Q \rangle \approx 0.66$. As a rule of thumb, it can be concluded that values $Q < 0.8$ indicate weak or no significant similarity of a given structure with the reference conformation. The statistical average of the overlap parameter is a particularly useful measure for the "distance" of a conformational phase from the native phase, if the ground-state conformation $\mathbf{X}^{(0)}$ is chosen as the reference structure.

9.2 Identification of characteristic folding channels

For the qualitative discussion of the folding behavior it is useful to consider the histogram of energy E and angular overlap parameter Q in the multicanonical ensemble as it is directly obtained from multicanonical simulations,

$$H_{\text{muca}}(E, Q) = \sum_t \delta_{E,E(\mathbf{X}_t)} \delta_{Q,Q(\mathbf{X}_t,\mathbf{X}^{(0)})}, \tag{9.5}$$

where the sum runs over all Monte Carlo sweeps t. In Figures 9.3(a)–9.3(c), the multi-canonical histograms $H_{\text{muca}}(E, Q)$ are plotted for three of the sequences listed in Table 8.1: 20.1, 20.3, and 20.4. Ideally, multicanonical sampling yields a constant energy distribution

$$h_{\text{muca}}(E) = \int_0^1 dQ\, H_{\text{muca}}(E, Q) = \text{const.} \tag{9.6}$$

In consequence, the distribution $H_{\text{muca}}(E, Q)$ can suitably be used to identify the folding channels, independently of temperature. This is more difficult with temperature-dependent canonical distributions $P^{\text{can}}(E, Q)$, which can, of course, be obtained from $H_{\text{muca}}(E, Q)$ by a simple reweighting procedure, $P^{\text{can}}(E, Q) \sim H_{\text{muca}}(E, Q)g(E)\exp(-E/k_B T)$. Nonetheless, it should be noted that, since there is a unique one-to-one correspondence between the average energy $\langle E \rangle$ and canonical temperature T, regions of changes in the monotonic behavior of $H_{\text{muca}}(E, Q)$ can also be assigned a temperature, where a conformational transition occurs.

Interpreting the ridges of the probability distributions in the left-hand panel of Fig. 9.3 as folding channels, it can clearly be seen that the heteropolymers exhibit noticeable differences in the folding behavior toward the native conformations (N). Considering natural proteins it would not be surprising that different sequences of amino acids in many cases not only cause different native folds but also vary in their folding behavior. Here we are considering, however, a highly minimalistic heteropolymer model and hitherto it is not clear whether it would be possible at all to separate characteristic folding channels in such a simple model, but, as Fig. 9.3 demonstrates, in fact, it is. For sequence 20.1, we identify in Fig. 9.3(a) a typical two-state characteristic. Approaching from high energies (or high temperatures), the conformations in the ensemble D have an angular overlap $Q \approx 0.7$, which means that there is no significant similarity with the reference structure, i.e., the ensemble D consists mainly of unfolded peptides. For energies $E < -30$ a second branch opens. This channel (N) leads to the native conformation (for which $Q = 1$ and $E_{\min} \approx -33.8$). The constant-energy distribution, where the main and native-fold channels D and N coexist, exhibits two peaks noticeably separated by a well. Therefore, the conformational transition between the channels looks

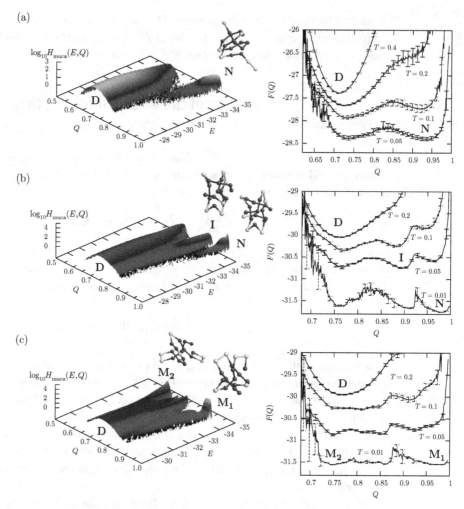

Fig. 9.3 Multicanonical histograms $H_{muca}(E, Q)$ of energy E and angular overlap parameter Q and free-energy landscapes $F(Q)$ at different temperatures for the three sequences (see Table 8.1) (a) 20.1, (b) 20.4, and (c) 20.3. The reference folds reside at $Q = 1$ and $E = E_{min}$. Pseudophases are symbolized by D (denatured states), N (native folds), I (intermediates), and M (metastable states). Representative conformations in intermediate and folded phases are also shown. After [46, 47].

first-order-like, which is typical for two-state folding. The main channel D contains the ensemble of unfolded conformations, whereas the native-fold channel N represents the folded states.

The two-state behavior is confirmed by analyzing the temperature dependence of the minima in the free-energy landscape. The free energy as a function of the "order" parameter Q at fixed temperature can be suitably defined as:

$$F(Q) = -k_B T \ln p(Q). \tag{9.7}$$

In this expression,

$$p(Q_0) = \int \mathcal{D}\mathbf{X} \, \delta(Q_0 - Q(\mathbf{X}, \mathbf{X}^{(0)})) \, e^{-E(\mathbf{X})/k_B T} \qquad (9.8)$$

is related to the probability of finding a conformation with a given value of Q in the canonical ensemble at temperature T. The formal integration runs over all possible conformations $\mathbf{X}$ and guarantees the normalization of the probability distribution p. In the right-hand panel of Fig. 9.3(a), the free-energy landscape at various temperatures is shown for sequence 20.1. At comparatively high temperatures ($T = 0.4$), only the unfolded states ($Q \approx 0.71$) in the main folding channel D dominate. Decreasing the temperature, the second (native-fold) channel N begins to form ($Q \approx 0.9$), but the global free-energy minimum is still associated with the main channel. Near $T \approx 0.1$, both free-energy minima have approximately the same value, the folding transition occurs. The discontinuous character of this conformational transition is manifest by the existence of the free-energy barrier between the two macrostates. For even lower temperatures, the native-fold-like conformations ($Q > 0.95$) dominate and fold smoothly toward the $Q = 1$ reference conformation, which is the lowest-energy conformation found in the simulation.

A significantly different folding behavior is noticed for the heteropolymer with sequence 20.4. The corresponding multicanonical histogram is shown in Fig. 9.3(b) and represents a folding event through an intermediate macrostate. The main channel D bifurcates and a side channel I branches off continuously. This branching is followed by the formation of a third channel N, which ends in the native fold. The characteristics of folding-through-intermediates is also confirmed by the free-energy landscapes as shown for this sequence in Fig. 9.3(b) at different temperatures. Approaching from high energies, the ensemble of denatured conformations D ($Q \approx 0.76$) is dominant. Close to the transition temperature $T \approx 0.05$, the intermediary phase I is reached. The overlap of these intermediary conformations with the native fold is about $Q \approx 0.9$. Decreasing the temperature further below the native-folding threshold close to $T = 0.01$, the hydrophobic-core formation is finished and stable native-fold-like conformations with $Q > 0.97$ dominate (N).

The most extreme behavior of the three exemplified sequences is exhibited by the heteropolymer 20.3. The main channel D does not decay in favor of a native-fold channel. In fact, we observe the formation of *two* separate native-fold channels M_1 and M_2. Channel M_1 advances toward the $Q = 1$ fold and M_2 ends up in a completely different conformation with approximately the same energy ($E \approx -33.51$). The spatial structures of these two conformations are noticeably different and their mutual overlap is correspondingly very small, $Q \approx 0.75$. It should also be noted that the lowest-energy conformations in the main channel D have only slightly larger energies than the two native folds. Thus, the folding of this heteropolymer is accompanied by very complex folding characteristics. In fact, this multiple-peak distribution near minimum energies is a strong indication for metastability. A native fold in the natural sense does not exist, the $Q = 1$ conformation is only a reference state but the folding toward this structure is not distinguished as it is in the folding characteristics of sequences 20.1 and 20.4. This explains also why the bond- and torsion-angle distribution in Fig. 9.2 possesses so many spikes: it represents the ensemble of amorphous conformations rather than a distinct footprint of a distinguished native fold. The amorphous folding behavior is also seen in the free-energy landscapes in Fig. 9.3(c). Above the folding transitions ($T = 0.2$) the typical sequence-independent denatured conformations with $\langle Q \rangle \approx 0.77$ dominate (D). Then, in the annealing process, several channels are formed and coexist. The two most prominent channels (to which the lowest-energy conformations belong that we found in the simulations) eventually lead for $T \approx 0.01$ to ensembles of macrostates with $Q > 0.97$ (M_1), and conformations with $Q < 0.75$ (M_2). The lowest-energy conformation found in this regime is structurally different but energetically degenerate compared with the reference conformation.

Now that we have shown that it is indeed possible to classify protein folding characteristics employing a mesoscopic model, we will consider kinetic properties of a particularly interesting folding behavior: two-state folding.

9.3 Gō kinetics of folding transitions

Spontaneous protein folding is a dynamic process, which starts after the creation of the amino acid sequence in the ribosome. It is in many cases finished when the functional conformation, the native fold, is formed. As this process takes microseconds to seconds, a dynamical computational analysis of an appropriate microscopic model, which could lead to a better understanding of the conformational transitions accompanying folding [203], is extremely demanding. Since protein folding is a thermodynamic process at finite temperature, a certain folding trajectory in the free-energy landscape is influenced by Brownian collisions with surrounding solvent molecules. Therefore, it is more favorable to study the kinetics of the folding process by averaging over an appropriate ensemble of trajectories.

A significant problem is that the complexity of detailed semiclassical microscopic models based on force-fields and solvent parameter sets (or explicit solvent) rules out molecular dynamics (MD) in many cases and, therefore, Markovian Monte Carlo (MC) dynamics is a frequently used method for such kinetic studies.[2] It is obvious, however, that the timescale provided by MC is not directly comparable with the timescale of the folding process. It is widely believed that the folding path of a protein is strongly correlated with contact ordering [204], i.e., the order of the successive contact formation between residues and, therefore, long-range correlations and memory effects can significantly influence the kinetics.

A few years ago, experimental evidence was found that classes of proteins show particular simple folding characteristics, single exponential and two-state folding [205, 206]. In the two-state folding process, which we will focus on in the following example, the peptide either is in an unfolded, denatured state or it possesses a native-like, folded structure. In contrast to the barrier-free single-exponential folding, there exists an unstable transition state to be passed in the two-state folding process. Due to the comparatively simple folding

---

[2] Another reason why MD is typically outperformed by MC if thermodynamics becomes relevant is the fact that it is difficult to obtain the correct statistical distributions with MD. The proper usage of thermostats in MD remains a notorious problem [76].

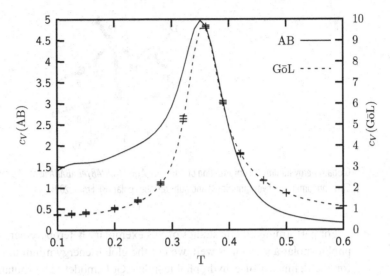

Fig. 9.4 Specific heats for the peptide with sequence 20.6 as obtained with the AB model [189] and the gauged GōL model as defined in Eq. (9.9). From [200].

characteristics, strongly simplified, effective models were established. Knowledge-based models of Gō-like type [207–210] were investigated in numerous recent studies [194, 211–219]. In Gō-like models the native fold must be known and is taken as input for the energy function. The energy of an actual conformation depends on its structural deviation from the native fold (e.g., by counting the number of already established native contacts). By definition, the energy is minimal, if conformation and native fold are identical in all degrees of freedom involved in the model. The simplicity of the model entails reduced computational complexity and also MD simulations, e.g., based on Langevin dynamics [217], can successfully be performed.

Here, we follow a different approach. We also study a Gō-like model, but it is based on a minimalistic coarse-grained hydrophobic–polar representation of the heteropolymer [200].

9.3.1 Coarse-grained Gō modeling

In the following, we compare the thermodynamics and kinetics of a physics-based model, the AB model [14], and a knowledge-based model of Gō type [207, 208, 212], which is referred to as the GōL (Gō-like) model [200] throughout the following discussion. We choose the AB heteropolymer sequence $A_3B_2AB_2ABAB_2ABABABA$ (sequence 20.6 in Table 8.1), because it possesses a pronounced first-order-like folding transition with two-state (folded/unfolded) folding characteristics. In the AB model, this heteropolymer experiences a hydrophobic collapse transition that is identified by the peak in the specific-heat curve plotted in Fig. 9.4. For such systems, it is known from model studies of realistic amino acid sequences that knowledge-based Gō type models reveal kinetic aspects of folding and unfolding processes reasonably well [212, 217].

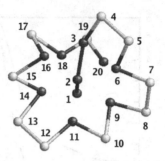

Fig. 9.5   Global-energy minimum conformation of sequence $A_3B_2AB_2ABAB_2ABABABA$ in the AB model with $E_{AB} \approx -19.3$. Dark monomers are hydrophobic (A) and light residues polar (B). From [200].

For performing kinetic studies of this exemplary heteropolymer in the simplified hydrophobic–polar approach as well, we use the global-energy minimum conformation as input for the definition of a hydrophobic–polar GōL model. The (putative) native conformation $\mathbf{X}^{(0)}$ is shown in Fig. 9.5 and its energy is $E_{AB} \approx -19.3$ in the units of the AB model (1.13). In GōL models, the "energy" of a given conformation is related to its similarity with the ground state. This means "energy" in the GōL model plays the role of a similarity or "order" parameter and, therefore, it is not a potential energy in the usual physical sense (as there is no physical force associated with it). Denoting the $(N-2)$ bending angles of the global-energy minimum conformation by $\vartheta_k^{(0)}$, its $(N-3)$ torsional angles as $\varphi_l^{(0)}$, and the inter-monomer distances by $r_{ij}^{(0)}$, we define the GōL model according to its form in microscopic representation [212] as:

$$E_{\text{GōL}}(\mathbf{X})/\varepsilon = K_\vartheta \sum_{k=1}^{N-2} \left(\vartheta_k - \vartheta_k^{(0)} \right)^2 + \sum_{n=1,3} \sum_{l=1}^{N-3} K_\varphi^{(n)} \left\{ 1 - \cos\left(n\left[\varphi_l - \varphi_l^{(0)} \right] \right) \right\}$$

$$+ \sum_{i<j-1}^{\text{native}} \left(5\left[\frac{r_{ij}^{(0)}}{r_{ij}} \right]^{12} - 6\left[\frac{r_{ij}^{(0)}}{r_{ij}} \right]^{10} \right) + \sum_{i<j-1}^{\text{non-native}} \frac{1}{r_{ij}^{12}}. \tag{9.9}$$

The last two sums run over all pairs of nonbonded monomers. In the case where the pair (i,j) forms a native contact, i.e., $r_{ij} < r_{\text{cut}}$ and $r_{ij}^{(0)} < r_{\text{cut}}$, the monomers experience a short-range 10-12 Lennard-Jones attraction, while for non-native contacts an overall repulsive $1/r^{12}$ contribution is taken into account. The constants K_ϑ, $K_\varphi^{(1)}$, and $K_\varphi^{(3)}$ weigh the relative strengths of the angular energy contributions.[3] The values are adjusted such as to have a reasonable coincidence of the peak temperature of the specific heat compared to the results obtained with the AB model [189] (see Fig. 9.4). The free overall energy scale ε is chosen such that $E_{AB}(\mathbf{X}^{(0)}) = E_{\text{GōL}}(\mathbf{X}^{(0)}) = -\varepsilon n_{\text{tot}}$, where n_{tot} is the total number of native contacts. The definition of a native contact requires the introduction of a cutoff radius, which for $r_{\text{cut}} = 1.14$ yields $n_{\text{tot}} = 20$ contacts in the native conformation $\mathbf{X}^{(0)}$, and therefore $\varepsilon \approx 0.966$. The choice of the cutoff radius is not very sensitive, provided it is not too small.

---

[3] The results presented in the following were obtained by setting $K_\vartheta = 20$, $K_\varphi^{(1)} = 0.5$, and $K_\varphi^{(3)} = 0.25$ [200].

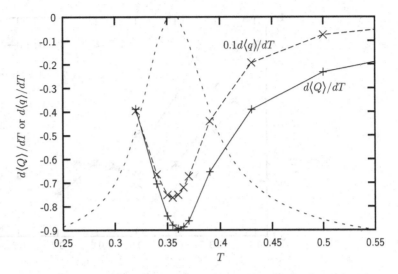

Fig. 9.6   Fluctuations of the similarity parameter q and the angular overlap parameter Q as functions of the temperature. For comparison, the specific heat (dotted line) is also plotted into this figure. From [200].

9.3.2 Thermodynamics

For the study of the folding transition, the introduction of an effective parameter that uniquely describes the macrostate of the ensemble of heteropolymer conformations is useful. A typical measure is the contact number $q(\mathbf{X})$, which for a given conformation $\mathbf{X}$ is simply defined as the fraction of the already formed native contacts $n(\mathbf{X})$ in conformation $\mathbf{X}$ and the total number of native contacts n_{tot} in the final fold, i.e., $q(\mathbf{X}) = n(\mathbf{X})/n_{\text{tot}}$. Then, the statistical ensemble average of this quantity $\langle q(\mathbf{X}) \rangle$ at a given temperature characterizes its macrostate. Roughly, for a two-state folder, if $\langle q(\mathbf{X}) \rangle > 0.5$, native-like conformations are dominating the statistical ensemble. If less than half the total number of contacts is formed, the heteropolymer tends to reside in the pseudophase of denatured conformations.

Another suitable parameter is the angular overlap parameter $Q(\mathbf{X})$ introduced in Section 9.1. The advantage is that it is particularly useful for classifying intermediate or metastable structures with stable but non-native contacts. In Fig. 9.6, the fluctuations of both parameters, i.e., $d\langle q \rangle/dT$ and $d\langle Q \rangle/dT$, respectively, are plotted as functions of the temperature for the GōL model of sequence 20.6. We see clearly that the temperature region of conformational activity as signaled by these two "order" parameters coincides with the thermally active region indicated by the peak of the specific heat, which is also shown for comparison. The folding temperature, i.e., the temperature of maximum activity, is $T_f \approx 0.36$.

The classification of the heteropolymer with sequence 20.6 as a two-state folder arises from the analysis of the free-energy landscape. We assume that q is a suitable parameter that describes the macrostate of the system adequately. Considering this parameter as a constraint, we can formally average out the conformational degrees of freedom, and the probability for a conformation in a macrostate with contact parameter q' reads

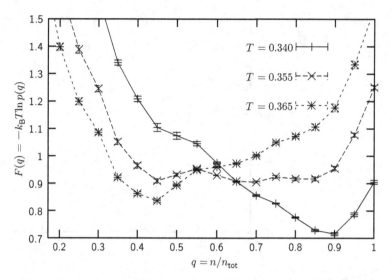

Fig. 9.7 Free-energy landscape $F(q)$ of sequence 20.6 for different temperatures close to the transition point. From [200].

$$p(q') = \langle \delta(q' - q(\mathbf{X})) \rangle = \frac{1}{Z} \int \mathcal{D}\mathbf{X}\, \delta(q' - q(\mathbf{X}))\, e^{-E_{\mathrm{G\bar{o}L}}(\mathbf{X})/k_B T}, \qquad (9.10)$$

where Z denotes the unrestricted partition function. The integral measure is simply $\mathcal{D}\mathbf{X} = \prod_{i=1}^{N}[d^3 r_i] \prod_{i=1}^{N-1}[\delta(r_{ii+1} - 1)]$. Expression (9.10) can be used to define a free energy as function of the q parameter by

$$F(q) = -k_B T \ln p(q). \qquad (9.11)$$

Since the value of q is a qualitative measure for the macrostate the system resides in, the minimum of the free energy at a given temperature T is related to the dominant macrostate in the canonical ensemble at this temperature. Actually, as can be seen in Fig. 9.7, the folding transition of the heteropolymer with sequence 20.6 is a phase-separation process, i.e., at the transition point close to $T_f \approx 0.36$, the folded and the denatured pseudophase coexist and the transition state barrier possesses a local maximum close to $q \approx 0.5$, as expected for a typical two-state folding characteristics. We will return to this point later, when we will shine light on this transition from the perspective of microcanonical analysis.

A useful measure of structure formation is the radial distribution function and its dependence on temperature. Due to translational invariance in three-dimensional space, three degrees of freedom contribute only a volume factor to the partition function. Therefore, we utilize this by fixing the position of the first monomer, $\mathbf{r}_1 = \mathbf{0}$. Thus, radial distances r of the other monomers are measured with respect to the first one and the radial distribution function can be defined by

$$G(r) = \frac{1}{4\pi r^2} \left\langle \sum_{i=3}^{N} \delta(|\mathbf{r}_i| - r) \right\rangle, \qquad (9.12)$$

where $i = 2$ is excluded as, by definition, the virtual covalent bonds are rigid and therefore $r_{12} = 1$. Actually, from our definition, $4\pi \int_0^\infty dr\, r^2 G(r) = N - 2$. In Fig. 9.8, the radial

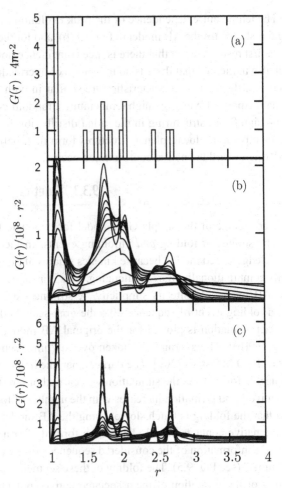

Fig. 9.8 Radial distribution functions of 20.6 (a) for the lowest-energy conformation (see Fig. 9.5), (b) at 16 temperatures in the interval $T \in [0.1, 1.5]$ (curves from top to bottom) employing the AB model, and (c) at 18 temperatures in the interval $T \in [0.1, 1.5]$ for the GōL model. From [200].

distribution functions are shown for sequence 20.6 for (a) the lowest-energy conformation and the ensembles at different temperatures in the (b) AB model and (c) GōL model. Although the global-energy minimum conformation shown in Fig. 9.5 obviously does not form a regular crystal structure, the Lennard-Jones-like interactions in combination with the rigid virtual bonds induce preferable substructures, favoring, e.g., bond angles of 60°, 90°, and 120° [47]. Therefore, the peaks of the radial distances from the first monomer for the global energy minimum conformation in Fig. 9.8(a) can be partly explained by these local segments. The first peak at $r = 2^{1/6} \approx 1.12$ is related to the minimum potential distance between two nonbonded A monomers (and the reference monomer at $\mathbf{r}_1 = \mathbf{0}$ is of type A). Also the location of other peaks can be deduced from Fig. 9.5 by similar geometric arguments.

The temperature dependence of the peak evolution in the radial distribution function of 20.6 is shown for the AB model in Fig. 9.8(b), and for the GōL model variant in Fig. 9.8(c). As a first result, we see that there is nice coincidence not only in the location of the peaks, but also in the fact that there is no indication of intermediary, weakly stable conformations. Actually, the folding characteristics are similar in both models, at least from the thermodynamic point of view. At high temperatures, random-coil conformations dominate: there is no significant structuring in the radial distribution function. Passing the folding transition temperature, local, planar structures form first, before the tertiary, three-dimensional ordering toward the native conformation occurs.

9.3.3 Kinetics

The advantage of the simple GōL model for two-state folding is that it enables likewise kinetic studies of folding *and* unfolding events. In fact, this is the main purpose of this knowledge-based model, because kinetics studies of physically motivated models are typically computationally more challenging. This is also the case for folding studies employing the AB model, despite its simplicity. A striking example is shown in Fig. 9.9, where for a folding event of sequence 20.6 the gray-scale coded average probability of native-contact formation is plotted for the original AB model [Fig. 9.9(a)] and the GōL model [Fig. 9.9(b)]. The average was taken over a "time" window $[t_{MC} - \Delta t, t_{MC} + \Delta t]$ with $\Delta t = 1000$ MC steps [200]. The darker a bar, the higher the probability that the associated contact is formed. As the simulation was carried out in the folding regime, the probability of native-bond formation increases with the number of total MC steps. It is, however, obvious that the folding is much slower using the AB model, whereas in the GōL model most of the native contacts already have been formed after a few ten thousand MC steps. The native conformation of the considered sequence possesses partly kind of zigzag structures, or "turns" (see Fig. 9.5). The folding of these segments is particularly simple and the probability of the formation of the associated native contacts of monomers i with monomers $i + 2$ or $i + 3$ increases much faster than for the other contacts. But even in this case, the GōL kinetics is unbeatable.

This example exhibits the dilemma of physics-based models in studying kinetic aspects of structure formation by means of computer simulations. It is a notoriously difficult problem, because the timescale of molecular dynamics is typically too small to see folding events, but also Markov chain Monte Carlo dynamics of physical models is typically too slow. Just for kinetic aspects, where absolute timescales are widely irrelevant, knowledge-based GōL models are an alternative that allow for increasing the sampling efficiency, at least for systems with simple folding characteristics (as, for example, two-state folding). Actually, for such models an adequate statistical sampling of ensembles close to the transition state can be achieved by sophisticated Monte Carlo methods where efficiency is typically gained by simulating generalized ensembles at the expense of an artificial dynamics. Therefore, it is widely believed that free-energy-driven dynamics, as it is relevant for protein folding, can be provided reasonably only by Boltzmann-Markov chains of conformational updates. One possibility to achieve this is the application of a conventional Metropolis Monte Carlo method. Although the overall timescale is left open, this method

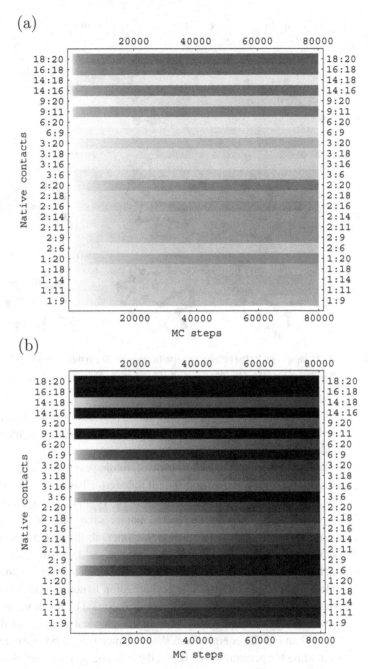

Fig. 9.9   Gray-scale coded averages of the probabilities for native-contact formation of sequence 20.6 as a function of Monte Carlo time for (a) AB model and (b) GōL model. The probability is calculated as a combined temporal average over 2000 MC steps in the ensemble of 1000 folding events. The darker the bars are, the higher is the probability that the associated contact has formed. The labels of the pair contacts refer to the numbering in the native conformation shown in Fig. 9.5. From [200].

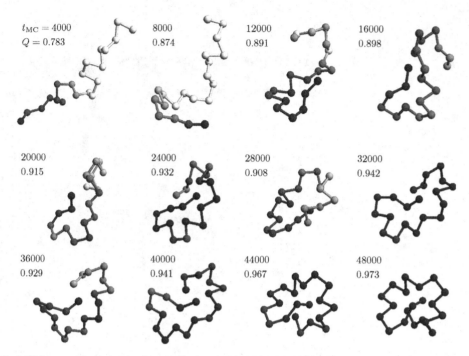

Fig. 9.10   Single folding event for the heteropolymer with sequence 20.6 in the GōL model at $T = 0.3 < T_f \approx 0.36$. The description of the gray-scale code is given in the text. From [200].

allows for comparative folding (unfolding) studies at various temperatures. The kinetics of the folding (unfolding) transitions is thereby obtained by averaging over sufficiently many folding (unfolding) trajectories.

In Fig. 9.10, snapshots of a single GōL folding event of the heteropolymer with sequence 20.6 at $T = 0.3$ are shown at different times t_{MC}. The gray scale of the monomers encodes the variance of the monomer positions, $\sigma_{\mathbf{r}_i}^2 = \overline{\mathbf{r}_i^2} - \overline{\mathbf{r}_i}^2$, averaged over the MC time interval $[t_{MC} - 1\,000, t_{MC} + 1\,000]$. The first monomer is fixed and therefore not moving. The higher the mobility of a monomer in this time interval, the brighter is its color. For $\sigma_{\mathbf{r}_i}^2 < 0.1$, the monomer is rendered in black, and for $\sigma_{\mathbf{r}_i}^2 > 1$ in white. The gray scales are linearly interpolated in between these boundaries. Although there are periods of relaxation and local unfolding, a stable intermediate conformation is not present and the folding process is a relatively "smooth" process. This is also confirmed by the more quantitative analysis of the same folding event in Fig. 9.11, where the temporal averages of the similarity parameters q and Q and of the energy E are shown. Since the temperature lies sufficiently far below the folding temperature ($T_f \approx 0.36$), the free-energy landscape does not exhibit substantial barriers which hinder the folding process.

Nonetheless, the chevron plot shown in Fig. 9.12 exhibits a rollover, which means that the folding characteristic is not perfectly of two-state type, in which case the folding (unfolding) branches would be almost linear [217]. In this plot, the temperature dependence of the mean first passage time τ_{MFP} is presented. We define τ_{MFP} as the average number of MC steps necessary to form at least 13 native contacts in the folding simulations,

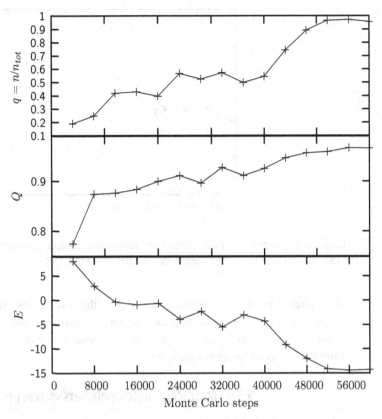

Fig. 9.11 Temporal averages of the native-contact and the overlap similarity parameters q and Q, respectively, and the energy for the folding event shown in Fig. 9.10. The temporal averages are calculated every 4000 MC steps over the time interval $[t_{MC} - 1\,000, t_{MC} + 1\,000]$. From [200].

starting from a random conformation. In the unfolding simulations, one starts from the native state and τ_{MFP} is the number of MC steps required to reach a conformation with fewer than 7 native contacts, i.e., 13 native contacts are broken. In all simulations performed at different temperatures, τ_{MFP} is averaged from the first passage times of a few hundred respective folding and unfolding trajectories. Assuming a linear dependence at least in the transition state region, τ_{MFP} is directly related to exponential folding and unfolding rates $k_{f,u} \approx 1/\tau_{MFP}^{f,u} \sim \exp(-\varepsilon_{f,u}/k_B T)$, respectively, where the constants $\varepsilon_{f,u}$ determine the kinetic folding (unfolding) propensities. The dashed lines in Fig. 9.12 are tangents to the logarithmic folding and unfolding curves at the transition state. The slopes are the folding (unfolding) propensities and have in this example values of $\varepsilon_f \approx -1.32$ and $\varepsilon_u \approx 5.0$.

In this variant of the chevron plot [217, 220], the temperature T mimics the effect of the denaturant concentration that is in experimental studies the more generic external control parameter. The hypothetic intersection point of the folding and unfolding branches defines the transition state. The transition state temperature estimated from this analysis coincides very nicely with the folding temperature $T_f \approx 0.36$ as identified in our earlier

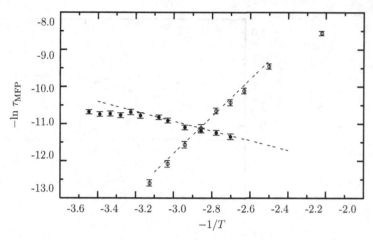

Fig. 9.12 Chevron plot of the mean-first passage times from folding (●) and unfolding (○) events at different temperatures. The hypothetic intersection point corresponds to the transition state. From [200].

discussion of the thermodynamic properties of the system. This result also demonstrates that the description of the folding and unfolding transitions from the kinetic point of view is not only qualitatively but even quantitatively consistent with the thermodynamic results from the canonical-ensemble analysis.

9.3.4 Mesoscopic heteropolymers vs. real proteins

In this context, it is not the purpose of the coarse-grained analysis to explain the two-state folding characteristics of a specific, real protein. Rather, we have shown that it can actually be useful to study thermodynamic and kinetic properties of mesoscopic peptide models without introducing atomic details. The main focus of such models is pointed toward general features of the folding transition (measured in terms of "order" parameters being specific for the corresponding transition, such as, e.g., the contact and overlap parameters investigated in this analysis) that are common to a number of proteins behaving qualitatively similarly. It is then furthermore assumed that these proteins can be grouped into classes of certain folding characteristics. The AB sequences, including sequence 20.6 considered here, are not obtained from a one-to-one hydrophobic–polar transcription of a real amino acid sequence. Such a mapping is not particularly useful. Rather, these heteropolymer models are considered as representatives for typical folding behaviors that are also found in real proteins. This implies, in general, that the classification of peptide folding behaviors is not necessarily connected with detailed atomic correlations and particular contact-ordering. It is rather an intrinsic property of protein-like heteropolymers and can thus already be discovered by employing models on mesoscopic scales [46, 47, 221].

Several proteins are known to be two-state folders, and their folding transitions exhibit the features we have also seen in the coarse-grained model study presented in this chapter. A famous example is *chymotrypsin inhibitor 2* (CI2) [205], one of the first proteins where two-state folding characteristics have been identified experimentally. Clear signals of a first-order-like folding–unfolding transition were also seen in computational Gō model analyses of that peptide [212, 217]. It is clear, however, that a precise characterization of the transition state ensemble, which is required for a better understanding of the folding (or unfolding) process of a specific peptide (e.g., secondary-structure formation or disruption in CI2 [216]), is not possible. In the particularly simple hydrophobic–polar models, for example, only tertiary folding aspects based on hydrophobic-core formation are considered as being relevant. If this hydrophobic-core collapse is the effective driving force in the folding process of a protein, it is likely that a simple model such as a mesoscopic Gō model exists that is capable of describing kinetic and thermodynamic aspects of the folding process of this protein very well. This is a direct consequence of *cooperativity* in the mechanical behavior of individual degrees of freedom during the folding process.

9.4 Microcanonical effects

The folded structure of a functional bioprotein is thermodynamically stable under physiological conditions, i.e., thermal fluctuations do not lead to significant globular conformational changes. To force tertiary unfolding requires an activation energy that is much larger than the energy of the thermal fluctuations. This activation barrier can be drastically reduced by the influence of other proteins like prions. The Creutzfeldt–Jakob disease is an example of the disastrous consequences prion-mediated degeneration of proteins can cause in the brain. The folded structure and the statistical ensemble of native-like structures, which are morphologically identical to the native fold, form a macrostate. It represents a conformational phase which is energy-dominated. Functionality of the native structure is only assured if entropic effects are of little relevance.

A significant change of the environmental conditions such as temperature, pH value, or, following the abovementioned example, the prion concentration, can destabilize the folded phase. Entropy becomes relevant, the entropic contribution to the free energy starts dominating over energy. Consequently, the hydrophobic core decays. This does not necessarily lead to a globular unfolding of the protein. A rather compact intermediate conformational phase can be stable. However, further imbalancing the conditions will finally lead to the phase of randomly unstructured coils. The latter transition is often called "folding/ unfolding transition," whereas the hydrophobic core formation is referred to as "glassy transition," as unresolvable competing energetic effects may result in frustration. The primary structure, i.e., the sequence of different amino acids lining up in proteins, is already sort of quenched disorder.

Simple examples for these transitions accompanying peptide folding processes have already been discussed in the analyses of Figures 8.2 and 8.6 for an exemplified lattice and an off-lattice heteropolymer, respectively. Obviously, the peak temperatures of the heat capacity C_V and the fluctuations of end-to-end distance and radius of gyration do not perfectly coincide, although all fluctuating quantities clearly signal the transitions. This is a typical indication for the finiteness of the system. There are no transition points in protein structure-formation processes, but rather transition regions (shaded areas in Fig. 8.2 and 8.6). This separates conformational transitions of finite-length polymers (pseudophase transitions) from thermodynamic phase transitions being considered in the thermodynamic limit.

The smallness of such systems can cause surprising side effects in nucleation processes that protein folding belongs to. Since the formation of the solvent-accessible hydrophilic surface and the bulky hydrophobic core is crucial for the whole tertiary folding process, the competition between surface and volume effects significantly influences the thermodynamics of nucleation. For this reason, it is not obvious at all which statistical ensemble represents the appropriate frame for the thermodynamic analysis of folding processes [222]. This is even more intricate than one may think. It is, for example, quite common to interpret phase transitions by means of fluctuating quantities calculated within the canonical formalism. Transition points are characterized by divergences in the fluctuations (second-order phase transitions) or entropy discontinuities (first-order transitions), occurring at unique transition temperatures. This standard analysis is based on the assumption that the temperature is a well-defined quantity, as it seems to be an easily accessible control parameter in experiments. This assumption is true for very large systems ($N \to \infty$) with vanishing surface/volume ratio in equilibrium, where surface fluctuations are irrelevant. The microcanonical entropy $S(E) = k_B \ln g(E)$ ($k_B = 1$ in the following), where $g(E)$ is the density of states, is a concave function and thus the microcanonical temperature, defined by the mapping $\mathcal{F} : E \mapsto T =: T(E) = [\partial S(E)/\partial E]^{-1}$, never decreases with increasing energy E. A discrimination of the parameter "temperature" in the canonical ensemble and the microcanonical (caloric) temperature $T(E)$ is not necessary, as energetic fluctuations vanish and thus the canonical and the microcanonical ensemble are equivalent in the thermodynamic limit.

But what if surface fluctuations are non-negligible? In this case, the canonical temperature can be a badly defined control parameter for studies of nucleation transitions with phase separation.[4] This becomes apparent in the following microcanonical folding analysis of the hydrophobic–polar heteropolymer sequence 20.6, whose canonical thermodynamic and kinetic properties have been investigated in detail in the previous section, by employing the AB model.

By means of generalized-ensemble computer simulations, an accurate estimate of the density of states $g(E)$ can be obtained. For this particular heteropolymer, it turns out that the entropy $S(E)$ exhibits a *convex* region, i.e., a double-tangent with two touching points, at E_{fold} and $E_{unf} > E_{fold}$, can be constructed. This so-called Gibbs hull is then parametrized by $H(E) = S(E_{fold}) + E/T_{fold}$, where $T_{fold} = [\partial H(E)/\partial E]^{-1} = [(\partial S(E)/\partial E)_{E_{fold},E_{unf}}]^{-1}$ is the microcanonically defined folding temperature, which is here $T_{fold} \approx 0.36$. As shown in Fig. 9.13, the difference $S(E) - H(E)$ has two zeros at E_{fold} and E_{unf}, and a noticeable well in between with the local minimum at E_{sep}. At this point the deviation $S(E_{sep}) - H(E_{sep})$ is called *surface entropy* as the convexity of the entropy in this region is caused by surface effects.

---

[4] Folding or "nucleation" processes of proteins are strongly dependent on the sequence of amino acids. Thus, folding is no generic phase transition and terms like "nucleation" should be used with some care.

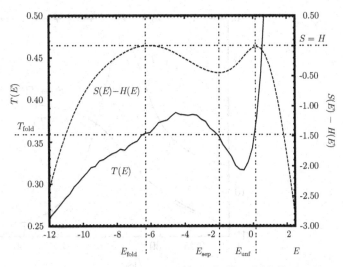

Folding transition as a phase-separation process for sequence 20.6 in the AB model. In the transition region, the caloric temperature $T(E)$ of the protein decreases with increasing total energy. Folded and unfolded conformations coexist at the folding transition temperature, where $S(E) = H(E)$, corresponding to $T_{fold} \approx 0.36$. The energetic transition region is bounded by the energies $E_{fold} \approx -6.3$ and $E_{unf} \approx 0.15$. Folding and unfolding regions are separated at $E_{sep} \approx -2.0$. From [222].

However, the most striking feature in Fig. 9.13 is the qualitative change of the micro-canonical temperature $T(E)$ in the transition region: approaching from small energies (folded phase), the curve passes the folding temperature T_{fold} and follows the overheating branch. It then *decreases* with *increasing energy* (passing again T_{fold}) before following the undercooling branch in reverse direction, crossing T_{fold} for the third time. In the unfolded phase, temperature and energy increase as expected. The unusual backbending of the caloric temperature curve within the transition region is a physical effect, as we have discussed in Section 2.7.[5]

In Fig. 9.14, results from the canonical calculations (mean energy $\langle E \rangle$ and specific heat c_V) are shown as functions of the temperature. The specific heat exhibits a clear peak near $T = 0.35$ which is close to the folding temperature T_{fold}, as defined before in the microcanonical analysis. The loss of information by the canonical averaging process is apparent by comparing $\langle E \rangle$ and the inverse, non-unique mapping $\mathcal{F}^{-1}$ of microcanonical temperature and energy. The temperature decrease in the transition region from the folded to the unfolded structures is unseen in the plot of $\langle E \rangle$.

---

[5] It is sometimes argued that proteins fold in solvent, where the solvent serves as a heat bath. This would provide a fixed canonical temperature such that the canonical interpretation is sufficient to understand the transition. However, the solvent-protein interaction is actually implicitly contained in the heteropolymer model and, nonetheless, the microcanonical analysis reveals this effect, which is simply "lost" by integrating out the energetic fluctuations in the canonical ensemble (see Fig. 9.14).

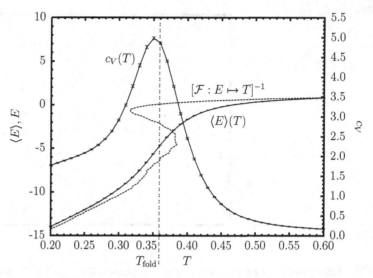

Fig. 9.14 Comparison of canonical and microcanonical analysis for the folding transition of 20.6: The fluctuations of energy as represented by the specific heat, $c_V(T)$, exhibit a sharp peak near $T \approx 0.35$. The canonical mean energy $\langle E \rangle(T)$ crosses over from the folded to the unfolded conformations. However, the canonical calculation averages out the overheating/undercooling branches and the backbending effect, which are clearly signaled by the microcanonical analysis of the (inverse) mapping between temperature and energy. The microcanonically defined "folding temperature" T_{fold} is close to the peak temperature of the specific heat. From [222].

Eventually, as we had already mentioned, there is no unique canonical folding temperature signaled by peaks of fluctuating quantities; rather, there are transition regions. Therefore, for small systems, the definition of transitions based on the canonical temperature is indeed of little use and the interpretation of these cooperative effects as thermodynamic phase transitions has noticeable limitations. This not only regards the impossibility to precisely identify definite transition points. It is even more fundamental to ask the question which of the typically used statistical ensembles provides the most comprehensive interpretation of finite-system structure formation processes. We have shown that the microcanonical analysis of folding thermodynamics is particularly advantageous, as, e.g., the remarkable temperature backbending effect is averaged out in the canonical ensembles, where the temperature is considered as an external control parameter, which seems to be questionable for small systems. As we will see in examples to be discussed later on, this also applies to molecular aggregation and adsorption transitions, where similar phenomena can occur.

Up to now, we have mainly investigated tertiary conformational transitions such as polymer collapse and crystallization, and also protein folding, which all require a cooperative behavior of the monomers on a length scale that corresponds to the chain length. We will now turn to effects on smaller (but still mesoscopic) scales that entail the formation of local symmetries. This class of symmetries is known as the secondary structures.

9.5 Two-state cooperativity in helix–coil transitions

Possibly the simplest representation for a two-state folding transition of a peptide is the crystallization into helical structures. Since most proteins possess helical segments, the helix–coil transition is perhaps the most prominent structure formation process of proteins. Helices form if the values of the dihedral angles ψ and ϕ (see Fig. 1.6) of amino acids adjacent in the sequence lie within a certain section of the Ramachandran map. The boundaries of this section are not well defined, but most amino acids in helical state are found with dihedral angles in the intervals $\phi \in (-90°, -30°)$ and $\psi \in (-77°, -17°)$ [223].

The simple sequence $(AAQAA)_3$ (A = alanine, Q = glutamine) is known to be a stable helix folder [224–226] and, therefore, it is a good starting point for the investigation of microcanonical properties of two-state folding processes associated with helix–coil formation. Figure 9.15(a) shows that only a single transition occurs in the entire energy region.[6] There is a nonzero entropic barrier ΔS in the transition region (between $E = 40\mathcal{E}$ and $E = 80\mathcal{E}$, where $\mathcal{E}$ is the artificial energy unit of the model used [227]. Thus, the transition energies are well separated and the energetic region in between is entropically depleted. It is a clear signal for a transition with coexisting structural phases, in which intermediate states are unstable. We have discussed this feature of a finite-size first-order-like transition, when we introduced the microcanonical statistical analysis method in Section 2.7. The inverse microcanonical temperature curve $T_{\mu c}^{-1}(E)$ responds to the entropic suppression of states by backbending, which is characteristic for this kind of transition. Useful order parameters of helix–coil transitions are the helicity, i.e., the probability $\theta(E)$ for an amino acid to be in helical state and the average number of helices in the ensemble $H(E)$. Both quantities are also shown for $(AAQAA)_3$ in the bottom figure of panel 9.15(a). Outside the transition region, $H(E) = 0$ in the coil phase ($E > 80\mathcal{E}$) and $H(E) = 1$ in the helix phase ($E < 40\mathcal{E}$), as expected. The helicity shows a similarly clear trend.

Increasing the number of segments, we observe that for $n = 7$ the first-order character of the helix–coil transition is still well sustained [Fig. 9.15(b)]. For $n = 10$ [Fig. 9.15(c)], however, the helix–coil transition is no longer a single-step process. The average number of helices $H(E)$ is noticeably larger than unity in the transition region, which clearly suggests a noncooperative but simultaneous folding of the coil into two independent helix segments. Since no side chain is available that stabilizes a tertiary fold formed by two aligned helices, the helices finally unite and form a single, large one. This completes the transition from the coil state into the helical conformation. In effect, the transition does not possess an apparent first-order character. There is no significant entropic barrier; it is virtually "flattened" by a sufficiently large number of intermediate states. This characteristic remains in effect

---

[6] The results discussed in this section were obtained by replica-exchange molecular dynamics simulations [225, 226] of a partially coarse-grained model [227] of C^β type. In a C^β model, the backbone of the protein is modeled in atomistic resolution, whereas the side chains are replaced by an effective interaction site "C^β" with few parameters that are specific to the individual amino acids. Such simplified models are particularly efficient and useful for the simulation of peptides with dominant secondary structures, where tertiary folding effects governed by the side chains are of less relevance.

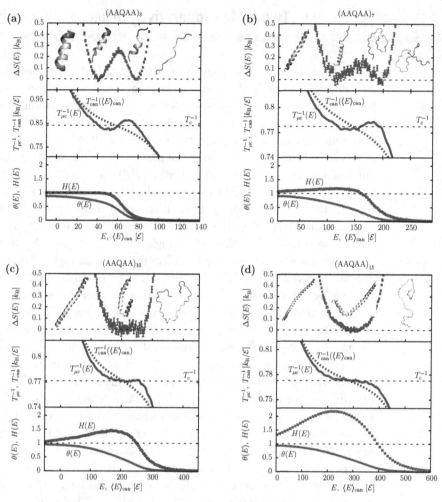

Fig. 9.15 Microcanonical quantities for sequences $(AAQAA)_n$ with (a) $n = 3$, (b) $n = 7$, (c) $n = 10$, and (d) $n = 15$. Shown are: (top figure in the panels) the entropic barrier $\Delta S(E)$, (center) the inverse microcanonical and canonical temperatures $T_{\mu c}^{-1}(E)$ and $T^{-1}(\langle E \rangle)$, and (bottom) the average helicity $\theta(E)$ and the average number of helices $H(E)$. After [226].

also for longer chains. Figure 9.15(c) shows the microcanonical helix-folding behavior for $n = 15$, and it appears to be a clear second-order-like structural transition. In this case, the ensemble in the transition region contains states with more than two uncooperatively folded helices that finally crystallize into a single one [225, 226].

The sequence $(AAQAA)_n$ is designed in a way that side-chain effects do not matter or at least do not substantially influence the folding into a single helix. It is certainly intriguing to study within the same context another example that exhibits stronger interactions among side chains of the amino acids in its sequence. The synthetic 73-residue helix bundle α3D [228] is a prominent representative of the group of proteins that fold into a structure

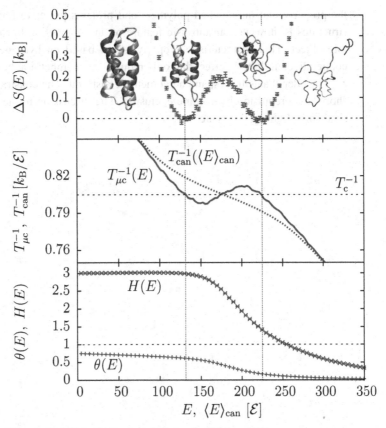

Fig. 9.16   Same quantities as in Fig. 9.15, but for the three-helix bundle α3D. From [226].

composed of individual helix segments that are aligned and stabilized by tertiary contacts.[7] Although this sounds like a multistep process, it reduces on the ensemble level to a two-state folding transition.[8] This is shown in Fig. 9.16. The insets in the top figure of the panel show characteristic structures. There is an entropic barrier that separates the unfolded from the tertiarily aligned helices. Helix formation and tertiary alignment are cooperative, i.e., these are not separate transitions here. The folded state is a three-helix bundle, and $H(E)$ shown in the bottom figure shows that the helix formation is a smooth, monotonic process in energy space. The entire secondary and tertiary folding process is a first-order-like two-state transition that is obviously more complex than the previously described purely secondary helix–coil transitions [225, 226].

It is commonplace that secondary structures require the formation of dipolar hydrogen bonds. Whereas hydrogen bonds are doubtless necessary to stabilize secondary structures,

---

[7] The sequence is MGSWA EFKQR LAAIK TRLQA LGGSE AELAA FEKEI AAFES ELQAY KGKGN PEVEA LRKEA AAIRD ELQAY RHN (the helical segments are underlined).

[8] Remember that we introduced "transitions" and their characteristics entirely on the basis of total system energy and the entropy. So, the cooperativity of a thermodynamic process is expressed by the combination of energy and entropy or, more precisely, by the change of entropy with respect to energy in the energetic transition interval (see also Section 2.7).

the property that unbranched polymers and proteins prefer to form helical or sheet-like structures is, however, an intrinsic feature of any interacting linear, flexible or semiflexible object. This characteristic property is not based on hydrogen bonds at all. It is a cooperative, multibody effect that only requires the competition between attractive forces, e.g., van der Waals forces, and a geometric constraint. The constraint is a consequence of short-range repulsion by volume exclusion. The discussion in the following chapter will shed some light on this issue.

10 Inducing generic secondary structures by constraints

10.1 The intrinsic nature of secondary structures

Resolving structural properties of single molecules is a fundamental issue as molecular functionality strongly depends on the capability of the molecules to form stable conformations. Experimentally, the identification of substructures is typically performed, for example, by means of single-molecule microscopy, X-ray analyses of polymer crystals, or NMR for polymers in solution. With these methods, structural details of *specific* molecules are identified, but frequently these can not be generalized systematically with respect to characteristic features being equally relevant for different polymers. Therefore, the identification of generic conformational properties of polymer classes is highly desirable. To date the most promising approach to attack this problem is to analyze polymer conformations by means of comparative computer simulations of polymer models on mesoscopic scales, i.e., by introducing relevant cooperative degrees of freedom and additional constraints. In these modeling approaches – we have already made use of it in the previous chapters – the linear polymer is considered as a chain of beads and springs. Monomeric properties are accumulated in an effective, specifically parametrized single interaction point of dimension zero ("united atom approach"). Noncovalent van der Waals interactions between pairs of such interaction points are typically modeled by Lennard-Jones (LJ) potentials. In such models, only the repulsive short-range part of the LJ potentials keeps the monomers *pairwisely* apart. Although such models have proven to be quite useful in identifying universal aspects of global structure formation processes, these approaches are less useful in this form to describe local symmetric arrangements of segments of the chain.

For the identification of underlying *secondary* structure segments like helices, strands, and turns, however, the modeling of volume exclusion by means of pure LJ pair potentials is not sufficient to form clearly distinct secondary structures enabling a classification scheme. Segments of such secondary structures were found, e.g., in dynamical LJ polymer studies [229] or as ground states in models with stiffness [230], explicit hydrogen bonding [231, 232], or explicit solvent particles [233, 234]. It could also be shown that helical structures form by introducing anisotropic monomer–monomer potentials in conjunction with a wormlike backbone model [235] or by combining excluded volume and torsional interactions [236].

The formation of secondary structures requires cooperative behavior of adjacent monomers, i.e., in addition to pairwise repulsion, information about the relative position of the monomers to each other in the chain is necessary to effectively model the competition between noncovalent monomeric attraction and short-range repulsion due to volume

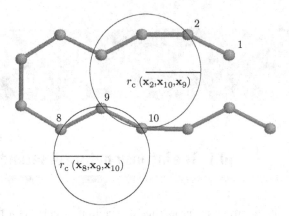

Fig. 10.1 Two examples of circles with different radii of curvature. The small circle is defined by the three consecutive monomers 8, 9, and 10, in which case its radius is called the *local* radius of curvature. From [243].

exclusion effects [237]. The simplest way to achieve this in a general, mesoscopic model is to introduce a hard single-parameter thickness constraint and, thus, to consider a polymer chain rather as a topologically three-dimensional tube-like object than as a one-dimensional, line-like string of monomers [238,239]. This approach differs from frequently studied cylindrical tube models [240], in which case the tube thickness rather mimics volume exclusion but not cooperativity such that explicit modeling of hydrogen bonds is required to generate and stabilize secondary structures.

In the following, we use the global radius of curvature as a quantity that effectively includes many-body interactions among the monomers. Defining a lower bound will allow for introducing a thickness constraint. Then, the polymer behaves like a tube. Variations of thickness and temperature will enable us to classify the geometric secondary phases classes of short polymers (or segments of larger polymers) can reside in [241–243].

10.2 Polymers with thickness constraint

10.2.1 Global radius of curvature

A natural choice for parametrizing the thickness of a polymer conformation with N monomers, $\mathbf{X} = (\mathbf{x}_1, \ldots, \mathbf{x}_N)$, is the global radius of curvature r_{gc} [244]. It is defined as the radius r_c of the smallest circle connecting any three different monomer positions $\mathbf{x}_i, \mathbf{x}_j, \mathbf{x}_k$ $(i, j, k = 1, \ldots, N)$:

$$r_{gc}(\mathbf{X}) = \min\{r_c(\mathbf{x}_i, \mathbf{x}_j, \mathbf{x}_k) \; \forall i, j, k \,|\, i \neq j \neq k\}. \tag{10.1}$$

In Fig. 10.1, two exemplified circles with their associated radii of curvature r_c are depicted.

The circumradius or local radius of curvature r_c can be easily calculated by the geometric relationships of angles in similar triangles in a circle with center $\mathbf{C}$ and radius r_c, as shown in Fig. 10.2. The triangles $\Delta(\mathbf{x}_i, \mathbf{x}_j, \mathbf{x}_k)$, $\Delta(\mathbf{x}_j, \mathbf{x}_k, \mathbf{D})$, and $\Delta(\mathbf{x}_j, \mathbf{x}_k, \mathbf{C})$ subtend the same

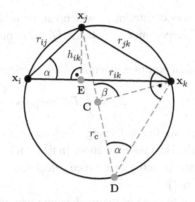

Fig. 10.2 Definition of the circumradius for the three coordinates x_i, x_j, and x_k.

arc (between $\mathbf{x}_j$ and $\mathbf{x}_k$). Hence, according to the theorem of inscribed angles, $\beta = 2\alpha$, and $\Delta(\mathbf{x}_j, \mathbf{x}_k, \mathbf{D}) \sim \Delta(\mathbf{x}_j, \mathbf{E}, \mathbf{x}_i)$, because of the respective right angles. Denoting the altitude with foot at $\mathbf{E}$ by h_{ik} and the distance between two points by $r_{ij} = |\mathbf{x}_i - \mathbf{x}_j|$, the similarity yields

$$\sin\alpha = \frac{h_{ik}}{r_{ij}} = \frac{r_{jk}}{2r_c}. \tag{10.2}$$

Since the area of $\Delta(\mathbf{x}_i, \mathbf{x}_j, \mathbf{x}_k)$ is given by $A_\Delta(\mathbf{x}_i, \mathbf{x}_j, \mathbf{x}_k) = h_{ik}r_{ik}/2$, the altitude h_{ik} in Eq. (10.2) can be replaced by the area. Then, the local radius of curvature can be calculated as

$$r_c(\mathbf{x}_i, \mathbf{x}_j, \mathbf{x}_k) = \frac{r_{ij}r_{jk}r_{ik}}{4A_\Delta(\mathbf{x}_i, \mathbf{x}_j, \mathbf{x}_k)}. \tag{10.3}$$

Eventually, with the definition of the global radius of curvature (10.1), the polymer tube $\mathbf{X}$ can be assigned the "thickness" (or diameter) $d(\mathbf{X}) = 2r_{gc}(\mathbf{X})$ [244, 245].

10.2.2 Modeling flexible polymers with constraints

In the following, we first want to consider linear, flexible polymers with stiff bonds of unit length ($r_{ii+1} = 1$) and pairwise interactions among nonbonded monomers are modeled by a standard LJ potential. Thus, the energy of a conformation $\mathbf{X}$ reads

$$E(\mathbf{X}) = \sum_{i,j>i+1} V_{LJ}(r_{ij}), \tag{10.4}$$

where as usual $V_{LJ}(r_{ij}) = 4\varepsilon[(\sigma/r_{ij})^{12} - (\sigma/r_{ij})^6]$. By setting $\sigma = 1$, $V_{LJ}(r_{ij})$ vanishes for $r_{ij} = 1$ and is minimal at $r_{ij}^{min} = 2^{1/6} \approx 1.122$.

Since we are interested in classifying conformational pseudophases of polymers with respect to their thickness, it is useful to introduce the restricted conformational space $\mathcal{R}_\rho = \{\mathbf{X} \mid r_{gc}(\mathbf{X}) > \rho\}$ of all conformations $\mathbf{X}$ with a global radius of curvature larger than a thickness constraint ρ, which can be understood as an effective measure for the extension of the polymer side chains. Given ρ, obviously only conformations with $r_{gc} \geq \rho$ can occur.

The canonical partition function of the restricted conformational space thus reads

$$Z_\rho = \int \mathcal{D}X \Theta(r_{gc}(\mathbf{X}) - \rho)e^{-E(\mathbf{X})/k_B T}, \tag{10.5}$$

where $k_B T$ is the thermal energy (we use units in which $\varepsilon = k_B = 1$ in the following) and $\Theta(z)$ is the Heaviside function. In this thickness-restricted space, canonical statistical averages of any quantity O are then calculated via $\langle O \rangle_\rho = Z_\rho^{-1} \int \mathcal{D}X O(\mathbf{X}) \Theta(r_{gc}(\mathbf{X}) - \rho)$ $\exp[-E(\mathbf{X})/k_B T]$.

For the investigation of thermodynamic features, we will focus in the following on the tube polymer with $N = 9$ monomers [241]. Results obtained for this system can be generalized and are thus also common to somewhat longer chains [242, 243]. However, it should already be noted here that this consideration does not make much sense for much longer chains, as large secondary structures in single molecules (such as proteins) often become unstable and a rearrangement of secondary-structure segments in favor of tertiary alignments occurs.

10.2.3 Thickness-dependent ground-state properties

Figure 10.3 shows the ground-state energy per monomer as a function of ρ. Also shown are lowest-energy conformations for exemplified values of ρ. The line-like 9-mer (i.e., $\rho = 0$) has the ground-state energy $E_{min}/N = -1.85$ in our chosen units. The thickness constraint becomes relevant, if ρ is larger than half the characteristic length scale r_{ij}^{min} of the LJ potential: in the interval $2^{-5/6} \approx 0.561 < \rho < \rho_\alpha \approx 0.686$ conformations are pre-helical. The nonbonded interaction distance is still allowed to be so small that structures are deformed. Nonetheless, the onset of helix formation is clearly visible. Optimal space-filling helical symmetry is reached when approaching ρ_α, where the ground-state conformation takes the perfect α-helical shape (see inset of Fig. 10.3). All torsional angles are identical (near $41.6°$) and also all local radii are constant; the number of monomers per winding is 3.6. Note that for proteins, where the effective distance between two C^α atoms is about $3.8\,\text{Å}$, ρ_α in our units corresponds indeed to a pitch of about $5.4\,\text{Å}$ as known from α-helices of proteins. Thus, an α-helix is a natural geometric shape for tube-like polymers. Hydrogen bonds stabilize these structures in nature – but are not a necessary prerequisite for forming such secondary structures. This is indeed a substantial result, as it shows that helical secondary structures form a subset of dominant basic geometries of string-like objects with thickness. In other words, we can expect that there is a helical phase that, depending on thickness and external parameters such as the temperature, is stable for a *large class* of polymers with different molecular compositions.

For larger values of ρ, helices unwind, i.e., the pitch gets larger and the number of monomers per winding increases. However, helical structures still dominate the ground-state conformations. It should be noted that the simple polymer model is energetically invariant under helicity reversal, i.e., left-handed helices or segments are not explicitly

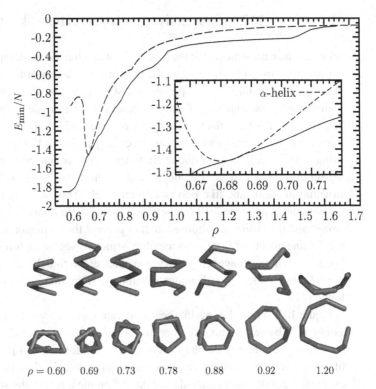

Fig. 10.3 Ground-state energy per monomer E_{min}/N of tube-like polymers with nine monomers as a function of the global radius of curvature constraint ρ (solid line). For comparison, the energy curve of the perfect α-helix is also plotted (dashed line). The inset shows that for a small interval around $\rho \approx 0.686$, the ground-state structure is perfectly α-helical. Also depicted are side and top views of putative ground-state conformations for various exemplified values of ρ. For the purpose of clarity, the conformations are not shown with their natural thickness. From [243].

disfavored and are, therefore, also equally present in the conformational space. In the interval $\rho_\alpha \lesssim \rho \lesssim 0.92$, there are also stable helical conformations in the vicinity of $\rho \approx 0.73$ (winding number ≈ 4.5) and $\rho \approx 0.78$ (winding number ≈ 5.0). Near $\rho \approx 0.92$, the final helical state has been reached. The thickness has increased in such a way that the most compact conformation is a helix with a single winding. After that, a topological change occurs and the ground-state conformations are getting flatter. The helix finally opens up and planar conformations with similarities to β-hairpins become dominant. These structures are still stabilized by nonbonded LJ interactions between pairs of monomers. Increasing the thickness further leads to a breaking of these contacts and ring-like conformations become relevant [245]. For values $r_{gc} \approx N/2\pi$, ground-state conformations are almost perfect circles with radius r_{gc}. The existence of ring-like conformations is a consequence of the long-range monomer–monomer attraction. Eventually, for $\rho \to \infty$, the effective stiffness increases, the end contacts disappear, and only thick rods are still present.

Now that we have completed these preparatory considerations of ground-state properties, we are going to discuss the thermodynamic behavior of the tube polymers.

10.2.4 Structural phase diagram of tube-like polymers

Based on the peak structure of the specific heat as a function of temperature T and thickness constraint ρ, we will now discuss the structure of the conformational ρ-T (pseudo)phase diagram. This requires a very precise statistical analysis. Even for such a small polymer with 9 monomers, hundreds of separate generalized-ensemble computer simulations had to be performed to obtain the results discussed here [241, 243]. The phase-diagram topology turns out to be general and valid for larger polymers, too [243]. Since even the expected shifts of the transition lines due to finite-length corrections are very small, one has good reason to assume that the pseudophase diagram of the 9-mer reflects the general phase structure of short tube-like polymers pretty well. This is partly due to the fact that the polymer thickness as defined via the global radius of curvature is a *length-independent constraint* and the chains are short enough to prevent the formation of tertiary structures (as, e.g., arrangements of different secondary-structure segments forming a tertiary domain). For longer chains, however, tertiary structures are definitely relevant. For the chain with $N = 13$ monomers, first indications of structure formation on globular length scales were found [243].

Figure 10.4 shows the specific-heat landscape $C_V(\rho, T) = (\langle E^2 \rangle_\rho - \langle E \rangle_\rho^2)/T^2$ for a 9-mer as obtained by reweighting the density of states for given thickness constraints ρ. This profile therefore represents the structure of the conformational phase diagram for a tube-like polymer in the perspective of secondary structures. Dark regions correspond to strong energetic fluctuations, i.e., the darker the region the larger is the specific-heat value. Data points ($+$) mark the peaks or ridges of the profile and indicate conformational activity and thus represent transitions between different conformational pseudophases. Error bars are not shown for clarity but are sufficiently small (for most data points smaller than symbol size), so that the identified phase boundaries are statistically significant.

Guided by the analyses of the ground-state properties, one can identify four principal pseudophases.[1] In region α, helical conformations are the most relevant structures. In particular, the α-helix resides in this pseudophase. Characteristic for the transition from pseudophase α to β is the unwinding of the helical structures that are getting more planar. Thus, region β is dominated by simple sheet-like structures. Since the 9-mer is rather short, the only sheet-like class of conformations is the hairpin. For longer chains, one also finds more complex sheets, e.g., lamellar structures [238, 242, 243]. A characteristic property of the hairpins is that these are still stabilized by nonbonded interactions. These break with larger thickness and higher temperature. Entering pseudophase γ, dominating structures possess ring-like shapes. Finally, region δ is the phase of random coils, which are getting stiffer for large thickness and eventually resemble rods. Representative polymer conformations dominating the pseudophases in the regions α to δ are depicted in Table 10.1 in different representations.

---

[1] We note that there are singular points in the parameter space corresponding to special geometric representations of secondary structures. For the chain with length $N = 8$ and $r_{gc} \approx 1/\sqrt{2}$, for example, the degenerate ground-state conformation exhibits an almost perfect alignment of the chain along the edges of a cube.

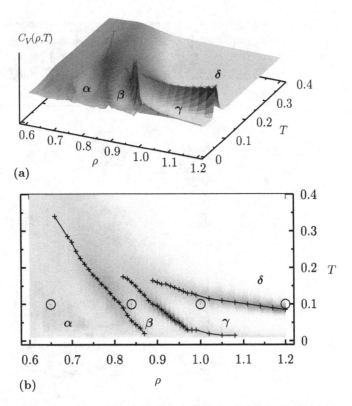

Fig. 10.4 (a) Perspective and (b) projected view of the specific-heat profile $C_V(\rho, T)$ for a 9-mer which is interpreted as a structural phase diagram of thermodynamically relevant tube polymer conformations in thickness–temperature parameter space. Dark regions and data points (+) indicate the ridges of the landscape and separate conformational phases. Helical or helix-like conformations dominate in region α, sheets in region β, rings in region γ, and stiff rods in pseudophase δ. Circles ($\odot$) indicate the locations where the exemplified conformations of Table 10.1 are relevant. The general structure of the phase diagram in the secondary-structure sector remains widely unchanged also for the longer polymers. From [241].

10.3 Secondary-structure phases of a hydrophobic–polar heteropolymer model

Proteins form the most prominent class of polymers, where different types of secondary structures occur, typically within the same molecule. It is therefore useful to extend the homopolymer model employed in the previous section by introducing two types of monomers which can be hydrophobic or polar. For this purpose, we will now combine the already introduced and for the line-like heteropolymers in Chapter 8 extensively studied AB model with the thickness constraint.

In order to get an impression of the effects caused by introducing a sequence-dependent model, we here consider an exemplified Fibonacci sequence $AB_2AB_2ABAB_2AB$ with 13

Table 10.1 Exemplified conformations being thermodynamically relevant in the respective pseudophases shown in Fig. 10.4, visualized in different representations. From [241].

phase	type	views of representative example
α	helix	
β	sheet	
γ	ring	
δ	rod	

monomers. The line-like case corresponds to the constraint $\rho = 0$ (no restriction of the conformational space). Figure 10.5 shows the structural phase diagram analogously to Fig. 10.4, as well as selected ground-state conformations. The general structure including the several separated structural subphases is similar to that for the homopolymer. The most noteworthy section of the phase diagram is the very stable β-sheet region in the interval $0.90 \leq \rho \leq 1.01$, as $T \to 0$. The dominant conformations are indeed "planar" (see low-energy conformations depicted in Fig. 10.5), and the qualitative properties do not change in the entire region. A quantitatively remarkable fact is the variation of the intra-monomer distances. We note that the interaction length between the opposite hydrophobic A monomers 1–12 ($r_{1,12} = 1.13$, see Fig. 10.5 for monomer numbering) and 4–9 ($r_{4,9} = 1.15$) in the sheet conformation does not change in the whole region of thickness constraint at all.

On the other hand, the distances of B monomer pairs 2–11 and 3–10 increase ($\Delta r_{2,11} = \Delta r_{3,10} = 0.27$) and decrease between the A monomers 1–4 and 9–12 ($\Delta r_{1,4} = \Delta r_{9,12} = -0.10$; cf. the conformations at $\rho = 0.9$ and $\rho = 1.0$). The van der Waals attraction between the A monomers is thus responsible for the stabilization of the β-sheet structures.

For smaller thickness constraints, structures with helical properties are found, which depends on the monomer sequence. We note here a very pronounced conformational

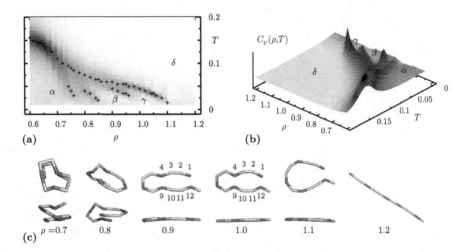

Fig. 10.5 Structural phase diagram for the $N = 13$ Fibonacci AB heteropolymer sequence $AB_2AB_2ABAB_2AB$. (a) Top view with marked peak positions of the specific heat for various parameters ρ; (b) qualitative view of the specific-heat landscape. Gray scales encode the value of the specific heat; (c) exemplified ground-state conformations shown from different viewpoints, A monomers are marked in dark gray, B monomers are white. From [243].

transition from random coils to native conformations in the temperature interval $0.1 \lesssim T \lesssim 0.15$. With increasing thickness, the ground-state conformation becomes ring-like and finally switches to a stretched rod.

From our discussion in this chapter, we can conclude that the tube constraint concept is well suited to mimic the volume exclusion of polymers with extended side chains like, for example, the residual groups of amino acids in proteins. The effective thickness of the polymer tubes is responsible for the formation of stable ordered, short-length structures that resemble secondary structures such as α-helices and β-strands. This is a remarkable result because short-range interactions such as hydrogen bonding are obviously not necessary to form secondary structures (however they are needed for their stabilization). The constraint investigated here represents an effective global multibody volume exclusion, which consequently, in combination with local attraction by pairwise van der Waals forces, causes long-range cooperative behavior of the monomers.

After the discussion of single-chain properties, we are now going to investigate thermodynamic properties of the formation of structures of several individual chains. In these systems, single-chain structuring processes will compete with cooperative effects among different chains, which finally leads to another type of structural transition – the aggregation of polymers.

11

Statistical analyses of aggregation processes

Beside receptor-ligand binding mechanisms, folding and aggregation of proteins belong to the biologically most relevant molecular structure formation processes. While the specific binding between receptors and ligands is not necessarily accompanied by global structural changes, protein folding and oligomerization of peptides are typically cooperative conformational transitions [246]. Proteins and their aggregates are comparatively small systems. A typical protein consists of a sequence of some hundred amino acids and aggregates are often formed by only a few peptides. A very prominent example is the extracellular aggregation of the $A\beta$ peptide, which is associated with Alzheimer's disease. Following the amyloid hypothesis, it is believed that these aggregates (which can also take fibrillar forms [247]) are neurotoxic, i.e., they are able to fuse into cell membranes of neurons and create pores that are penetrable to calcium ions. It is known that extracellular Ca^{2+} ions intruding into a neuron can promote its degeneration [248–250].

In this chapter, we will investigate thermodynamic properties of aggregation transitions of polymers and peptides from different perspectives of statistical analysis.

11.1 Pseudophase separation in nucleation processes of polymers

We have already discussed why conformational transitions polymers experience in structure formation processes are not phase transitions in the strict thermodynamic sense. This will be similar for the aggregation of a finite number of finitely long polymers, because surface effects are also not negligible in these cases. The statistical analysis of any kind of transition is traditionally based on studies of signals exposed by energetic and structural fluctuations, as well as system-specific "order" parameters. In these cases, the temperature T is considered as an adjustable, external control parameter and, for the analysis of the pseudophase transitions, the peak structure of quantities such as the specific heat and the fluctuations of the gyration tensor components or any "order" parameter as functions of the temperature are investigated. The natural ensemble for this kind of analysis is the canonical ensemble, where the possible states of the system with energies E are distributed according to the Boltzmann probability $\exp(-E/k_BT)$, where k_B is the Boltzmann constant. However, phase separation processes of small systems as, e.g., droplet condensation, are accompanied by surface effects at the interface between the pseudophases [251, 252]. As discussed in detail in Section 2.7, the influence of the surface effects is reflected by the behavior of the microcanonical entropy $S(E)$, which can exhibit *convex* monotony in the transition region. Consequences are the backbending of the caloric temperature $T(E) = (\partial S/\partial E)^{-1}$, i.e., the *decrease* of temperature with increasing system energy, and the negativity of the micro-canonical specific heat $C_V(E) = (\partial T(E)/\partial E)^{-1} = -(\partial S/\partial E)^2/(\partial^2 S/\partial E^2)$. The physical reason is that the free-energy balance in phase equilibrium requires the minimization of the interfacial surface and, therefore, the loss of entropy [253–255]. A reduction of the entropy can, however, only be achieved by transferring energy into the system.

It is a surprising fact that this so-called backbending effect is indeed observed in transitions with phase separation. Although this phenomenon has already been known for a long time from astrophysical systems [256], it has been widely ignored since then as a somehow "exotic" effect. However, experimental evidence was also found from melting studies of sodium clusters by photo-fragmentation [257]. Bimodality and negative specific heats are also known from nuclei fragmentation experiments and models [258, 259], as well as from spin models on finite lattices that experience first-order transitions in the thermodynamic limit [260, 261]. This phenomenon is also observed in a large number of other isolated finite model systems for evaporation and melting effects [262, 263]. We have discussed methods to analyze the microcanonical finite-size effects in detail in Section 2.7.

These phenomena can also be expected to occur in the aggregation process of polymers. To this end, we will now investigate the aggregation behavior of a mesoscopic hydrophobic–polar heteropolymer model for aggregation [254, 255], which is based on the simple AB model [14] that we have already discussed in the context of tertiary folding behavior of proteins from a generic, coarse-grained point of view.

11.2   Mesoscopic hydrophobic–polar aggregation model

As for modeling heteropolymer folding, we will assume in the following that the tertiary folding process of the individual chains is governed by hydrophobic-core formation in an aqueous environment. For systems of more than one chain, we further take into account that the interaction strengths between nonbonded residues are independent of the individual properties of the chains the residues belong to. Therefore, we use the same parameter sets as in the AB model [14] for the pairwise interactions between residues of different chains. Therefore, a simple aggregation model can be introduced by [254, 255]

$$E = \sum_{\mu} E_{AB}^{(\mu)} + \sum_{\mu<\nu} \sum_{i_\mu, j_\nu} \Phi(r_{i_\mu j_\nu}; \sigma_{i_\mu}, \sigma_{j_\nu}), \tag{11.1}$$

where μ, ν label the M polymers interacting with each other, and i_μ, j_ν index the $N_{\mu,\nu}$ monomers of the respective μth and νth polymer. The intrinsic single-chain energy of the μth polymer is given by [cf. Eq. (1.13)]

$$E_{AB}^{(\mu)} = \frac{1}{4} \sum_{i_\mu} (1 - \cos \vartheta_{i_\mu}) + \sum_{j_\mu > i_\mu + 1} \Phi(r_{i_\mu j_\mu}; \sigma_{i_\mu}, \sigma_{j_\mu}), \tag{11.2}$$

with $0 \leq \vartheta_{i_\mu} \leq \pi$ denoting the bending angle between monomers i_μ, $i_\mu + 1$, and $i_\mu + 2$. The nonbonded inter-residue pair potential

$$\Phi(r_{i_\mu j_\nu}; \sigma_{i_\mu}, \sigma_{j_\nu}) = 4 \left[r_{i_\mu j_\nu}^{-12} - C(\sigma_{i_\mu}, \sigma_{j_\nu}) r_{i_\mu j_\nu}^{-6} \right] \tag{11.3}$$

depends on the distance $r_{i_\mu j_\nu}$ between the residues, and on their type, $\sigma_{i_\mu} = A, B$. The long-range behavior is attractive for like pairs of residues $[C(A, A) = 1, C(B, B) = 0.5]$ and repulsive otherwise $[C(A, B) = C(B, A) = -0.5]$. The lengths of all virtual peptide bonds are set to unity.

We now discuss the statistical mechanics of the aggregation processes of short polymers, governed by model (11.1). As an example, we choose again the heteropolymer with the Fibonacci sequence $AB_2AB_2ABAB_2AB$, which we will simply denote "F1" from now on. In the following, we are going to study the thermodynamics of systems with up to four chains of this sequence over the whole energy and temperature regime.[1]

11.3 Order parameter of aggregation and fluctuations

In order to distinguish between the fragmented and the aggregated regime, it is useful to introduce the "order" parameter [255]

$$\Gamma^2 = \frac{1}{2M^2} \sum_{\mu, \nu = 1}^{M} \mathbf{d}_{\text{per}}^2 \left(\mathbf{r}_{\text{COM}, \mu}, \mathbf{r}_{\text{COM}, \nu} \right). \tag{11.4}$$

The summations are taken over the minimum distances $\mathbf{d}_{\text{per}} = (d_{\text{per}}^{(1)}, d_{\text{per}}^{(2)}, d_{\text{per}}^{(3)})$ of the respective centers of mass of the chains (or their periodic continuations). The center of mass of the μth chain in a box with periodic boundary conditions is defined as $\mathbf{r}_{\text{COM}, \mu} = \sum_{i_\mu = 1}^{N_\mu} [\mathbf{d}_{\text{per}}(\mathbf{r}_{i_\mu}, \mathbf{r}_{1_\mu}) + \mathbf{r}_{1_\mu}] / N_\mu$, where $\mathbf{r}_{1_\mu}$ is the coordinate vector of the first monomer and serves as a reference coordinate in a local coordinate system.

This aggregation parameter must be considered as a qualitative measure; roughly, fragmentation corresponds to large values of Γ, whereas aggregation requires the centers of masses to be in close distance, in which case Γ is comparatively small. Despite its qualitative nature, it turns out to be a surprisingly manifest indicator for the aggregation transition and allows even a clear discrimination of different aggregation pathways, as will be seen later on.

According to the Boltzmann distribution, we define canonical expectation values of any observable O by

$$\langle O \rangle(T) = \frac{1}{Z_{\text{can}}(T)} \prod_{\mu = 1}^{M} \left[\int \mathcal{D}\mathbf{X}_\mu \right] O(\{\mathbf{X}_\mu\}) e^{-E(\{\mathbf{X}_\mu\})/k_B T}, \tag{11.5}$$

where the canonical partition function Z_{can} is given by

---

[1] In the actual simulations, a typical multicanonical run contained on the order of 10^{10} single updates. The polymer chains were embedded into a cubic box with edge lengths $L = 40$ and periodic boundary conditions were used [254, 255].

$$Z_{can}(T) = \prod_{\mu=1}^{M} \left[\int \mathcal{D}\mathbf{X}_\mu \right] e^{-E(\{\mathbf{X}_\mu\})/k_B T}.$$ (11.6)

Formally, the integrations are performed over all possible conformations $\mathbf{X}_\mu$ of the M chains.

Similar to the specific heat per monomer[2]

$$c_V(T) = \frac{1}{N_{tot}} \frac{d\langle E \rangle}{dT} = \frac{1}{N_{tot} k_B T^2} \left(\langle E^2 \rangle - \langle E \rangle^2 \right)$$ (11.7)

(with $N_{tot} = \sum_{\mu=1}^{M} N_\mu$), which expresses the thermal fluctuations of energy, the temperature derivative of $\langle \Gamma \rangle$ per monomer,

$$\frac{1}{N_{tot}} \frac{d\langle \Gamma \rangle}{dT} = \frac{1}{N_{tot} k_B T^2} \left(\langle \Gamma E \rangle - \langle \Gamma \rangle \langle E \rangle \right),$$ (11.8)

is a useful indicator for cooperative behavior of the multiple-chain system. Since the system size is small – the number of monomers N_{tot} as well as the number of chains M – aggregation transitions, if any, are expected to be signaled by the peak structure of the fluctuating quantities as functions of the temperature. This requires the temperature to be a unique external control parameter, which is a natural choice in the canonical statistical ensemble. Furthermore, this is a typically easily adjustable and, therefore, convenient parameter in experiments. However, aggregation is a phase separation process and, since the system is small, there is no uniform mapping between temperature and energy [254, 255]. For this reason, the total system energy is the more appropriate external parameter. Thus, the microcanonical interpretation will turn out to be the more favorable description, at least in the transition region. We will discuss this in detail in the following section.

11.4 Statistical analysis in various ensembles

For the qualitative description of the aggregation and the accompanied conformational cooperativity within the whole system, it is sufficient to consider a very small system. We will first focus on a heteropolymer system consisting of two identical chains with the sequence F1. This system will be denoted as $2\times$F1 in the following discussion of its aggregation behavior from the multicanonical, the canonical, and the microcanonical points of view.

11.4.1 Multicanonical results

In a multicanonical simulation, the phase space is sampled in such a way that the energy distribution gets as flat as possible. Thermodynamically, this means that the sampling of the

---

[2] As in the previous chapters, we denote by "monomer" a single chemical unit within a single chain. However, in the aggregation literature, the notion "monomer" is often also used for entire chains.

phase space is performed for all temperatures within a single simulation [94, 95, 264]; see also the detailed description of the method in Section 4.6.3. The desired information for the thermodynamic behavior of the system at a certain temperature is then obtained by simply reweighting the multicanonical into the respective canonical distribution, according to

$$p_{\text{can}}(E) \sim W_{\text{muca}}^{-1}(E; T) p_{\text{muca}}(E) e^{-E/k_{\text{B}}T}. \tag{11.9}$$

Since the multicanonical ensemble contains all thermodynamic information, including the conformational transitions, it is quite useful to measure within the simulation the multicanonical histogram (cf. also Section 9.2)

$$H_{\text{muca}}(E_0, \Gamma_0) = \sum_t \delta_{E_t, E_0} \delta_{\Gamma_t, \Gamma_0} \propto \langle \delta(E_t - E_0) \delta(\Gamma_t - \Gamma_0) \rangle_{\text{muca}}, \tag{11.10}$$

where t labels the Monte Carlo "time" steps. More formally, this distribution can be expressed as a conformation-space integral

$$\langle \delta(E - E_0) \delta(\Gamma - \Gamma_0) \rangle_{\text{muca}} = \frac{1}{Z_{\text{muca}}} \prod_{\mu=1}^{M} \left[\int \mathcal{D}\mathbf{X}_\mu \right] \delta(E(\{\mathbf{X}_\mu\}) - E_0) \delta(\Gamma(\{\mathbf{X}_\mu\}) - \Gamma_0)$$
$$\times \exp\left[-\mathcal{H}_{\text{muca}}(E(\{\mathbf{X}_\mu\}))/k_{\text{B}}T \right] \propto e^{-\mathcal{F}_{\text{muca}}(E_0, \Gamma_0)/k_{\text{B}}T} \tag{11.11}$$

with the multicanonical energy $\mathcal{H}_{\text{muca}}(E) = E - k_{\text{B}}T \ln W_{\text{muca}}(E; T)$, which is independent of temperature.

The multicanonical partition function is also trivially constant with respect to temperature,

$$Z_{\text{muca}} = \prod_{\mu=1}^{M} \left[\int \mathcal{D}\mathbf{X}_\mu \right] e^{-\mathcal{H}_{\text{muca}}(E(\{\mathbf{X}_\mu\}))/k_{\text{B}}T} = \text{const}_T. \tag{11.12}$$

It is obvious that integrating $H_{\text{muca}}(E, \Gamma)$ over Γ recovers the uniform multicanonical energy distribution:

$$\int_0^\infty d\Gamma \, H_{\text{muca}}(E, \Gamma) \sim h_{\text{muca}}(E). \tag{11.13}$$

The canonical distribution of energy and Γ parameter at temperature T can be retained, similar to inverting Eq. (11.9), by performing the simple reweighting

$$H_{\text{can}}(E, \Gamma; T) = H_{\text{muca}}(E, \Gamma) W_{\text{muca}}^{-1}(E; T), \tag{11.14}$$

which is, due to the restriction to a certain temperature, less favorable to gain an overall impression of the phase behavior (i.e., the transition pathway) of the system, compared to the multicanonical analog $H_{\text{muca}}(E, \Gamma)$.

In Fig. 11.1(a), $H_{\text{muca}}(E, \Gamma)$ is shown for the two-peptide system 2×F1 as a density plot in the E-Γ plane, which is the direct output obtained in a multicanonical simulation. Qualitatively, we observe two separate main branches (which are "channels" in the corresponding free-energy landscape), between which a noticeable transition occurs. In the vicinity of the energy $E_{\text{sep}} \approx -3.15$, both channels overlap, i.e., the associated macrostates

(a)

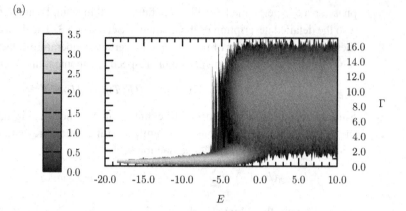

(b)

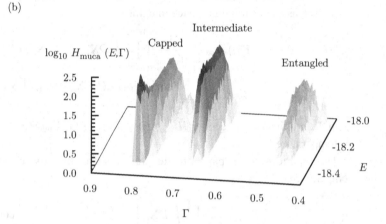

Fig. 11.1 (a) Multicanonical histogram $\log_{10} H_{\mathrm{muca}}$ as a function of energy E and aggregation parameter Γ, (b) section of $\log_{10} H_{\mathrm{muca}}$ in the low-energy tail. From [255].

coexist. Since Γ is an effective measure for the spatial distance between the two pep-tides, it is obvious that conformations with separated or fragmented peptides belong to the dominating channel in the regime of high energies and large Γ values, whereas the aggregates are accumulated in the narrow low-energy and small-Γ channel. Thus, the main observation from the multicanonical, comprising point of view is that the aggregation tran-sition is a phase-separation process which, even for this small system, already appears in a surprisingly clear fashion.

The typically high precision of results obtained by multicanonical sampling allows us to even reveal further details in the lowest-energy aggregation regime, which is usually a notoriously difficult sampling problem. Figure 11.1(b) shows that the tight aggregation channel splits into three separate, almost degenerate subchannels at lowest energies. From the analysis of the conformations in this region, we find that representative conformations with smallest Γ values, $\Gamma \approx 0.45$, are typically entangled, while those with $\Gamma \approx 0.8$ have a spherically capped shape. This is the subchannel connected to the lowest-energy states. Examples are shown in Fig. 11.2. The also highly compact conformations belonging to

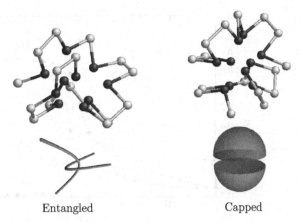

Fig. 11.2   Representatives and schematic characteristics of entangled and spherically capped conformations dominating the lowest-energy branches in the multicanonical histogram shown in Fig. 11.1(b). Dark spheres correspond to hydrophobic (*A*), light ones to polar (*B*) residues. From [255].

the intermediate subphase do not exhibit such characteristic features and rather resemble globules without noticeable internal symmetries. In all cases, the aggregates contain a single compact core of hydrophobic residues. Thus, for the system discussed here, the aggregation is not a simple docking process of two prefolded peptides, but a complex cooperative folding-binding process. This is a consequence of the energetically favored hydrophobic inter-residue contacts which, as the results show, overcompensate the entropic steric constraints. The story is, however, even more interesting, as non-negligible surface effects also come into play. After the following standard canonical analysis, this will be discussed in more detail in the subsequent microcanonical interpretation of these features.

11.4.2   Canonical perspective

Phase transitions are typically described in the canonical ensemble with the temperature kept fixed. This is also natural from an experimentalist's point of view, since the temperature is a convenient external control parameter. The macrostates are weighted according to the Boltzmann distribution $p_{can}(E) \sim g(E) \exp(-E/k_B T)$. A nice feature of the canonical ensemble is that the temperature dependence of fluctuations of thermodynamic quantities is usually a very useful indicator for phase or pseudophase transitions. This cooperative thermodynamic activity is usually indicated by peaks or, in the thermodynamic limit (if it exists), by divergences of these fluctuations. Even for small systems, peak temperatures can frequently be identified with transition temperatures. Although in these cases peak temperatures typically depend on the fluctuating quantities considered, in most cases associated pseudophase transitions are doubtless manifest.

For the $2 \times F1$ system, the canonical analysis reveals a surprisingly clear picture of the aggregation transition. Figure 11.3(a) shows the canonical mean energy $\langle E \rangle$ and the specific heat c_V, plotted as functions of the temperature T. In Fig. 11.3(b), the temperature dependence of the mean aggregation order parameter $\langle \Gamma \rangle$ and the fluctuations of Γ are

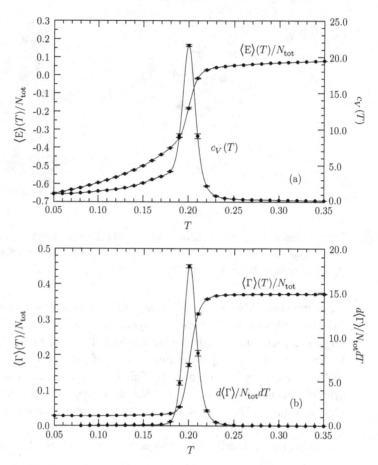

Fig. 11.3 (a) Mean energy $\langle E \rangle / N_{tot}$ and specific heat per monomer c_V, and (b) $\langle \Gamma \rangle / N_{tot}$ and $(1/N_{tot})d\langle \Gamma \rangle / dT$ as functions of the temperature. From [255].

shown. The aggregation transition is indicated by very sharp peaks, and from both figures we read peak temperatures close to $T_{agg} \approx 0.20$. The aggregation of the two peptides is a single-step process, in which the formation of the aggregate with a common compact hydrophobic core governs the folding behavior of the individual chains. Folding and binding are not separate processes. This is a consequence of the small size of the individual chains and the fact that the energy scales of intra-monomer and inter-monomer interaction were chosen to be identical.[3]

The dominance of the inter-chain binding interaction can also be seen by considering the lowest-energy structure. The energy of this conformation, which is shown in Fig. 11.4, is $E_{min} \approx -18.4$ in our energy units. The peptide–peptide binding energy [i.e., the second term in Eq. (11.1)] is with $E_{AB,min}^{(1,2)} \approx -11.4$ much stronger than the intrinsic single-chain

---

[3] Intra-monomer interaction refers to the interaction of monomers that belong to the same polymer chain, whereas monomer–monomer interaction between different chains is called inter-chain interaction.

Fig. 11.4 The minimum-energy $2 \times F1$ complex with $E_{min} \approx -18.4$ is a capped aggregate. From [255].

energies $E^{(1)}_{AB,min} \approx -3.2$ and $E^{(2)}_{AB,min} \approx -3.8$, respectively. The single-chain minimum energy is with $E^{single}_{AB,min} \approx -5.0$ [189] noticeably smaller.

The comparatively strong inter-chain interaction and the strength of the aggregation transition despite the smallness of the system lead to the conclusion that surface effects are of essential importance for the aggregation of the peptides. This is actually confirmed by the detailed microcanonical statistical analysis that will be performed next.

11.4.3 Microcanonical interpretation: The backbending effect

We introduced the microcanonical analysis in Section 2.7 and found that the density of states $g(E)$ already contains all relevant information about the phases of the system. Alternatively, one can also use the phase space volume $\Delta G(E)$ of the energetic shell that represents the macrostate in the microcanonical ensemble in the energetic interval $(E, E + \Delta E)$ with ΔE being sufficiently small to satisfy $\Delta G(E) = g(E)\Delta E$. In the limit $\Delta E \to 0$, the total phase space volume up to the energy E can thus be expressed as $G(E) = \int_{E_{min}}^{E} dE' g(E')$. Since $g(E)$ is positive for all E, $G(E)$ is a monotonically increasing function and this quantity is suitably related to the microcanonical entropy $\mathcal{S}(E)$ of the system. In the definition of Hertz,

$$\mathcal{S}(E) = k_B \ln G(E), \qquad (11.15)$$

which is similar to the definition $S(E) = k_B \ln g(E)$ that is already familiar to us from Section 2.7. Since the density of states exhibits a decrease much faster than exponential toward the low-energy states, the phase-space volume at energy E is strongly dominated by the number of states in the energy shell ΔE. Thus $G(E) \approx \Delta G(E) \sim g(E)$ is directly related to the density of states, i.e., both definitions virtually coincide in the most interesting section of the energetic space, where transitions occur [254, 255]. This virtual identity breaks down in the higher-energy region, where $\ln g(E)$ is getting flat – but this is far above the energetic regions being relevant for the discussion of transitions. The (reciprocal) slope of the microcanonical entropy fixes the temperature scale and the corresponding caloric temperature is then defined via $T(E) = (\partial \mathcal{S}(E)/\partial E)^{-1}$ for fixed volume V and particle number N_{tot}.

As long as the mapping between the caloric temperature T and the system energy E is bijective, the canonical analysis of crossover and phase transitions is suitable since the temperature can be treated as external control parameter. For systems where this condition

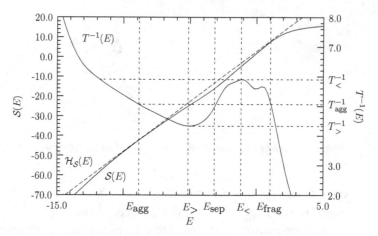

Fig. 11.5 Microcanonical Hertz entropy $\mathcal{S}(E)$ of the $2\times$F1 system, concave Gibbs hull $\mathcal{H}_\mathcal{S}(E)$, and inverse caloric temperature $T^{-1}(E)$ as functions of energy. The phase separation regime ranges from E_{agg} to E_{frag}. Between $T_<^{-1}$ and $T_>^{-1}$, the temperature is not a suitable external control parameter and the canonical interpretation is not useful: the inverse caloric temperature $T^{-1}(E)$ exhibits an obvious backbending in the transition region. Note the second, less-pronounced backbending in the energy range $E_< < E < E_{\text{frag}}$. From [255].

is not satisfied, however, in a standard canonical analysis one may easily miss a physical effect accompanying condensation processes: due to surface effects (the formation of the contact surface between the peptides requires a rearrangement of monomers in the surfaces of the individual peptides), additional energy does not necessarily lead to an increase in temperature of the condensate. Actually, the aggregate can even become colder. The supply of additional energy supports the fragmentation of parts of the aggregate, but this is over-compensated by cooperative processes of the particles aiming to reduce the surface tension. Condensation processes are phase-separation processes and as such aggregated and frag-mented phases coexist. Since in this phase-separation region T and E are not bijective, the same "backbending effect" that we have already discussed in crystallization and fold-ing processes of polymers and proteins in the previous chapters occurs. Perhaps the most important class of systems exhibiting this effect is characterized by the smallness and the capability of these systems to form aggregates, depending on the interaction range. The fact that this effect could be indirectly observed in sodium clustering experiments [257] gives rise to the hope that backbending could also be observed in aggregation processes of small peptides.

In Fig. 11.5 the microcanonical entropy $\mathcal{S}(E)$ is plotted as a function of the system energy. It shows the characteristic backbending feature and thus the aggrega-tion/fragmentation transition in the $2\times$F1 system apparently is a phase-separation process. The phase-separation region of aggregated and fragmented conformations lies between $E_{\text{agg}} \approx -8.85$ and $E_{\text{frag}} \approx 1.05$. Constructing the concave Gibbs hull $\mathcal{H}_\mathcal{S}(E)$ by linearly connecting $\mathcal{S}(E_{\text{agg}})$ and $\mathcal{S}(E_{\text{frag}})$ (straight dashed line in Fig. 11.5), the entropic devia-tion due to surface effects is simply $\Delta\mathcal{S}(E) = \mathcal{H}_\mathcal{S}(E) - \mathcal{S}(E)$. The deviation is maximal for $E = E_{\text{sep}}$ and $\Delta\mathcal{S}(E_{\text{sep}}) \equiv \Delta\mathcal{S}_{\text{surf}}$ is the surface entropy. The Gibbs hull also defines

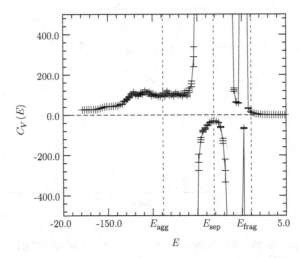

Fig. 11.6   Microcanonical specific heat $C_V(E)$ for the 2×F1 complex. Note the negativity in the backbending regions. From [254].

the aggregation transition temperature $T_{\text{agg}} = (\partial \mathcal{H}_S(E)/\partial E)^{-1}$. For the 2×F1 system, we find $T_{\text{agg}} \approx 0.198$, which is virtually identical to the peak temperatures of the fluctuating quantities discussed in Section 11.4.2.

The inverse caloric temperature $T^{-1}(E)$ is also plotted in Fig. 11.5. For a fixed temperature in the interval $T_< < T < T_>$ ($T_< \approx 0.169$ and $T_> \approx 0.231$), different energetic macrostates coexist. This is a consequence of the backbending effect. Within the backbending region, the temperature decreases with increasing system energy. The horizontal line at $T_{\text{agg}}^{-1} \approx 5.04$ is the Maxwell construction, i.e., the slope of the Gibbs hull $\mathcal{H}_S(E)$. Although the transition seems to have similarities with the van der Waals description of the condensation/evaporation transition of gases – the "overheating" of the aggregate between T_{agg} and $T_>$ (within the energy interval $E_{\text{agg}} < E < E_> \approx -5.13$) is as apparent as the "undercooling" of the fragments between $T_<$ and T_{agg} (in the energy interval $E_{\text{frag}} > E > E_< \approx -1.13$) – it is important to notice that in contrast to the van der Waals picture, the backbending effect in between is a real physical effect.[4] We also conclude that in the transition region the temperature is not a suitable external control parameter, as the macrostate of the system cannot be adjusted by fixing the temperature. Another direct consequence of the energetic ambiguity for a fixed temperature between $T_<$ and $T_>$ is that the canonical interpretation is not suitable for detecting the backbending phenomenon.

The negativity of the specific heat of the system in the backbending region, as shown in Fig. 11.6, is also a remarkable side effect. Negative specific heat values in the phase separation regime are due to the non-extensivity of the energy of the two subsystems resulting from the interaction between the polymers. "Heating" a *large* aggregate would lead to the stretching of monomer–monomer contact distances, i.e., the potential energy of an exemplified pair of monomers increases, while kinetic energy and, therefore, temperature remain widely constant. In a comparatively *small* aggregate, additional energy leads to cooperative

---

[4] See also the discussion about Maxwell constructions in both situations in Sections 2.7.2 and 2.7.3, respectively.

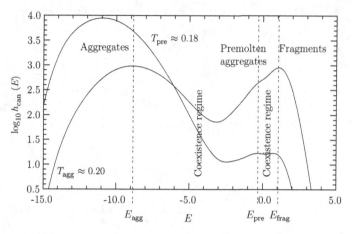

Fig. 11.7 Logarithmic plots of the canonical energy histograms (not normalized) at $T \approx 0.18$ and $T \approx 0.20$, respectively. From [255].

rearrangements of monomers in the aggregate in order to reduce surface tension, i.e., the formation of molten globular aggregates is suppressed. In consequence, additional energy is transferred into potential energy and, effectively, the temperature decreases. In this regime, the aggregate becomes colder, although the total energy increases [254]. Further increasing the energy leads to fragmentation. Translational entropy dominates over conformational entropy and the temperature increases steadily as a function of energy above the fragmentation threshold E_{frag}.

The precise microcanonical analysis also reveals an additional detail of the aggregation transition. Close to $E_{\mathrm{pre}} \approx -0.32$, the T^{-1} curve in Fig. 11.5 exhibits another "backbending," which signalizes a second but unstable transition of the same type. The associated transition temperature $T_{\mathrm{pre}} \approx 0.18$ is smaller than T_{agg}, but this transition occurs in the energetic region where fragmented states dominate. Hence, this transition can be interpreted as the premelting of aggregates by forming intermediate states. These intermediate structures are rather weakly stable: the population of the premolten aggregates never dominates. In particular, at T_{pre}, where premolten aggregates and fragments coexist, the population of compact aggregates is much larger. This is nicely evident in the canonical energy histograms at these temperatures plotted in Fig. 11.7, where the second backbending is indicated only by a small cusp in the coexistence region. Since both transitions are phase-separation processes, structure formation is accompanied by releasing latent heat, which can be defined as the energetic widths of the phase coexistence regimes, i.e., $\Delta Q_{\mathrm{agg}} = E_{\mathrm{frag}} - E_{\mathrm{agg}} = T_{\mathrm{agg}}[\mathcal{S}(E_{\mathrm{frag}}) - \mathcal{S}(E_{\mathrm{agg}})] \approx 9.90$ and $\Delta Q_{\mathrm{pre}} = E_{\mathrm{frag}} - E_{\mathrm{pre}} = T_{\mathrm{pre}}[\mathcal{S}(E_{\mathrm{frag}}) - \mathcal{S}(E_{\mathrm{pre}})] \approx 1.37$. Obviously, the energy required to melt the premolten aggregate is much smaller than the energy necessary to dissolve a compact (solid) aggregate.

The difference of the microcanonical surface entropies in the transition regime defines the surface entropy. In the case of the aggregation transition, it is given by the maximum separation between the Gibbs hull and the entropy curve as shown in Fig. 11.5: $\Delta \mathcal{S}_{\mathrm{surf}}^{\mathrm{agg}} \approx \Delta S_{\mathrm{surf}}^{\mathrm{agg}} = H_S(E_{\mathrm{sep}}) - S(E_{\mathrm{sep}})$, where $H_S(E) \approx \mathcal{H}_{\mathcal{S}}(E)$ is the concave Gibbs hull of $S(E)$. Since $H_S(E_{sep}) = H_S(E_{frag}) - (E_{frag} - E_{sep})/T_{agg}$ and $H_S(E_{frag}) = S(E_{frag})$, the surface entropy is

$$\Delta S_{surf}^{agg} = S(E_{frag}) - S(E_{sep}) - \frac{1}{T_{agg}}(E_{frag} - E_{sep}). \qquad (11.16)$$

Yet utilizing that with Eq. (11.14), the canonical distribution $h_{can}(E) = \int d\Gamma\, H_{can}(E, \Gamma; T)$ at T_{agg} (shown in Fig. 11.7) is $h_{can}(E) \sim g(E)\exp(-E/k_B T_{agg})$, the surface entropy can be written in the simple and computationally convenient form [260]:

$$\Delta S_{surf}^{agg} = k_B \ln \frac{h_{can}(E_{frag})}{h_{can}(E_{sep})}. \qquad (11.17)$$

A similar expression is valid for the coexistence of premolten and fragmented states at T_{pre}. The corresponding canonical distribution is also shown in Fig. 11.7. Thus, the surface entropy (in units of k_B) of the aggregation transition in this example is $\Delta S_{surf}^{agg} \approx 2.48$ and for the premelting $\Delta S_{surf}^{pre} \approx 0.04$, confirming the weakness of the interface between premolten aggregates and fragmented states.

11.5 Aggregation transition in larger heteropolymer systems

The microcanonical signatures of the $2 \times F1$ discussed in the previous section are also, in general, valid for larger systems. We will investigate this in the following for systems consisting of three (in the following referred to as $3 \times F1$) and four ($4 \times F1$) identical peptides with sequence F1 [255].

Although the formation of compact hydrophobic cores is more complex in larger compounds of our exemplified sequence F1, the aggregation transition is little influenced by this. This is nicely seen in Figs. 11.8(a) and 11.8(b), where the temperature dependence of the canonical expectation values of Γ and E, as well as for their fluctuations, are shown for the $3 \times F1$ system. For comparison, results for the $4 \times F1$ system are plotted into the same figures.[5]

As has already been discussed for the $2 \times F1$ system, there are also for the larger systems no obvious signals for separate aggregation and hydrophobic-core formation processes.

---

[5] Note that in the actual simulations the simulation box size has been kept constant (with edge lengths $L = 40$) for computational reasons [255], which particularly affects the $4 \times F1$ system, because L is smaller than the successive arrangement of four straight chains with 13 monomers. This influences primarily the translational entropy in the high-energy regime far above the aggregation transition energy, which is irrelevant for the microcanonical analysis of the aggregation transition. Nonetheless, in the canonical interpretation, it acts back on the transition as undesired states (chain ends overlapping due to the periodic boundary conditions) are (weakly) populated at the transition temperature, whereas others are suppressed. A more detailed analysis of the box size dependence reveals that the canonical transition temperature scales slightly, but noticeably with the box size [255]. Thus, the results obtained by *canonical* statistics for the $4 \times F1$ system should not quantitatively be compared to the canonical results for the $2 \times F1$ and $3 \times F1$ systems.

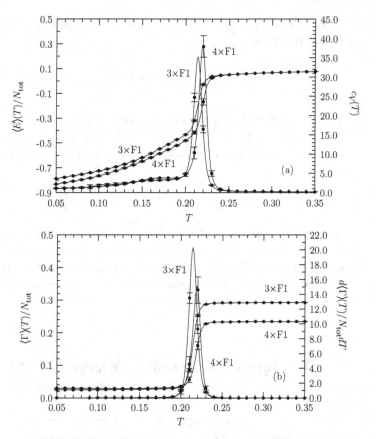

Fig. 11.8 (a) Mean energy $\langle E \rangle / N_{\text{tot}}$ and specific heat c_V, (b) mean aggregation parameter $\langle \Gamma \rangle / N_{\text{tot}}$ and its fluctuations $d\langle \Gamma \rangle / N_{\text{tot}} dT$ as functions of the temperature for the $3 \times$F1 and $4 \times$F1 heteropolymer systems. From [255].

Only weak activity in the energy fluctuations in the temperature region below the aggregation transition temperature indicates that local restructuring processes of little cooperativity (comparable with the discussion of the premolten aggregates in the discussion of the $2 \times$F1 system) are still happening. The strength of the aggregation transition is also documented by the fact that the peak temperatures of energetic *and* aggregation parameter fluctuations are virtually identical for the $3 \times$F1 system, i.e., the aggregation temperature is $T_{\text{agg}} \approx 0.21$ (for $4 \times$F1 $T_{\text{agg}} \approx 0.22$).

For homogeneous multiple-chain systems, two variants of thermodynamic limits are of particular interest: (i) $M \to \infty$, while $N_\mu = \text{const}$, (ii) $N_\mu \to \infty$ with $M = \text{const}$; both limits considered for constant polymer density. Since for proteins the sequence of amino acids is fixed, in this case only (i) is relevant and it is possible to perform a scaling analysis for multiple-peptide systems in this limit. A particularly interesting question is to what extent remnants of the finite-system effects survive in the limit of an infinite number of chains, dependent of the peptide density. Since we have focused our study on the precise analysis of systems of few peptides for all energies and temperatures, it is computationally

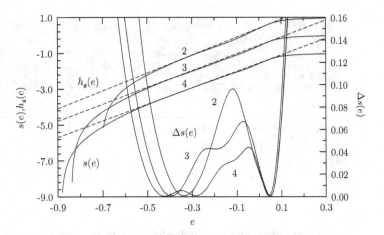

Microcanonical entropies per monomer $s(e)$, respective Gibbs constructions $h_s(e)$ (left-hand scale), and deviations $\Delta s(e) = h_s(e) - s(e)$ (right-hand scale) for $2 \times$F1 (labeled as 2), $3 \times$F1 (3), and $4 \times$F1 (4) as functions of the energy per monomer e. From [255].

inevitable to restrict oneself to small systems, for which a scaling analysis is not very useful. Nonetheless, it is instructive to devote a few interesting remarks to the comparison, once more, of microcanonical aspects of the aggregation transition in dependence of the system size.

In Fig. 11.9, the microcanonical entropies per monomer $s(e) = \mathcal{S}(e)/N_{\text{tot}}$ (shifted by an unimportant constant for clearer visibility) and the corresponding Gibbs hulls $h_s(e) = \mathcal{H}_{\mathcal{S}}(e)/N_{\text{tot}}$ are shown for $2 \times$F1 (in the figure denoted by "2"), $3 \times$F1 ("3"), and $4 \times$F1 ("4"), respectively, as functions of the energy per monomer $e = E/N_{\text{tot}}$. Although the convex entropic "intruder" is apparent for larger systems as well, its relative strength decreases with increasing number of chains. As usual, the slopes of the respective Gibbs constructions determine the aggregation temperatures, which are found to be $T_{\text{agg}}^{3 \times \text{F1}} \approx 0.212$ and $T_{\text{agg}}^{4 \times \text{F1}} \approx 0.217$ confirming the peak temperatures of the fluctuation quantities plotted in Fig. 11.8.

The existence of the interfacial boundary entails a transition barrier whose strength is characterized by the surface entropy $\Delta \mathcal{S}_{\text{surf}}$. In Fig. 11.9, the individual entropic deviations per monomer, $\Delta s(e) = \Delta \mathcal{S}(e)/N_{\text{tot}}$, are also shown and the maximum deviations, i.e., the surface entropies $\Delta \mathcal{S}_{\text{surf}}$ and relative surface entropies per monomer $\Delta s_{\text{surf}} = \Delta \mathcal{S}_{\text{surf}}/N_{\text{tot}}$ are listed in Table 11.1. There is no apparent difference between the values of $\Delta \mathcal{S}_{\text{surf}}$ that would indicate a trend for a vanishing of the *absolute* surface barrier in larger systems. However, the *relative* surface entropy Δs_{surf} obviously decreases. In the thermodynamic limit it disappears completely, because the surface-to-volume ratio converges to zero in this limit, i.e., entropic effects at the surface are not relevant for the transition behavior. Remember the analogous discussion in Section 6.2, where we had found that in icosahedral clusters the number of monomers at the surface scales like $\sim N^{2/3}$, whereas in bulk it grows like $\sim N$.

Table 11.1 Aggregation temperatures T_{agg}, surface entropies $\Delta \mathcal{S}_{surf}$, relative surface entropies per monomer Δs_{surf}, relative aggregation and fragmentation energies per monomer, e_{agg} and e_{frag}, respectively, latent heat per monomer Δq, and phase-separation entropy per monomer $\Delta q / T_{agg}$. All quantities for systems consisting of two, three, and four 13-mers with AB sequence F1.

system	T_{agg}	$\Delta \mathcal{S}_{surf}$	Δs_{surf}	e_{agg}	e_{frag}	Δq	$\Delta q / T_{agg}$
$2 \times$F1	0.198	2.48	0.10	-0.34	0.04	0.38	1.92
$3 \times$F1	0.212	2.60	0.07	-0.40	0.05	0.45	2.12
$4 \times$F1	0.217	2.30	0.04	-0.43	0.05	0.48	2.21

It is also interesting that subleading effects increase and the double-well form found for $2 \times$F1 changes by higher-order effects, and it seems that for larger systems the almost single-step aggregation of $2 \times$F1 is replaced by a multiple-step process.

Not surprisingly, the fragmented phase is hardly influenced by side effects and the rightmost minimum in Fig. 11.9 lies well at $e_{frag} = E_{frag}/N_{tot} \approx 0.04 - 0.05$. Since the Gibbs construction covers the whole convex region of $s(e)$, the aggregation energy per monomer $e_{agg} = E_{agg}/N_{tot}$ corresponds to the leftmost minimum and its value changes noticeably with the number of chains. In consequence, the latent heat per monomer $\Delta q = \Delta Q/N_{tot} = T_{agg}[\mathcal{S}(E_{frag}) - \mathcal{S}(E_{agg})]/N_{tot}$ that is required to fragment the aggregate increases from two to four chains in the system (see Table 11.1). Although the systems under consideration are too small to extrapolate phase transition properties in the thermodynamic limit, it is obvious that the aggregation–fragmentation transition exhibits strong similarities to condensation–evaporation transitions of colloidal systems. Given that, the entropic transition barrier $\Delta q / T_{agg}$, which we see increasing with the number of chains (cf. the values in Table 11.1), would survive in the thermodynamic limit and the transition was first-order-like. It should be noted that protein aggregates forming in biological systems often consist of only a few peptides and are definitely of small size. Thus, the surface effects are responsible for structure formation and are not unimportant side effects. One should keep in mind that standard thermodynamics and the thermodynamic limit are somewhat theoretical constructs valid only for very large systems. The increasing interest in physical properties of small systems, in particular in conformational transitions in molecular systems, requires in part a revision of dogmatic thermodynamic views. Indeed, by means of today's chemo-analytical and experimental equipment, effects like those described in this chapter should actually experimentally be verifiable as these are real physical effects [257].

In this chapter, we have discussed the aggregation of peptides with an exemplary sequence. The question remains whether the effects discussed qualitatively depend on this particular choice, or whether these features are rather generic signatures of aggregation processes. For this reason, we will first study another striking example, aggregation of semiflexible homopolymers, before we return to peptide aggregation in the context of microcanonical hierarchies in aggregation processes.

Hierarchical nature of phase transitions

12.1 Aggregation of semiflexible polymers

In the following, we will discuss the aggregation of interacting semiflexible polymers by analyzing the order and hierarchy of subphase transitions that accompany the aggregation transition.

Cluster formation and crystallization of polymers are processes that are interesting for technological applications, e.g., for the design of new materials with certain mechanical properties or nanoelectronic organic devices and polymeric solar cells. From a biophysical point of view, the understanding of oligomerization, but also the (de)fragmentation in semiflexible biopolymer systems like actin networks is of substantial relevance. This requires a systematic analysis of the basic properties of the polymeric cluster formation processes, in particular, for small polymer complexes on the nanoscale, where surface effects are competing noticeably with structure-formation processes in the interior of the aggregate.

A further motivation for investigating the aggregation transition of semiflexible *homo*polymer chains derives from the intriguing results of the similar aggregation process for peptides [254, 255] discussed in Chapter 11, which were modeled as *hetero*polymers with a sequence of two types of monomers, hydrophobic (A) and hydrophilic (B). By specializing the previously employed heteropolymer model to the apparently simpler homopolymer case, we now, by comparison, aim at isolating those properties that were driven mainly by the sequence properties of heteropolymers. In fact, while in both cases the aggregation transition is a first-order-like phase-coexistence process, it will turn out that for the homopolymer model considered in the following, aggregation and crystallization (if any) are separate conformational transitions, if the bending rigidity of the interacting homopolymers is sufficiently small. This was different in the example of heteropolymer aggregates that we discussed in the previous chapter, where these transitions were found to coincide [254, 255]. As we will also see, the aggregation of semiflexible polymers turns out to be a process in which the constituents experience strong structural fluctuations, which is similar to the behavior of peptides in coupled folding-binding cluster formation processes.

Rather stiff semiflexible polymers are likely to behave like liquid crystals. In analogy to the heteropolymer case, the first-order-like aggregation transition of the complexes can be expected to be accompanied by strong system-size-dependent surface effects. In consequence, polymer aggregation will also be a phase-coexistence process with entropy reduction [265].

Because of its unique perspective, we will explain the physical origin causing these differences by microcanonical analysis, which proves to be particularly suitable for this type of problem.

12.2 Structural transitions of semiflexible polymers with different bending rigidities

We consider the same model as in Section 11.2, but here we assume that all monomers $i_\mu = 1, \ldots, N^{(\mu)}$ of the μth chain ($\mu = 1, \ldots, M$) at positions $\mathbf{x}_{i_\mu}$ are hydrophobic (A). The bonds between adjacent monomers are taken to be rigid (bead-stick model) and pairwise interactions among nonbonded monomers are modeled by a Lennard-Jones potential

$$V_{\mathrm{LJ}}(r_{i_\mu j_\nu}) = 4\left[r_{i_\mu j_\nu}^{-12} - r_{i_\mu j_\nu}^{-6} \right], \tag{12.1}$$

where $r_{i_\mu j_\nu} = |\mathbf{x}_{i_\mu} - \mathbf{x}_{j_\nu}|$ is the distance between monomers i_μ and j_ν of the μth and νth chain, respectively. Intra-chain ($\mu = \nu$) and inter-chain ($\mu \neq \nu$) contacts are not distinguished energetically. The semiflexible nature of the individual chains is reflected by a nonzero bending energy

$$E_{\mathrm{bend}}^{(\mu)} = \kappa \sum_{i_\mu} (1 - \cos\vartheta_{i_\mu}), \tag{12.2}$$

where $0 \leq \vartheta_{i_\mu} \leq \pi$ is the bending angle formed by the monomers i_μ, $i_\mu + 1$, and $i_\mu + 2$. For the comparison with the heteropolymer aggregation, we consider a bending rigidity $\kappa = 0.25$, which is at the rather floppy end of semiflexibility. Thus, the single-chain energy reads

$$E^{(\mu)} = E_{\mathrm{bend}}^{(\mu)} + \sum_{j_\mu > i_\mu + 1} V_{\mathrm{LJ}}(r_{i_\mu j_\mu}) \tag{12.3}$$

and the total energy of the polymer system is given by

$$E = \sum_\mu E^{(\mu)} + \sum_{\mu < \nu} \sum_{i_\mu j_\nu} V_{\mathrm{LJ}}(r_{i_\mu j_\nu}). \tag{12.4}$$

All chains are assumed to have the same degree of polymerization, i.e., the same number of monomers, $N^{(\mu)} = N$, $\mu = 1, \ldots, M$.[1]

In Fig. 12.1(a), the specific-heat curve for a system of two identical semiflexible polymers ($2 \times A_{13}$) is compared with the energetic fluctuations of a single chain ($1 \times A_{13}$). The single chain exhibits a very weak coil–globule collapse transition (shoulder near $T \approx 0.88$), whereas the crystallization near $T \approx 0.24$ is a pronounced, separate process. The thermodynamic phase behavior of single semiflexible polymers in solvent has already been discussed in Chapter 7. The first result for the semiflexible *multiple*-chain system obtained from

---

[1] The following results were obtained in multicanonical computer simulations that yielded precise estimates for the density of states $g(E)$, on which the subsequent microcanonical analysis is based [265].

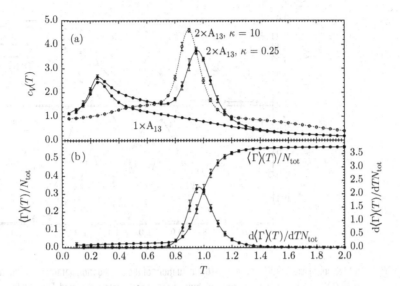

(a) Specific heat per monomer as a function of temperature for a single semiflexible homopolymer with 13 monomers ($1 \times A_{13}$, $\kappa = 0.25$) and for systems of two such chains ($2 \times A_{13}$) with different bending rigidities. (b) Canonical expectation value $\langle \Gamma \rangle$ and fluctuation $d\langle \Gamma \rangle/dT$ of the aggregation parameter Γ for the two-chain system with $\kappa = 0.25$. From [265].

Fig. 12.1(a) is that aggregation and collapse are not separate processes (near $T \approx 0.97$), similar to the corresponding heteropolymer system; compare Fig. 11.3(a).

At about $T \approx 0.24$ (close to the single-chain freezing temperature), the multiple-chain homopolymer complex crystallizes in a separate process. This is in strong contrast to the heteropolymer systems, where aggregation, collapse, *and* crystallization (hydrophobic-core formation) is a single-step process (cf. Fig. 11.3) [254, 255]. In Fig. 12.1(a), we compare with a two-chain system with much stronger bending rigidity $\kappa = 10$. As expected, there is a single aggregation transition near the temperature where the system of less stiff semiflexible polymers with $\kappa = 0.25$ collapses. However, a further crystallization process at lower temperatures does not occur: there is no globular (liquid) pseudophase of defragmented relatively stiff semiflexible polymers.

In Fig. 12.1(b), the canonical expectation value of the aggregation parameter Γ as defined in Section 11.3, $\langle \Gamma \rangle$, and its fluctuation $d\langle \Gamma \rangle/dT$ are plotted. The peak position of $d\langle \Gamma \rangle/dT$ coincides nicely with the corresponding peak temperature of the specific heat and thus signals the aggregation transition.

In Fig. 12.2, the aggregation parameter and its fluctuations are shown for different system sizes of up to four chains. The peak shifts toward higher temperatures and gets sharper with an increasing number of chains, since intra-chain and inter-chain monomer–monomer contacts are not energetically distinguished. The hypothetic maximal total number of intrinsic contacts is

$$n_{\text{intra}}^{\text{max}} = M(N-2)(N-1)/2 \sim MN^2 = NN_{\text{tot}}, \qquad (12.5)$$

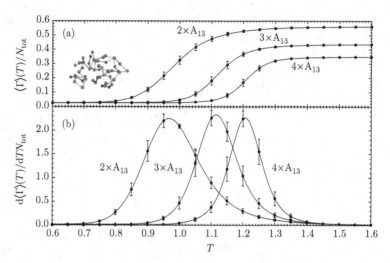

Fig. 12.2 (a) Canonical expectation values and (b) fluctuations of the aggregation parameter Γ for multi-chain systems with $\kappa = 0.25$. In (a), also an exemplified globular $4 \times A_{13}$ aggregate is depicted. From [265].

and for the maximally possible number of inter-chain contacts one finds

$$n_{\text{inter}}^{\max} = M(M-1)N^2/2 \sim M^2N^2 = N_{\text{tot}}^2. \tag{12.6}$$

Due to excluded volume and optimal space-filling constraints, these numbers need to be scaled by about q/N_{tot}, where q is the effective coordination number (which is, e.g., $q \approx 10.0$ for a perfect fcc crystal). For large M, the relative fraction r_{inter} of the inter-chain contacts,

$$r_{\text{inter}} = \frac{n_{\text{inter}}^{\max}}{n_{\text{intra}}^{\max} + n_{\text{inter}}^{\max}} = 1 - \frac{1}{M}\left(1 - \frac{1}{N}\right)\left(1 - \frac{2}{N}\right) + \mathcal{O}\left(\frac{1}{M^2N}\right), \tag{12.7}$$

behaves like $r_{\text{inter}} \sim 1 - M^{-1}$, i.e., the relative influence of inter-chain contacts increases rapidly toward unity with the number of chains. In consequence, aggregation dominates over collapse of the individual chains. Even for the two-chain system $2 \times A_{13}$ this estimate is reasonable: the energy of the lowest-energy conformation found numerically is $E^{(\min)} \approx -83.61$ and the contribution of the inter-chain contacts is $E_{\text{inter}}^{(\min)} \approx -50.20 \approx r_{\text{inter}}^{(\min)} E^{(\min)}$ with $r_{\text{inter}}^{(\min)} \approx 0.60$. This coincides nicely with the corresponding value of the above contact ratio, $r_{\text{inter}} \approx 0.56$. In fact, considering energetic and structural fluctuations of the larger systems with three and four chains, there are no indications for an additional collapse transition at temperatures higher than the aggregation transition.

Below the aggregation temperature, the entropic loss of the individual chains is over-compensated by the energetic gain of forming a joint globular aggregate. However, the entropic change while passing the aggregation transition is noticeably smaller than what was found for heteropolymer systems, where no separate freezing transition occurs. In the intermediate fluid "globular" pseudophase, the aggregate of the homopolymers thus behaves like a single chain of length MN. Consequently, reducing the temperature, the aggregate optimizes the monomer arrangements in order to maximize energetic contacts.

Indicated by the peak in the specific heat near $T \approx 0.2$ (see Fig. 12.1), the small globular aggregates freeze into spherical amorphous structures with a maximum number of inter-chain contacts. For rather stiff semiflexible polymers (as our example with $\kappa = 10$), however, a separate freezing transition does not occur and the peak of c_V near $T \approx 0.9$ indicates a single-step transition from rod-like coils to a liquid-crystal-like phase.

In the studies of heteropolymer aggregation in Section 11.4, we also observed a pronounced single transition of a different nature. In this case, the formation of a heteropolymer complex consisting of different chains with compact hydrophobic core was roughly a single-step process, because hydrophobic-core formation ("freezing") favors conformations with very small entropy. For semiflexible homopolymer systems, we find in the rather floppy limit that the *freezing* temperature ($T \approx 0.2$, for single chains or globular aggregates with $\kappa = 0.25$) is almost identical to the *aggregation* temperature for heteropolymer systems (compare Figs. 11.3 and 12.1). This coincidence in the behavior of these different systems is due to the formation of a single, very compact hydrophobic domain (the monomers of the interacting homopolymers are as hydrophobic as the A-type monomers of the heteropolymer), which maximizes the number of energetic contacts. Stiff homopolymers, on the other hand, cannot form a maximally compact hydrophobic core and hence do not crystallize in a separate transition.

The investigation of stiffness effects upon aggregation is certainly of interest, even more so if the knowledge of the single-chain behavior (see Chapter 7) is taken into account. However, we will now focus on the even more exciting feature of transition hierarchies that we have already observed in the case of heteropolymer aggregation in the last chapter, without having it discussed in detail there.

12.3 Hierarchies of subphase transitions

For a deeper understanding of the aggregation transition, we now analyze entropic effects accompanying the aggregation transition in the microcanonical ensemble for an isolated system of multiple polymer chains. The microcanonical entropies per monomer, $s = S/N_{tot}$, are shown in Fig. 12.3(a) as functions of the total energy per monomer, $e = E/N_{tot}$. The reciprocal caloric temperature, $T^{-1}(E) = \partial S/\partial E$, is plotted in Fig. 12.3(b). The plots reveal that increasing the energy entails a reduction of temperature in the transition region, i.e., the backbending effect, which was already discussed in Section 11.4.3.

In the study of the heteropolymer aggregates, we used the microcanonical view to identify the peptide aggregation process as a "coupled binding–folding transition," i.e., the individual heteropolymer chains refold during the binding process and the finally formed aggregate possesses a single compact hydrophobic domain. A closer inspection of Fig. 12.3(b) reveals for the homopolymer system a *hierarchical substructure* caused by these surface effects. The frequency of oscillations of the curves, increasing with system size, reveals that the aggregation transition is actually a composition of different subprocesses, each of which is an individual phase-separation process. The amplitude of

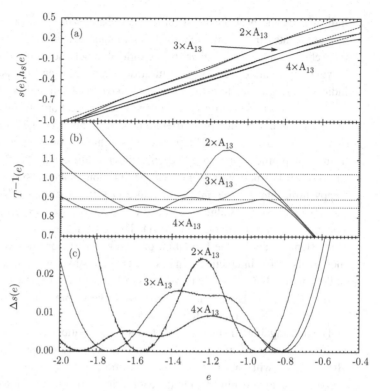

Fig. 12.3   (a) Microcanonical entropy per monomer $s(e)$ and the Gibbs constructions $h_s(e)$ (dashed lines), (b) reciprocal caloric temperatures $T^{-1}(e)$ and Maxwell constructions, and (c) relative surface entropies per monomer $\Delta s_{\mathrm{surf}}(e)$. From [265].

these oscillations decreases with system size, showing that these subprocesses comprise a smaller surface-entropic barrier [see Fig. 12.3(c)].

The $2 \times A_{13}$ system exhibits a single backbending effect as only two chains aggregate. For the three-chain system $3 \times A_{13}$, a different scenario is apparent. In the higher-energy regime, first two chains stick together, and the formation of the three-chain globule is a separate process at lower energies. This hierarchical procedure continues for larger systems as, e.g., for $4 \times A_{13}$. However, the impact of the individual backbending effects is getting weaker and, in the thermodynamic limit, these effects are expected to disappear asymptotically, whereas the first-order character of this transition remains. The horizontal lines in Fig. 12.3(b) are the respective Maxwell constructions and define the aggregation temperatures T_{agg}. In the following, we denote the leftmost (rightmost) energy where $T^{-1}(e) = T_{\mathrm{agg}}^{-1}$ as e_{agg} (e_{frag}). In the entropy curves, the Maxwell constructions correspond to concave hulls $h_s(e)$ with $\partial h_s(e)/\partial e = T_{\mathrm{agg}}^{-1}$ (Gibbs constructions) in the transition regime [see Fig. 12.3(a)], where entropy is reduced due to surface effects.

For a quantitative analysis, the deviation $\Delta s(e) = h_s(e) - s(e)$ is plotted in Fig. 12.3(c). Within the transition region (i.e., for $e_{\mathrm{agg}} \leq e \leq e_{\mathrm{frag}}$), the peak height defines the surface

Table 12.1 Aggregation temperatures T_{agg}, relative surface entropies per monomer Δs_{surf}, relative aggregation and fragmentation energies per monomer, e_{agg} and e_{frag}, respectively, latent heat per monomer Δq, and phase-separation entropy per monomer $\Delta q / T_{\mathrm{agg}}$.

System	T_{agg}	Δs_{surf}	e_{agg}	e_{frag}	Δq	$\Delta q / T_{\mathrm{agg}}$
$2 \times A_{13}$	0.973	0.024	-1.566	-0.944	0.622	0.639
$3 \times A_{13}$	1.118	0.016	-1.730	-0.831	0.899	0.804
$4 \times A_{13}$	1.172	0.009	-1.892	-0.799	1.093	0.932

entropy per monomer. The energetic width of the phase-coexistence regime is the latent heat per monomer,

$$\Delta s_{\mathrm{surf}} = \max_{e_{\mathrm{agg}} \leq e \leq e_{\mathrm{frag}}} \Delta s(e) \tag{12.8}$$

$$\Delta q = e_{\mathrm{frag}} - e_{\mathrm{agg}} = T_{\mathrm{agg}}[s(e_{\mathrm{frag}}) - s(e_{\mathrm{agg}})]. \tag{12.9}$$

Thus the entropic phase separation barrier $\Delta q / T_{\mathrm{agg}}$ should survive in the thermodynamic limit, if the aggregation of semiflexible polymers is a first-order-like phase-separation process with coexistence of aggregates and fragments.

Values for these quantities are listed in Table 12.1 for the three polymer systems considered here. We see that with increasing system size the surface entropies Δs_{surf} decrease and thus the influence of surface effects is getting weaker. On the other hand, the latent heat per monomer increases, supporting the first-order character of the aggregation transition.

We have seen in this analysis that the aggregation of interacting semiflexible polymers is a first-order phase-separation process. It is accompanied by sequentially ordered, i.e., hierarchical, subphase transitions of aggregation. Without the microcanonical analysis, it would have been difficult to reveal this hierarchical character of the transition. This analysis indeed sheds new light on the general nature of phase transitions, which appear to be composed of separate subprocesses. Thus, the cooperative interplay of small parts of the system that undergoes a phase transition is hierarchical and not a single-step process. This is probably a generic feature of all phase transitions and thus of fundamental importance for understanding the onset of phase transitions in general.

12.4 Hierarchical peptide aggregation processes

Another very important example for a nucleation process that exhibits hierarchical subphase transitions is the aggregation of proteins. For this purpose, we return to our discussion of aggregation properties of the hydrophobic–polar peptide with the Fibonacci sequence F1 ($AB_2AB_2ABAB_2AB$) in the previous chapter.

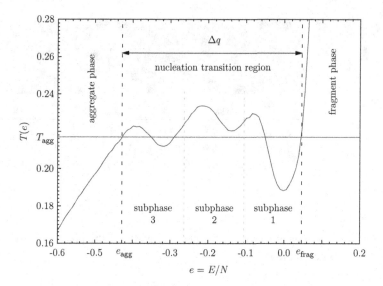

Fig. 12.4   Microcanonical temperature $T(e)$ as a function of energy per monomer, $e = E/N$, where N is the total number of all monomers in the system. The horizontal Maxwell line marks the aggregation temperature T_{agg}, obtained by Gibbs construction. Vertical dashed lines separate the different phases and subphases, respectively. From [266].

The surface entropy curves $\Delta s(e)$ for various numbers of aggregating chains shown in Fig. 11.9 resemble the plots for the homopolymers in Fig. 12.3(c). Therefore, we expect that peptide aggregation processes are also hierarchical [266].

In order to better understand the subphases, we discuss in the following the inverse derivative of the entropy, the microcanonical temperature $T(e)$, for the system containing four F1 chains. It is plotted in Fig. 12.4. The slope of the Gibbs tangent corresponds to the Maxwell line in Fig. 12.4 at the aggregation temperature $T_{agg} \approx 0.217$. Therefore, the intersection points of the Maxwell line and the temperature curve define the energetic phase boundaries e_{agg} and e_{frag}, respectively, as both phases coexist at T_{agg}.

For energies $e < e_{agg} \approx -0.43$, conformations of a single aggregate, composed of all four chains, dominate. On the other hand, conformations with $e > e_{frag} \approx 0.05$ are mainly entirely fragmented, i.e., all chains can form individual conformations, almost independently of each other. The entropy is governed by the contributions of the individual translational entropies of the chains, outperforming the conformational entropies. The translational entropies are limited only by the volume that corresponds to the simulation box size. The energetic difference $\Delta q = e_{frag} - e_{agg}$, serving as an estimator for the latent heat, is obviously larger than zero, $\Delta q \approx 0.48$. It thus corresponds to the total energy necessary to entirely melt the aggregate into fragments at the aggregation (or melting) temperature.

The relation between energy and microcanonical temperature in the aggregate phase and in the fragment phase is as intuitively expected: with increasing energy, $T(e)$ also increases. However, much more interesting is the behavior of the system in the energy interval $e_{agg} < e < e_{frag}$, which represents the energetic nucleation transition region.

fragment

subphase 1

subphase 2

subphase 3

aggregate

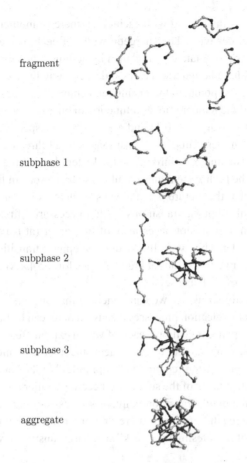

Fig. 12.5   Representative conformations in the different structural phases. From [266].

Figure 12.4 clearly shows that in our example $T(e)$ changes the monotonic behavior three times in this regime [266]. Representative conformations in the different structural phases are shown in Fig. 12.5. The surface entropy per monomer will vanish in the thermodynamic limit, as we expected from our analysis in Section 11.5.

If two chains aggregate (subphase 1 in Fig. 12.4), the total translational entropy of the individual chains is reduced by $k_B \ln V$, where V is the volume (corresponding to the simulation box size), whereas the energy of the aggregate is much smaller than the total energy of the system with the individual chains separated. Thus, the energy associated to the interaction between different chains, i.e., the cooperative formation of inter-chain contacts between monomers of different chains, is highly relevant here. This causes the latent heat between the completely fragmented and the two-chain aggregate phase to be nonzero, which signals a first-order-like transition. This procedure continues when an additional chain joins a two-chain cluster. Energetically, the system enters subphase 2. Qualitatively, the energetic and entropic reasons for the transition into this subphase are the same as explained before, since it is the same kind of nucleation process. In our example of four chains interacting with each other, there is another subphase 3 which also shows the described behavior. The energetic width of each of the subphase transitions corresponds to the respective latent heat gained by adding another chain to the already formed cluster. The subphase boundaries (vertical dashed, gray lines in Fig. 12.4) have been defined to be the inflection points in the raising temperature branches, thus enclosing a complete "oscillation" of the temperature as a function of energy. The energetic subphase transition points are located at $e_{12} \approx -0.11$ and $e_{23} \approx -0.26$, respectively. Therefore, the latent heat associated with these subphase transitions is in all three cases about $\Delta q_{ij} \approx 0.16$ ($i, j = 1, 2, 3$, $i \neq j$), thus being one-third of the total latent heat of the complete nucleation process. This reflects the high systematics of subphase transitions in first-order nucleation processes.

Note that the picture is not always as clear as it is in the examples discussed here. The individual subphase transitions are not necessarily first-order-like, and also the complete transition itself is not necessarily of first order (at least in finitely large systems), but of second order. This typically occurs if subphase transitions are energetically close to each other [the two transitions at high energies for the $4 \times A_{13}$ system, shown in Fig. 12.3(b) are examples of such a case].

Generally speaking, we can conclude that any first-order phase transition, such as the molecular nucleation processes discussed here, can be understood as a composite of hierarchical subphase transitions, each of which exhibits features of individual transitions. Since with increasing number of constituents the microcanonical entropy converges to the Gibbs hull in the transition region, the "amplitudes" of the backbending oscillations and the individual latent heats of the subphases become smaller and smaller, whereas the overall latent heat of the entire transition converges to its limiting value. Thus, in the thermodynamic limit, aggregation transitions are first-order nucleation processes composed of an infinite number of infinitesimally "weak" subphase transitions, which span a finite range in energy space.

12.5 Hierarchical aggregation of GNNQQNY

The question remains whether the hierarchical features of aggregation transitions are a special feature of coarse-grained models that might be irrelevant in more detailed representations of aggregating systems and thus also for realistic systems. The answer is no, although in systems, where folding and aggregation are well-separated processes, the identification of hierarchies can be difficult (which, in fact, is also true for coarse-grained models of such systems).

The heptapeptide GNNQQNY is a polar fragment of the yeast protein Sup35 and has become a popular subject of simulational studies, because it possesses rather simple aggregation pathways that are hardly influenced by folding effects [267–270]. Aggregates are structurally stable β sheets. Another candidate for basic aggregation studies is the hydrophobic peptide KLVFFAE, which is a fragment of the infamous Aβ peptide that is associated with Alzheimer's disease. KLVFFAE exhibits more complicated aggregation pathways and can form various morphologies. This is due to strong hydrophobic

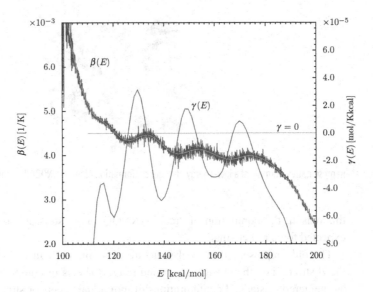

Fig. 12.6 Inflection-point analysis of the aggregation process of four GNNQQNY peptides. Shown are the inverse temperature curve $\beta(E)$ (original data and Bézier smoothed curve, see Section 4.3) and its derivative $\gamma(E)$ (obtained by Bézier smoothing).

interactions among the side chains, which can overcompensate the energetic gain by the formation of planar (parallel or anti-parallel) β sheets that are caused by orderly hydrogen-bond registration along the backbones of the peptides. GNNQQNY is dominated by polar side chains so that the interaction of side chains with each other is of less relevance for the aggregation process than backbone–backbone interaction. The consequence is that β-sheet formation and aggregation is a single-step process, i.e., disordered aggregates are not representative. This explains the rather simple and clear aggregation process and boosts our expectation that subphases and transitions between these should be easy to identify in a microcanonical statistical analysis.

Figure 12.6 shows the inverse temperature β as a function of the total energy E for a system of four GNNQQNY peptides and its derivative $\gamma(E) = \partial\beta/\partial E$.[2] The inverse temperature curve $\beta(E)$ exhibits transition signatures similar to the generic coarse-grained approach that led to the hierarchical behavior shown in Fig. 12.4. The inflection points of $\beta(E)$ can be clearly identified from the derivative $\gamma(E)$. The peak locations correspond to the energetic transition points that separate the subphases. These "fluctuations" are caused by individual chains (or by already formed separate clusters) that lose translational entropy by attaching themselves to other chains or clusters, respectively. This "loss" of translational entropy is accompanied by a reduction of the overall system energy. Note that the individual microcanonical subphase aggregation temperatures lie between 210 K and 250 K. For

---

[2] The results were obtained by microcanonical inflection-point analysis of data generated in extensive replica-exchange Monte Carlo simulations of the implicit all-atom force-field peptide model introduced in Refs. [223, 271, 272].

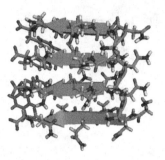

Fig. 12.7   Representative example of a low-energy β-sheet conformation of four GNNQQNY peptides in the aggregate phase.

this reason, the population of "real" GNNQQNY clusters at room temperature must be expected to be rather small.

The only representative local structure of a single chain in the aggregation region is the β strand, i.e., the clusters are found to be β sheets or compositions of β sheets (such as staggered or stacked configurations of individual sheets or strands). The lowest-energy conformation is, as expected, found to be a planar β sheet of the four peptides. It is shown in Fig. 12.7. However, it shall be emphasized again that not all aggregation processes are so smooth as in this example. If individual folding features influence the aggregation process or if intermediate amorphous aggregates or different aggregate shapes (such as barrels) can form, the microcanonical features of the overall aggregation transition can be much more complex and the identification of individual processes can be more difficult if different subphases are energetically degenerate.

With this we conclude the investigation of folding properties of individual polymer or peptide chains and the aggregation of multiple chains, and turn our interest to another large class of structure-formation processes that accompany the adsorption of polymers or peptides at solid substrates. Hybrid systems composed of organic and inorganic matter are of particular relevance for micro-scale technological applications.

13 Adsorption of polymers at solid substrates

13.1 Structure formation at hybrid interfaces of soft and solid matter

The requirement of higher integration scales in electronic circuits, the design of nanosensory applications in biomedicine, and also the fascinating capabilities of modern experimental setup with its enormous potential in polymer and surface research, have produced an increased interest in macromolecular structural behavior at hybrid interfaces of organic and inorganic matter [273–278]. Relevant processes include, e.g., wetting [279–281], pattern recognition [282–284], protein–ligand binding and docking [285–287], effects specific to polyelectrolytes at charged substrates [288] as well as electrophoretic polymer deposition and growth at surfaces [289].

The recent developments of experimental single-molecule techniques at the nanometer scale, e.g., by means of atomic force microscopy (AFM) [290, 291] and optical tweezers [292, 293], allow for a more detailed exploration of structural properties of polymers in the vicinity of adsorbing substrates [278]. The possibility to perform such studies is of essential biological and technological significance. From the biological point of view, the understanding of the binding and docking mechanisms of proteins at cell membranes is important for the reconstruction of biological cell processes. Similarly, specificity of peptides and binding affinity to selected substrates are relevant features that need to be taken into account for potential applications in nanoelectronics and pattern recognition nanosensory devices on a molecular basis [282].

In this chapter, theoretical approaches to the modeling of hybrid systems of soft and solid matter will be discussed. A comprehensive analysis of conformational transitions experienced by polymers in the binding process, of the phase-diagram structure, and of dominant polymer conformations in these phases can only be performed efficiently by means of computer simulations of hybrid physisorption models on mesoscopic scales. The influence of adhesion and steric hindrance for polymers grafted to a flat substrate [294–300], conformational pseudophase transitions for nongrafted polymers and peptides in a cavity with attractive substrate [301–309], adsorption dynamics and substructure formation of adhered polymers [310], polymer shape response to pulling forces [311, 312] or external fields [313], and biomembrane response to structural changes of grafted polymers [314] are subjects of computer simulations and analytical approaches of different such models [315, 316]. Proteins exhibit a strong specificity as the affinity of peptides to adsorb at surfaces depends on the amino acid sequence, solvent properties, substrate shape, and crystal orientation, among others. Differences in binding behavior are known for peptide-metal [273, 317] and peptide-semiconductor [274–276] interfaces. The problem of coupled

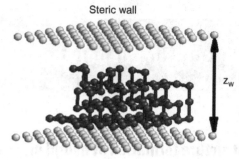

Steric wall

z_w

Attractive substrate

Fig. 13.1 Cavity model used for studies of the adsorption of single polymers at solid substrates on the simple-cubic lattice. The lower of the two parallel surfaces is attractive to the polymer, the upper is steric only. The distance between the surfaces is z_w lattice units. From [302].

binding/folding or simple docking properties of proteins at an adsorbing surface is a topic in its own right [287].

So the sheer number of possible scenarios and hybrid composites is certainly overwhelming. Yet the same question as in the discussions in the preceding chapters applies here, too: is it possible to reveal generic features in adsorption processes and, in the same context, what can we learn about the specificity of hybrid interfaces?

13.2 Minimalistic modeling and simulation of hybrid interfaces

The simplest, yet very instructive, model of a hybrid system is shown in Fig. 13.1. As in the discussion of general folding properties of polymers and proteins, employing a minimalistic simple-cubic (sc) lattice model [39] allows for a systematic analysis of the conformational phases experienced by a nongrafted polymer in a cavity with one adhesive surface. The polymer can move between the two infinitely extended parallel planar walls, separated by a distance z_w (represented in lattice units). The substrate is short-range attractive to the monomers of the polymer chain, while the influence of the other wall is purely steric.

Denoting the number of nearest-neighbor, but nonadjacent monomer–monomer contacts by n_m and the number of nearest-neighbor monomer–substrate contacts by n_s, the energy of the hybrid system can be expressed in the simplest model [295] as

$$E_s(n_s, n_m) = -\varepsilon_s n_s - \varepsilon_m n_m, \tag{13.1}$$

where ε_s and ε_m are the respective contact energy scales, which are left open in the following. For simplicity, we perform a simple rescaling and set $\varepsilon_s = \varepsilon_0$ and $\varepsilon_m = \varepsilon_0 s$. By doing so, we have introduced the overall energy scale ε_0 and the dimensionless reciprocal solubility s that controls the quality of the implicit solvent surrounding the polymer (the larger the value of s, the worse the solvent quality). Since contacts with the substrate usually entail a reduction of monomer–monomer contacts, there are two competing forces (rated against each other by the energy scales) affecting the formation of intrinsic and surface contacts.

In the following, we will focus mainly on the conformational transitions the polymer experiences under different environmental conditions [301, 302]. More specifically, we are interested in how energetic and structural quantities depend on temperature T and reciprocal solubility s in equilibrium. The probability (per unit area) for a conformation with n_s surface and n_m monomer–monomer contacts at temperature T and reciprocal solubility s is given by

$$p_{T,s}(n_s, n_m) = \frac{1}{Z} g_{n_s n_m} e^{\varepsilon_0(n_s + s n_m)/k_B T}, \qquad (13.2)$$

where $g_{n_s n_m} = \delta_{n_s 0} g_{n_m}^{u} + (1 - \delta_{n_s 0}) g_{n_s n_m}^{b}$ is the contact density and Z the partition sum. In this decomposition, $g_{n_m}^{u}$ stands for the density of unbound conformations, whereas $g_{n_s n_m}^{b}$ is the density of surface and intrinsic contacts of all conformations bound to the substrate. Obviously, the number of the conformations without contact to the attractive substrate, $g_{n_m}^{u}$, depends on the distance z_w between the cavity walls. For a sufficiently large distance z_w from the substrate, the influence of the neutral surface on the unbound polymer is small. For $z_w \to \infty$, however, $g_{n_m}^{u}$ formally diverges. Therefore, the nonadhesive, impenetrable steric wall is necessary for the regularization of the translational entropy of the polymer in a desorbed ("free") state.

The partition sum of the system as a function of the two parameters T and s is simply $Z = \sum_{n_s, n_m} g_{n_s n_m} \exp[\varepsilon_0(n_s + s n_m)/k_B T]$ and the statistical average of any function $O(n_s, n_m)$ is given by the formula

$$\langle O \rangle(T, s) = \sum_{n_s, n_m} O(n_s, n_m) p_{T,s}(n_s, n_m), \qquad (13.3)$$

which is very convenient since it only requires to estimate the contact density $g_{n_s n_m}$ in the simulation. Denoting contact correlation matrix elements as $M_{xy}(T, s) = \langle xy \rangle_c = \langle xy \rangle - \langle x \rangle \langle y \rangle$ with $x, y = n_s, n_m$, the heat capacity can be written as

$$C_V(T, s) = k_B \left(\frac{\varepsilon_0}{k_B T} \right)^2 (1, s) M(T, s) \binom{1}{s} \qquad (13.4)$$

and can be imagined as kind of a "landscape" in T–s parameter space. This will enable a very convenient way of interpreting the conformational phase structure in this parameter space. Eventually, all quantities depending only on the contact numbers n_m and n_s can simply be calculated from the estimate of the contact density $g_{n_s n_m}$.

Although the two contact parameters are sufficient to describe the macrostate of the system and their fluctuations characterize the main pseudophase transition lines, it is often useful also to introduce non-energetic quantities such as the end-to-end distance and the gyration tensor for gaining more detailed structural information of the polymer. For our specific problem at hand, it is particularly useful to study the structural anisotropy of the adsorbed polymer in the different phases. To this end, we define the general gyration tensor for a polymer chain of N beads with the components

$$R_{ij}^2 = \frac{1}{N} \sum_{n=1}^{N} \left(x_i^{(n)} - \overline{x}_i \right) \left(x_j^{(n)} - \overline{x}_j \right), \tag{13.5}$$

where $x_i^{(n)}$, $i = 1, 2, 3$, is the ith Cartesian coordinate of the nth monomer and $\overline{x}_i = \sum_{n=1}^{N} x_i^{(n)}/N$ is the center of mass with respect to the ith coordinate. Anisotropy in the polymer fluctuations is connected with the system's geometry and therefore it will be sufficient to study the components of the gyration tensor parallel (x, y components) and perpendicular (in z direction) to the planar walls,

$$R_{\parallel}^2 = \frac{1}{N} \sum_{n=1}^{N} \left[\left(x^{(n)} - \overline{x} \right)^2 + \left(y^{(n)} - \overline{y} \right)^2 \right] \tag{13.6}$$

and

$$R_{\perp}^2 = \frac{1}{N} \sum_{n=1}^{N} \left(z^{(n)} - \overline{z} \right)^2. \tag{13.7}$$

The gyration radius is then simply the trace of the gyration tensor, $R_{\mathrm{gyr}}^2 = \mathrm{Tr}\, R^2 = \sum_{i=1}^{3} R_{ii}^2 = R_{\parallel}^2 + R_{\perp}^2$. The calculation of statistical averages for quantities R that are not necessarily functions of the contact numbers n_s and n_m cannot be performed via Eq. (13.3). In this case only the more general relation $\langle R \rangle = \sum_{\mathbf{X}} R(\mathbf{X}) \exp\{-E_s(\mathbf{X})/k_B T\}/Z$ holds, where the sum runs over all polymer conformations $\mathbf{X}$. Introducing the accumulated density $R_{\mathrm{acc}}(n_s', n_m') = \sum_{\mathbf{X}} R(\mathbf{X}) \delta_{n_s(\mathbf{X})\, n_s'} \delta_{n_m(\mathbf{X})\, n_m'}/g_{n_s' n_m'}$, where δ_{ij} is the Kronecker symbol, the expectation value can be expressed, however, in a form similar to Eq. (13.3):

$$\langle R \rangle = \sum_{n_s, n_m} R_{\mathrm{acc}}(n_s, n_m) p_{T,s}(n_s, n_m). \tag{13.8}$$

The quantity $R_{\mathrm{acc}}(n_s, n_m)$ can easily be measured in simulations with the contact density chain-growth algorithm described in the next section. In the following, natural units are used as usual for brevity, i.e., we set $k_B = \varepsilon_0 \equiv 1$.

13.3 Contact-density chain-growth algorithm

Lattice models of hybrid systems with decoupled energy scales as in Eq. (13.1) are most efficiently simulated by means of variants of generalized-ensemble chain-growth methods such as the multicanonical chain-growth algorithm [39,40,318] introduced in Section 4.11 or flat-histogram PERM [103].[1] The main advantage of the improved methods is that these directly sample the contact density $g_{n_s n_m}$ and do not require energy scales to be set before

---

[1] Conventional, generalized-ensemble Monte Carlo methods can also be employed, of course, but require sophisticated Monte Carlo updates to be efficient [306,309].

the simulation is started. The contact-density chain growth method [318] generalizes the ordinary multicanonical version [39,40] which samples the density of states, i.e., the number of states for given energy. Here, the two independent energy scales ε_m and ε_s or their ratio s, respectively, can still be set to any value *after* the simulation has finished. This allows to introduce the reciprocal solubility s as a second external environmental parameter in addition to the temperature T and offers maximum flexibility for reweighting the ensemble to all possible s, T values by using Eqs. (13.2) and (13.8).

The determination of the contact density $g_{n_m n_s}$ follows similar lines as the multicanonical chain-growth method for the estimation of the density of states. In fact, the only change in the algorithm described in Section 4.11 is that the multicanonical chain-growth weights, defined in Eq. (4.161), are replaced by

$$W_{n,\alpha_j}^{ss,is} = W_{n-1}^{ss,is} \frac{m_n}{k \binom{m_n}{k} p_A} \frac{W_n^{\text{flat},(i)}\left(n_s^{(n,\alpha_j)}, n_m^{(n,\alpha_j)}\right)}{W_{n-1}^{\text{flat},(i)}\left(n_s^{(n-1)}, n_m^{(n-1)}\right)}, \tag{13.9}$$

where the multi-contact weights $W_n^{\text{flat},(i)}(n_s^{(n,\alpha_j)}, n_m^{(n,\alpha_j)}) \sim 1/g_{n_m^{(n,\alpha_j)}, n_s^{(n,\alpha_j)}}$ again have to be determined recursively.

The extension of this method incorporating more than two system parameters is straightforward, but the efficiency of flattening the high-dimensional histograms at all levels of the growth process decreases, whereas the duration of the runs to maintain sufficient statistics and the storage requirements for these fields rapidly increase.

13.4 Pseudophase diagram of a flexible polymer near an attractive substrate

For an exemplary study of a hybrid system on a simple-cubic (sc) lattice in equilibrium, we choose a polymer with 179 monomers. Since its length is a prime number, the polymer is unable to form perfect cuboid conformations on the sc lattice, as it is, e.g., the case for a 100-mer [301]. In that case, two low-temperature subphases, dominated by the same $4 \times 5 \times 5$ cuboid, are present. In one subphase it had 20 surface contacts, while in the other the cuboid is simply rotated, entailing 25 surface contacts. This is a typical example of how the exact number of monomers in the linear chain is directly connected to the occurrence of such specific pseudophases, although in this case it is only an uninteresting lattice effect. Nonetheless, the enormous progress in high-resolution experimental structure analyses and in the technological equipment for precise polymer deposition, as well as the natural finite length of classes of polymers (e.g., peptides and proteins), explain the growing interest in pseudophases and the conformational transitions between them. In the following, we will focus on the expected thermodynamic phase transitions [295, 296, 302] and low-temperature higher-order layering pseudophase transitions [298, 302].

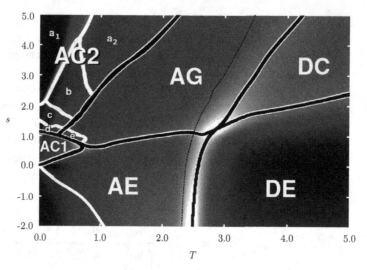

Fig. 13.2 Solubility-temperature pseudophase diagram of a lattice homopolymer with 179 monomers. The gray scales encode the specific heat as a function of reciprocal solubility s and temperature T – the brighter it is, the larger its value. Drawn lines emphasize the ridges of the profile and indicate transitions between the different conformational phases. Black lines mark expected thermodynamical phase transitions, while white lines belong to pseudo-transitions specific to finite-length polymers. Along the dashed black line coexisting desorbed and adsorbed conformations are equally probable. From [302].

13.4.1 Solubility–temperature pseudophase diagram

As has been outlined earlier, finite-size systems usually exhibit a zoo of crossover- or pseudo-transitions, most of which disappear in the thermodynamic limit. These peculiarities of finite systems are also relevant for the polymer adsorption problem considered here.

The structural phase diagram of the lattice homopolymer with 179 monomers in a cavity with $z_w = 200$, i.e., the projection of the specific heat profile $c_V(T, s) = C_V(T, s)/N$ onto the solubility–temperature plane [302], is plotted in Fig. 13.2. Canonical structural phase diagrams can be obtained by reweighting fluctuating response quantities based on the contact density $g_{n_s n_m}$ and accumulated densities like $R_{\mathrm{acc}}(n_s, n_m)$, as obtained in the contact-density chain-growth simulation, with respect to the external parameters temperature T and reciprocal solubility s by utilizing Eqs. (13.2) and (13.8).

The gray scales in Fig. 13.2 encode the value of the specific heat and the brighter the shading, the larger the value of c_V. Black and white lines emphasize the ridges of the profile. The specific heat profile typically is a reasonable quantity for the identification of pseudophases and, therefore, these ridges mark pseudophase boundaries. As expected, the pseudophase diagram is divided into two main parts – the phases of adsorption and desorption. The two desorbed pseudophases DC (desorbed-compact conformations) and DE (desorbed-expanded structures) are separated by the collapse transition line that corresponds to the Θ transition of the infinite-length polymer which is allowed to extend into the three spatial dimensions.[2]

The region of the adsorbed pseudophases is much more complex, and separate major structural phases of adsorbed-expanded (AE), adsorbed globules (AG), and adsorbed-compact (AC1, AC2) conformations can be identified. These regions are divided by the black lines indicating the transitions between them. These are expected to be transitions in the thermodynamic meaning; only the precise location of the transition lines will still change with increasing length of the polymer. The transition between the compact phases AC1 and AC2 is a layering transition between film-like and dual-layer conformations [298].

Furthermore, highlighted by white lines, there are transitions in between the major phases. These subphases are dominated by finite-size effects and do not survive in this form in the thermodynamic limit. This concerns, e.g., the higher-order layering transitions among the compact pseudophases AC2a$_{1,2}$-d. In the following sections we will analyze the properties of the structural phases in detail.

13.4.2 Contact-number fluctuations

The contact numbers n_s and n_m can be considered as system parameters appropriately describing the state of the system and are therefore useful to identify the pseudophases. Peaks and dips in the external-parameter dependence of self-correlations $\langle n_s^2 \rangle_c$, $\langle n_m^2 \rangle_c$ and cross-correlations $\langle n_s n_m \rangle_c$ indicate activity in the contact-number fluctuations and, analyzing the expectation values $\langle n_s \rangle$ and $\langle n_m \rangle$ in these active regions of the external parameters T and s, allow for an interpretation of the respective conformational transitions between the structural phases.

In Fig. 13.3, these quantities are plotted for the 179-mer and, for comparison, the heat capacity, as functions of the temperature T at a fixed solvent parameter $s = 1$. This example is quite illustrative as the system experiences several conformational transitions when increasing the temperature starting from $T = 0$ (see Fig. 13.2). At temperatures very close to $T = 0$ (pseudophase AC1) all 179 monomers have contact with the substrate and 153 monomer–monomer contacts are formed. This is the most compact contact set being possible for *topologically two-dimensional*, film-like conformations. It should be noted, however, that approximately 2×10^{18} conformations (self-avoiding walks) belong to this contact set.[3] This high degeneracy is an artifact of the minimalistic lattice polymer model used. It is remarkable that the conformations with the highest number of total contacts $n = n_s + n_m$ are film-like compact ($n = 332$). All other conformations possess fewer

---

[2] In very poor solvent (i.e., for very large values of s), the DC phase may degenerate, at least for flexible finite-length polymers, into separate globular and crystalline subphases (see the discussion in Section 5 and, e.g., in Refs. [127, 128, 137]).

[3] In Chapter 4, it has already been stated that it is an advantage of the simple-sampling algorithms based on Rosenbluth sampling [33], compared to importance-sampling methods, that they allow for the approximation of the degeneracy ("density") of states *absolutely*, i.e., free energy and entropy can be explicitly determined.

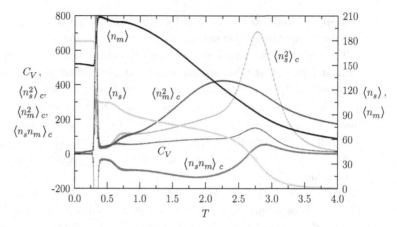

Fig. 13.3 Expectation values, self- and cross-correlations of the contact numbers n_s and n_m as functions of the temperature T in comparison with the heat capacity for a 179-mer in solvent with $s = 1$. From [302].

contacts, even the most compact contact set that dominates the five-layer pseudophase AC2a₁ ($n_s = 36$, $n_m = 263$, i.e., $n = 299$). The reason is that for low temperatures, those macrostates are formed that are energetically favored. Entropy is not yet relevant – for the $s = 1$ example $\langle n_s \rangle$ drops to 149 only up to $T \approx 0.3$. Increasing the temperature further, the situation changes dramatically, as can be seen in Fig. 13.3. In a highly cooperative process, the average number of intrinsic contacts $\langle n_m \rangle$ significantly increases (to ≈ 208) at the expense of surface contacts ($\langle n_s \rangle$ drops to approximately 104). Consequently, the strong fluctuations $\langle n_{s,m}^2 \rangle_c$ signal a conformational transition, and the anti-correlation indicated by $\langle n_s n_m \rangle_c$ confirms that surface contacts turn into intrinsic contacts, which indirectly leads to the conclusion that the film-like structure is given up in favor of layered, spatially three-dimensional conformations. The system has entered subphase AGe, which is the part of the phase AG where two-layer conformations dominate. The subphase transition near $T \approx 0.7$ from the two-layer (AGe) to the bulky regime of AG is due to the ongoing, rather unstructured expansion of the polymer into the z direction by forming so-called surface-attached globules [296]. This is accompanied by a further reduction of surface contacts, while the number of intrinsic contacts changes weakly. Approaching $T \approx 2.0$, the situation is just vice versa. Intrinsic contacts dissolve and the system experiences a conformational phase transition from globular conformations in AG to random strands in AE. Crossing this transition line, the system enters the good-solvent regime. Eventually, close to $T \approx 2.8$, the polymer unbinds off the substrate. A clear signal is observed in the fluctuations of n_s, i.e., the number of average surface contacts rapidly decreases. The expanded polymer is "free" and the influence of *both* walls is effectively steric. This phase (AE) is closely related to the typical random-coil phase of entirely free and dissolved polymers in good solvent.

This example shows that a study of the contact number fluctuations is indeed sufficient to qualitatively identify and describe the conformational transitions between the pseudophases of the hybrid system. For this reason, n_s and n_m are adequate system parameters, which play a similar role to order parameters in thermodynamic phase transitions.

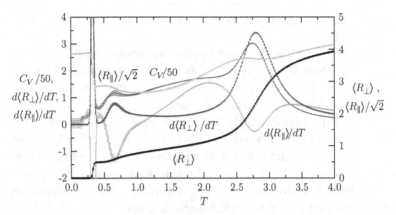

Fig. 13.4 Anisotropic behavior of gyration tensor components parallel and perpendicular to the substrate and their fluctuations as functions of the temperature T for a 179-mer at $s = 1$. For comparison, we have also plotted the associated heat capacity curve. From [302].

13.4.3 Anisotropic behavior of gyration tensor components

One of the most interesting structural quantities in studies of polymer phase transitions is the gyration tensor (13.5). For the hybrid system considered here, it is expected that the respective components parallel (13.6) and perpendicular (13.7) to the substrate will behave differently when the polymer passes transition lines. This anisotropy is obvious from Fig. 13.4, where the expectation values, $\langle R_{\parallel,\perp}\rangle$, and the fluctuations of these two components, $d\langle R_{\parallel,\perp}\rangle/dT$ are plotted, again for the polymer in solvent with $s = 1$. For interpreting the peaks of the fluctuations, the heat capacity curve is included for comparison. The immediate observation is that the temperatures, where one or both gyration tensor components exhibit peaks, almost perfectly coincide with those of the specific heat. This is a strong confirmation for the phase diagram in Fig. 13.2, which is based on the specific heat. Obviously, even for the rather short polymer with 179 monomers, we encounter the onset of fluctuation collapse near the (pseudo)phase transitions [302].

At very low temperatures, i.e., in pseudophase AC1, we have argued in the previous section that the dominant polymer conformation is the most compact single-layer film. This is confirmed by the behavior of $\langle R_{\parallel}\rangle$ and $\langle R_{\perp}\rangle$, the latter being zero in this phase. A simple argument that the structure is indeed maximally compact is as follows. It is well known that the most compact shape in the two-dimensional continuous space is the circle. For n monomers residing in it, $n \approx \pi r^2$, where r is the (dimensionless) radius of this circle. The usual squared gyration radius is

$$R_{\text{gyr}}^{\text{circ}\,2}(\approx R_{\parallel}^2) = \frac{1}{\pi r^2}\int\limits_{r' \leq r} d^2 r'\, r'^2 = \frac{1}{2}r^2 \tag{13.10}$$

and therefore $R_{\text{gyr}}^{\text{circ}} \approx \sqrt{n/2\pi} \approx 5.34$ for $n = N = 179$. Indeed, this is close to the value $R_{\parallel} \approx 5.46$ of the ground-state conformation identified in phase AC1. Note that the most compact shape in this simple lattice polymer model is a square and not a discretized circle.[4]

Near $T \approx 0.3$, the strong layering transition from AC1 to AGe is accompanied by an immediate decrease of $\langle R_\parallel \rangle$, while $\langle R_\perp \rangle$ rapidly increases from zero to about 0.5, which is exactly the gyration radius (perpendicular to the layers) of a two-layer system, where both layers cover approximately the same area. Note that the single layers are still compact, but not maximally. Applying the same approximation as in Eq. (13.10), the planar gyration radius for each of the two layers is (with $n \approx N/2$) $R_{\mathrm{gyr}}^{\mathrm{circ}} \approx 3.77$, while the measured value in this phase (AGe) is $R_\parallel \approx 4.05$. This separates the subphase AGe from the other two-layer pseudophase AC2d in Fig. 13.2, where the dominating conformation has perfect two-layer (lattice) structure with $R_\parallel \approx 3.85$ (this is the same 2% difference between continuous and lattice calculation for perfect shapes as above).

We will discuss the conformational peculiarities in the following in more detail. The subphase transition from AGe to AG near $T \approx 0.7$ is accompanied by a further decrease of $\langle R_\parallel \rangle$ whereas $\langle R_\perp \rangle$ increases, i.e., the height of the surface-adsorbed globule increases at the expense of the width. This tendency is stopped when approaching the transition ($T \approx 2.0$) from the globular regime AG to the phase of expanded but still adsorbed conformations. While $\langle R_\perp \rangle$ remains widely constant (the fluctuation does not signalize any transition), the polymer strongly extends in the directions parallel to the substrate, as indicated by the peak of $d\langle R_\parallel \rangle/dT$. After unbinding from the substrate, parallel and perpendicular gyration radii behave widely isotropically ($\langle R_\perp \rangle^2 \approx \langle R_\parallel \rangle^2/2 \approx \langle R_{\mathrm{gyr}} \rangle^2/3$) as the influence of the isotropy-disturbing walls is weak in this regime.

13.5 Alternative view: The free-energy landscape

It was shown in Section 13.4.2 that the contact numbers n_s and n_m are unique system parameters for the pseudophase identification of the hybrid system. We define the restricted partition sum for a macrostate with n_s surface contacts and n_m monomer–monomer contacts by

$$Z_{T,s}(n_s, n_m) = \sum_{n_s', n_m'} \delta_{n_s' n_s} \delta_{n_m' n_m} g_{n_s' n_m'} e^{-E_s(n_s', n_m')/k_B T} = g_{n_s n_m} e^{-E_s(n_s, n_m)/k_B T}, \qquad (13.11)$$

such that $Z = \sum_{n_s, n_m} Z_{T,s}(n_s, n_m)$. Assuming as usual that the dominating macrostate is given by the minimum of the free energy as a function of appropriate system parameters, it is useful to define the specific contact free energy as a function of the contact numbers n_s and n_m,

$$F_{T,s}(n_s, n_m) = -k_B T \ln g_{n_s n_m} e^{-E_s(n_s, n_m)/k_B T} = E_s(n_s, n_m) - TS(n_s, n_m), \qquad (13.12)$$

where $S(n_s, n_m) \equiv k_B \ln g_{n_s n_m}$ is the contact entropy. For given external parameters T and s, this relation can be used to determine the minimum of the contact free energy and,

---

[4] This is also true for most compact three-dimensional structures that are cube-like. Therefore, globules in the lattice model are cubes rather than shapes that would more closely resemble a sphere.

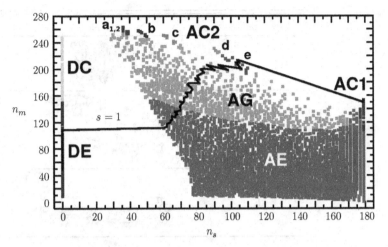

Fig. 13.5 Map of all minima of the contact free energy $F_{T,s}(n_s, n_m)$ in the parameter intervals $T \in [0, 10]$ and $s \in [-2, 10]$ for the 179-mer. The solid line connects the free-energy minima taken by the polymer in solvent with $s = 1$ by increasing the temperature from $T = 0$ to $T = 5$ and thus symbolizes its "path" through the free-energy landscape. The solid line is only a guide to the eyes. From [302].

therefore, it enables the identification of the dominant macrostate with respect to the contact numbers. In turn, this quantity allows for an alternative representation of the pseudophase diagram, complementary to the one shown in Fig. 13.2 in that it is related to the contact numbers n_s and n_m. This is done by determining for (in principle) all values of the external parameters T and s the minima of the contact free energy (13.12). Then the pair of values n_s and n_m of the minimum contact free-energy state are marked in an n_s-n_m phase diagram. This is shown in Fig. 13.5, where all free-energy minima of the 179-mer for the parameter set $T \in [0, 10]$ and $s \in [-2, 10]$ are included and, based on the arguments of the previous section, differently shaded according to the pseudophase they belong to. A nice aspect of this representation is that it allows for the differentiation of continuous and discontinuous pseudophase transitions.

The first important observation is that the diagram is divided into two separate regions, the pseudophases of desorbed conformations (DC and DE) and the remaining different phases of adsorption. The "space" in-between is blank, i.e., none of these (possible) conformations represents a free-energy minimum state. This shows that transitions between the adsorbed and desorbed pseudophases are first-order-like in this finite system. It should be noted that the regime of contact pairs (n_s, n_m) lying *above* the shown compact phases is forbidden, i.e., conformations with such contact numbers do not exist on the sc lattice.

The second remarkable result is that the pseudophases DC, DE, AE, and AG are "bulky," while all AC subphases are highly localized in the plot of the free-energy minima. Comparing with Fig. 13.2, the conclusion is that conformations in the AC phases are energetically favored (more explicitly, for $s/T > 0.8$ in AC1 and $s/T > 2.2$ in the AC2 subphases), while the behavior in the other pseudophases is entropy-dominated: the number of conformations with similar contact numbers in the globular or expanded regime is higher

pseudophase	example	n_s	n_m
DC		0	219
DE	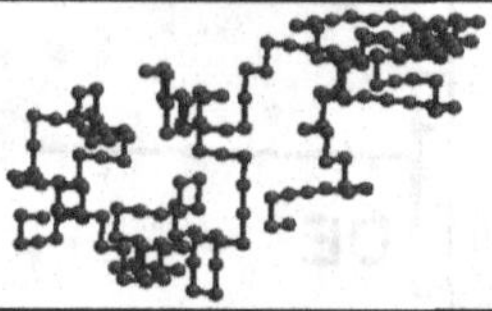	0	50
AE	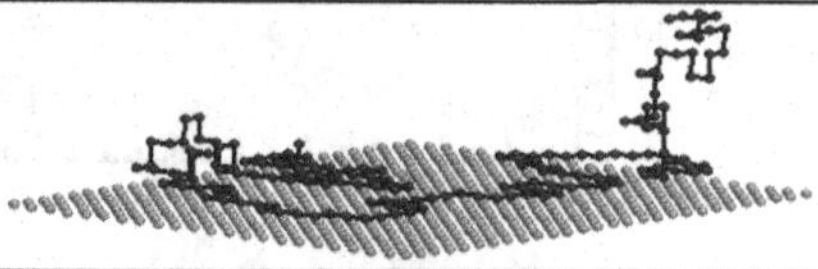	135	33
AG	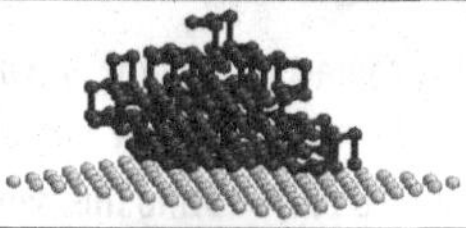	49	227
$AC2a_1$	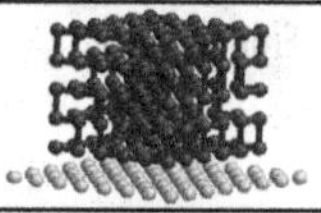	36	263
$AC2a_2$	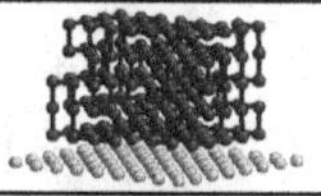	39	256
AC2b	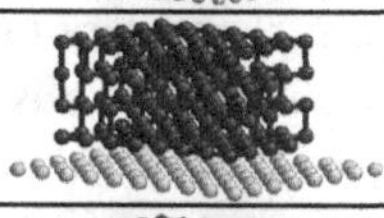	46	257
AC2c		60	251
AC2d		90	231
AGe	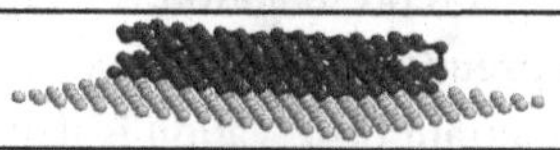	103	207
AC1	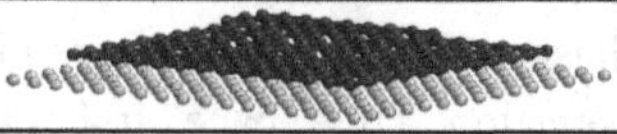	179	153

Fig. 13.6 Representative minimum free-energy examples of conformations in the different pseudophases of a 179-mer in a cavity. The substrate is shaded in light gray. From [302].

than the rather exceptional conformations in the compact phases, i.e., for sufficiently small s/T ratios the entropic effect overcompensates the energetic contribution to the free energy.

The subphases $AC2a_{1,2}$-d are strongly localized, thorn-like "peninsulas" that stand out from the AG regime. The discrete number and their separation leads to the conclusion that they have related structures. Indeed, as can be seen in Fig. 13.6, where representative conformations are listed for all structural phases, the few conformations dominating these subphases exhibit compact layered structures. The most compact three-dimensional conformation with 263 monomer–monomer contacts and 36 surface contacts is favored in subphase $AC2a_1$ and possesses five layers. Starting from this subphase and increasing the temperature, two things may happen. A rather small change is accompanied by the transition to $AC2a_2$, where the number of intrinsic contacts is reduced but the global five-layer structure remains. On the other hand, passing the transition line toward AC2b, the monomers prefer to arrange in compact four-layer conformations. Advancing toward AC2d, the typical conformations reduce layer by layer in order to increase the number of surface contacts. In AC2d, there are still two layers lying almost perfectly on top of each other. This is similar in subphase AGe, where also two-layer but less compact conformations dominate. In pseudophase AC1 only the film-like surface layer remains. The reason for the differentiation of the phases AC1 and AC2 of layered conformations is that the transition from single- to double-layer conformations is expected to be a real phase transition, while the transitions between the higher-layer AC2 subphases are assumed to disappear in the thermodynamic limit [298].

As can be seen in Fig. 13.2, a transition between AC1 and the phase of adsorbed, expanded conformations, AE, is possible. Since these two phases are connected in Fig. 13.5, the transition between these phases is second-order-like. Indeed, this transition is strongly related with the *two-dimensional* Θ transition since, close to the transition line, all monomers form a planar (surface-)layer. Similarly, there is also a second-order-like transition line $s_0(T)$ between AG and AE, which separates the regions of poor (AG: $s > s_0$) and good (AE: $s < s_0$) solvent. Also, the transition between the desorbed compact (DC) and expanded (DE) conformations is second-order-like: this transition resembles the well-known Θ transition in three dimensions [119]. Eventually, the transitions from the layer phases $AC2a_2$, AC2b, AC2c, and AGe to the globular pseudophase AG as well as transitions between pseudophases dominated by the same layer type (i.e., between the two-layer subphases AC2d and AGe, and between the five-layer subphases $AC2a_1$ and $AC2a_2$) are expected to be continuous.

On the other hand, the transitions among the energetically caused compact low-temperature pseudophases are rather first-order-like, due to their noticeable localization in the map of free-energy minima (Fig. 13.5). The possible transitions (see Fig. 13.2) are $AC2a_{1,2}$–AC2b, AC2b–AC2c, and AC2c–AC2d, respectively. Even more interesting, however, are the transitions from the single-layer pseudophase AC1 to the double-layer subphases AC2d and AGe. We discussed this transition previously for the special choice $s = 1$, where near $T \approx 0.3$ the fluctuations of the contact numbers and the components of the gyration tensor exhibit a strong activity. Also included in Fig. 13.5 is the "path" of macrostates the system passes by increasing the temperature from $T = 0$ to $T = 5$. At $T = 0$, the system is in a film-like, single-layer state. Near $T \approx 0.3$, it rearranges into two layers and enters subphase AGe in a single step. The probability distribution $p_{T,s}(n_s, n_m)$ for $s = 1$ and $T = 0.34$ is shown in Fig. 13.7(a), and it can clearly be seen that two distinguished macrostates coexist. By further increasing the temperature, the system experiences the continuous transitions from AGe via AG until the polymer unfolds when entering pseudophase AE. However, it is still in contact with the substrate. Close to a temperature $T \approx 2.4$, however,

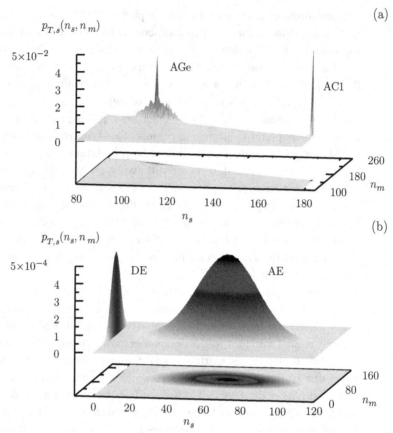

Fig. 13.7 Probability distributions $p_{T,s}(n_s, n_m)$ for the 179-mer in solvent with $s = 1$ (a) near the layering transition from AC1 to AGe at $T \approx 0.34$ and (b) near the adsorption–desorption transition from AE1 to DE at $T \approx 2.44$. Both transitions are expected to be real phase transitions in the thermodynamic limit and look first-order-like. From [302].

the unbinding of the polymer off the substrate occurs (from AE to DE). Comparing Figures 13.5 and 13.7(b), where the probability distribution at $T = 2.44$ is shown, we also see a clear indication for a discontinuous transition. We consider here the transition state where the two minima of the free energy coincide[5] (see also the black dashed line in Fig. 13.2) and not the point where the width of the distribution, i.e., the specific heat, is maximal. Since the system is finite, the transition temperature ($T \approx 2.8$), as signaled by the fluctuations studied in the previous sections, deviates slightly from the transition-state temperature.

---

[5] In the canonical statistical analysis, the coexistence point is typically defined by the temperature where the weights under the two peaks are equal, see, e.g., Refs. [319, 320].

13.6 Continuum model of adsorption

In order to get rid of undesired lattice effects like the almost cuboid form of the most compact adsorbed conformations in the subphases of AC2 in the phase diagram of a polymer on a simple-cubic lattice (see Fig. 13.2), we now investigate the structure of conformational phases of a generic off-lattice polymer near an attractive substrate [304, 307, 308].

13.6.1 Off-lattice modeling

As on the lattice, we assume that adjacent monomers are connected by rigid covalent bonds. Thus, the distance is fixed and set to unity. Bond and torsional angles are free to rotate. The energy function consists of three terms,

$$E = V_{\text{bend}} + V_{\text{LJ}} + V_{\text{sur}}, \tag{13.13}$$

associated with the bending stiffness (V_{bend}), monomer–monomer Lennard-Jones interaction (V_{LJ}), and monomer–surface attraction (V_{sur}). The Lennard-Jones potential of non-bonded monomers is of standard form, $V_{\text{LJ}} = 4 \sum_{i=1}^{N-2} \sum_{j=i+2}^{N} (r_{ij}^{-12} - r_{ij}^{-6})$ where $r_{ij} = |\mathbf{r}_j - \mathbf{r}_i|$ is the distance between the monomers i and j. The bending energy is $V_{\text{bend}} = \sum_{i=1}^{N-2} [1 - \cos(\vartheta_i)]/4$, where ϑ_i is the bending angle in the interval $[0, \pi)$. The bending energy can be considered as a penalty for successive bonds deviating from a straight arrangement.

For the interaction with the substrate we assume that each monomer interacts with each atom of the substrate via a Lennard-Jones potential. Considering for simplicity the attractive substrate as a continuum with atomic density n in the half-space $z < 0$ (see Fig. 13.8), the surface interaction potential of the ith monomer in a distance z_i from the surface can be written as

$$v_{\text{sur}}(z_i) = 4n \int_0^{2\pi} d\phi \int_0^\infty dr\, r \int_{-\infty}^{z_i} dz' \left(\left[z'^2 + r^2 \right]^{-6} - \left[z'^2 + r^2 \right]^{-3} \right)$$
$$= \frac{2\pi n}{3} \left(\frac{2}{15} z_i^{-9} - z_i^{-3} \right) \tag{13.14}$$

and thus $V_{\text{sur}} = \sum_{i=1}^{N} v_{\text{sur}}(z_i) = \epsilon_s \sum_{i=1}^{N} (2z_i^{-9}/15 - z_i^{-3})$, where $\epsilon_s = 2\pi n/3$ defines the surface attraction strength. As such it weighs the energy scales of intrinsic monomer–monomer attraction and monomer–surface attraction.

In order to prevent the molecule from escaping, the upper half-space $z > 0$ is once more regularized by a steric wall, which is placed a distance $z = L_z$ away from the attractive surface.[6]

---

[6] The following results for a chain with $N = 20$ monomers were obtained in multicanonical simulations [304, 307, 308].

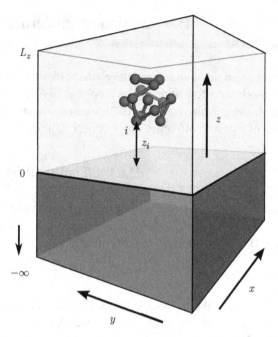

Fig. 13.8   Polymer interacting with a continuum substrate. The translational freedom of the polymer in z direction is limited by the presence of the attractive substrate at $z = 0$ and a steric wall at $z = L_z$.

13.6.2 Energetic and structural quantities for phase characterization by canonical statistical analysis

In order to understand as much as possible about the canonical equilibrium behavior of a polymer adsorbed at a substrate, the following suitable quantities O will be considered in the following. Besides the canonical expectation values $\langle O \rangle$, the fluctuations about these averages, as represented by the temperature derivative $d\langle O \rangle / dT = (\langle OE \rangle - \langle O \rangle \langle E \rangle)/T^2$ ($k_B \equiv 1$), are required to identify transition points. Apart from energetic fluctuations such as the specific heat $c_V = (1/N) d \langle E \rangle / dT$ and fluctuations of structural quantities like the parallel and perpendicular components of the gyration tensor, $R_\parallel$ and $R_\perp$, respectively [cf. Eqs. (13.6) and (13.7)], clear evidence that the polymer is moving "freely" in the box, or whether it is very close to the surface, can be provided by the average distance of the center of mass of the polymer from the surface, $\langle z_{\mathrm{cm}} \rangle = \sum_{i=1}^{N} \langle z_i \rangle / N$.

Another useful quantity is the mean number of monomers docked to the surface. A single-layer structure is formed if all monomers are attached at the substrate; if none is attached, the polymer is quasi-free (desorbed) in solvent. The surface potential is a continuous potential and in order to distinguish monomers docked to the substrate from those not being docked, it is necessary to introduce a cutoff z_c. A monomer i is defined to be "docked" if $z_i < z_c$.[7] The corresponding measured quantity is the average ratio $\langle n_s \rangle$ of

---

[7] In the actual simulations, $z_c = 1.2$ was chosen [304].

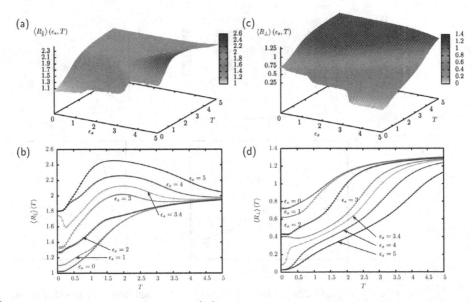

Fig. 13.9 (a) Parallel component of the radius of gyration $\langle R_\parallel \rangle$ as a function of temperature T and adsorption strength ϵ_s for a 20-mer; (b) $\langle R_\parallel \rangle$ for selected values of ϵ_s; (c) and (d) the same for the perpendicular component $\langle R_\perp \rangle$. After [304].

monomers docked to the surface and the total number of monomers. This can be expressed as $n_s = N_s/N$ with $N_s = \sum_{i=1}^{N} \Theta(z_c - z_i)$, where $\Theta(z)$ is the Heaviside step function.

13.6.3 Comparative discussion of structural fluctuations

In our previous discussion of the lattice model of adsorption, we have already made use of the components of the radius of gyration as excellent measures for the globular compactness of polymer conformations parallel [see Figs. 13.9(a,b)] and perpendicular [Figs. 13.9(c,d)] to the surface, respectively. These components are particularly helpful for the identification of structural changes induced by the presence of an attractive substrate.

For example, for $\epsilon_s \geq 3.4$, $\langle R_\perp \rangle$ vanishes at low temperatures, while $\langle R_\parallel \rangle$ attains small values at lower attraction strengths ϵ_s. The vanishing of $\langle R_\perp \rangle$ corresponds to conformations, where the polymer is spread out flat on the surface without any extension into the third dimension. The associated pseudophases are the adsorbed compact (AC1) and adsorbed expanded (AE1) phases. The phases AC1 and AE1 are separated by a freezing transition. Polymer structures in AC1 are maximally compact at lower temperatures, while AE1 conformations are less compact and more flexible but still lie rather planar at the surface.

In order to verify that conformations in AC1 are indeed maximally compact single layers, we can use the same simple argument as in Section 13.4.3. The most compact shape in the two-dimensional (2D) continuous space is the circular disk. Thus one can calculate $\langle R_\parallel \rangle$ for a disk and compare it with the value obtained in the simulation of the adsorption model. Assuming, for simplicity, N monomers (each assumed to occupy an area $a = 1$ in the

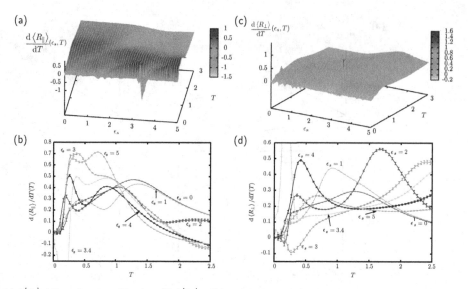

Fig. 13.10 (a) $d \langle R_{\parallel} \rangle /dT$ as a function of T and ϵ_s; (b) $d \langle R_{\parallel} \rangle /dT$ for selected values of ϵ_s; (c) and (d) the same for $d \langle R_{\perp} \rangle /dT$. After [304].

dimensionless length units used here) to be distributed uniformly in the disk, Eq. (13.10) yields $R_{\parallel} \approx 1.78$ for a polymer with $N = 20$ monomers, which is indeed close to the "true" value $R_{\parallel} \approx 1.81$ [304]. Most compact planar conformations in phase AC1 indeed assume disk-like shapes.

The most pronounced transition is the strong layering transition at $\epsilon_s \approx 3.4$ that separates regions of planar conformations (AC1, AE1) from the region of stable double-layer structures (AC2b) and adsorbed globules (AG), below and above the freezing transition respectively. For high surface attraction strengths ϵ_s, it is energetically favorable to form as many surface contacts as possible. In the layering-transition region, a higher number of monomer–monomer contacts causes the double-layer structures to have just the same energy as single-layer structures. For lower ϵ_s-values, the double-layer structures possess the lowest energies. Hence, this transition is a sharp energetic transition.

Although for the considered short chain no higher-layer structures exist, the components $\langle R_{\parallel,\perp} \rangle$ indicate some activity for low surface attraction strengths. For $N = 20$, $\epsilon_s \approx 1.4$ is the lowest attraction strength, where still stable double-layer conformations are found below the freezing transition. What follows for lower ϵ_s values after a seemingly continuous transition is a low-temperature subphase of surface attached compact conformations that we call AC2a. AC2a conformations occur if the monomer–surface attraction is not strong enough to induce layering in compact attached structures. The characterization of structures in this subphase requires some care, as system-size effects are dominant. Although the surface attraction is sufficiently strong to enable polymer–substrate contacts, compact desorbed polymer conformations below the Θ-transition are not expected to change much. Thus, layering effects do not occur.

In the wetting transition, elastic polymers with stretchable bonds can form perfectly icosahedral morphologies. This would additionally stabilize the polymer conformation

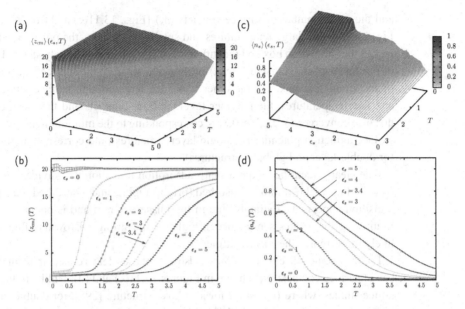

Fig. 13.11 (a) Mean center-of-mass coordinate $\langle z_{cm} \rangle$ as a function of T and ϵ_s; (b) $\langle z_{cm} \rangle$ for selected values of ϵ_s; (c) and (d) the same for the mean number of surface contacts $\langle n_s \rangle$. After [304].

and is already known from studies of atomic clusters. The smallest icosahedron with characteristic fivefold symmetry is formed by 13 atoms. Thus, it is plausible that the wetting transition is accompanied by strong finite-size effects and morphologies of adsorbed crystalline structures depend on the precise length of the polymer.

Raising the temperature above the freezing temperature, polymers form adsorbed and still rather compact conformations that look like globular, unstructured drops on the surface (AG: surface-attached globules [296]). We are already familiar with this structural phase from our discussion on the lattice-polymer adsorption model. At even higher temperatures, two scenarios can be distinguished by taking into account the relative strengths of monomer–monomer and monomer–substrate interactions. In the first case, the polymer first desorbs from the surface [from AG to the desorbed globular (DG) bulk phase] and disentangles at even higher temperatures [from DG to the desorbed expanded bulk phase (DE)]. In the latter case, the polymer expands while it is still located at the surface (from AG to AE2) and desorbs at higher temperatures (from AE2 to DE). Due to the higher relative number of monomer–monomer contacts in compact bulk conformations of longer chains, the Θ-temperature increases with N. The same holds true for the surface attraction strength ϵ_s associated with the layering transition.

13.6.4 Adsorption parameters

The adsorption transition can be discussed best when looking at the distance of the mean center-of-mass distance $\langle z_{cm} \rangle$ of the polymer from the surface [Figs. 13.11(a) and (b)] and the mean number of surface contacts $\langle n_s \rangle$ [Figs. 13.11(c) and (d)]. As can be seen in Fig. 13.11(a), for large temperatures and small values of ϵ_s, the polymer can move freely within the simulation box and the influence of the substrate is purely steric. Thus, the mean of the center-of-mass distance $\langle z_{cm} \rangle$ of the polymer from the surface is just half the height of the simulation box. On the other hand, for large enough surface attraction strengths and low temperatures, the polymer favors surface contacts and the mean center-of-mass distance converges to $\langle z_{cm} \rangle \approx 0.858$, corresponding to the minimum-energy distance of the surface attraction potential for single-layer structures (and correspondingly larger values for double-layer and globular structures).

One clearly identifies a quite sharp adsorption transition that divides the projection of $\langle z_{cm} \rangle$ in Fig. 13.11(a) into an adsorbed (bright) regime and a desorbed (dark) regime. This transition appears as a straight line in the phase diagram and is parametrized by $\epsilon_s \propto T$. Intuitively, this makes sense since at higher T the stronger Brownian fluctuation is more likely to overcome the surface attraction.

For the detailed discussion of these adsorbed phases, let us concentrate on the mean number of surface contacts $\langle n_s \rangle$ shown in Figures 13.11(c) and (d). Unlike in the simple-cubic lattice studies, where $\langle n_s \rangle \approx 1/l$ for an l-layer structure [298], for double-layer structures in the continuum model $\langle n_s \rangle > 1/2$. The reason is that most compact multilayer structures are cuboids on the lattice, while in the continuum case, "layered" conformations correspond to semispherical shapes, where – for optimization of the surface of the compact shape – the surface layer contains more monomers than the upper layers. Since this only pertains to the outer part of the layers, the difference is more pronounced the shorter the chain is. Further layering transitions are not observed. When the double-layer structure gets unstable at lower values of ϵ_s, $\langle n_s \rangle$ starts to decrease again. The conformations in AC and AC2a thus do not exhibit a pronounced number of surface contacts, and $\langle n_s \rangle$ varies with ϵ_s. To conclude, the single- to double-layer "layering transition" is a topological transition from two-dimensional to three-dimensional polymer conformations adsorbed at the substrate. The solvent-exposed part of the adsorbed compact polymer structure, which is not in direct contact with the substrate, reduces under poor solvent conditions the contact surface to the solvent. Due to the larger number of degrees of freedom for the off-lattice polymer, layered structures are not favored in this case. Thus, higher-order layering transitions are not present (which in part is also due to the short length of the chain), but are also not expected in pronounced form.

13.6.5 The pseudophase diagram of the hybrid system in continuum

To summarize all the information gained from the different observables, we construct the approximate boundaries of different regimes in the T-ϵ_s plane. The pseudophase diagram is displayed in Fig. 13.12 and the different pseudophases are denoted by the abbreviations introduced in the previous subsections. Transitions between conformational phases are indicated by stripes. It must be noted that their thickness is due to the different locations of peaks when considering different fluctuating quantities. This is a typical feature of a finite system, where different indicators of transitions (e.g., transition temperatures) do

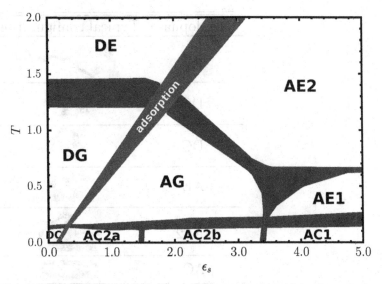

Fig. 13.12 Structural phase diagram of an off-lattice homopolymer with 20 monomers interacting with a continuous attractive substrate. The diagram, which is parametrized by the surface attraction strength ϵ_s and the temperature T, was constructed by means of the information gained from the quantities discussed previously. Stripes separate the individual conformational phases. The thickness reflects the (fundamental) uncertainty in the canonical estimation of the location of the transition points and arises from the input of the different quantities investigated for a small system. The conformational phases are discussed in the text.

not necessarily coincide. Since this uncertainty is of fundamental nature, it again questions the usefulness of a quantitative discussion of (pseudo)phase transitions of a finite system by a canonical-ensemble analysis (see also the discussion of nucleation transitions in Chapters 11 and 12 in this context). The transition lines in the pseudophase diagram (Fig. 13.12), which also still vary with chain length N, represent the best compromise of all canonical quantities analyzed separately. Only in the thermodynamic limit of infinitely long chains, most of the identified pseudophase transitions are expected to occur at sharp values of the parameters ϵ_s and T for all observables. Taking that into account, the pseudophase diagram gives a good qualitative overview of the behavior of polymers near attractive substrates in dependence of environmental parameters such as solvent quality and temperature. The locations of the phase boundaries should be considered as rough guidelines.

For the exemplified 20-mer, the following pseudophases can be associated to the different regions in Fig. 13.12. Exemplified conformations representative for the different phases are shown in Fig. 13.13. In the hybrid model discussed here, where the chain is not grafted at the substrate, the polymer can completely desorb from the surface and thus the typical polymer bulk phases are present in the phase diagram, too. These phases are denoted as DE (desorbed expanded), which corresponds to the random-coil phase of the quasi-free desorbed polymer, DG (desorbed globular) representing the globular phase of the desorbed chain, and DC (desorbed compact) for the maximally compact, spherically shaped crystalline structures that dominate this desorption phase below the freezing-transition temperature.

Pseudophase	Typical configuration
DE	
DG	
DC	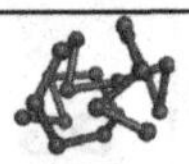
AE1	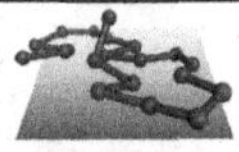
AE2	
AC1	
AG	
AC2a	
AC2b	

Fig. 13.13 Representative examples of conformations for a 20-mer in the different regions of the T-ϵ_s pseudophase diagram, shown in Fig. 13.12. DE, DG, and DC represent bulk "phases", where the polymer is preferably desorbed. In regions AE1, AE2, AC1, AG, AC2a, and AC2b conformations are favorably adsorbed. From [304].

In the adsorption part of the phase diagram, expanded, compact, and globular phases are also found, which, however, differ from their desorption counterparts. Two phases of adsorbed expanded conformations are distinguished – AE1 (adsorbed expanded single layer), which labels the phase of expanded, rather planar, but little compact random-coil conformations; and AE2 (adsorbed expanded three-dimensional conformations), associated with adsorbed, unstructured random-coil-like expanded conformations with typically more than half of the monomers in contact with the attractive substrate. Highly compact structures are found in AC1 (adsorbed compact single layer), where adsorbed circularly compact film-like (i.e., two-dimensional) conformations dominate, and in AC2a,b – two subphases of adsorbed compact three-dimensional conformations. AC2a corresponds to adsorbed compact, semi-spherically shaped crystalline conformations, whereas AC2b (adsorbed compact double layers) is the subphase of adsorbed, compact "double-layer" conformations, where the occupation of the surface layer is slightly larger than that of the other layer. For larger systems, droplet-shape conformations with more layers can form [308]. AC2a and AC2b are subphases in the regime of the phase diagram, where adsorbed compact and topologically three-dimensional conformations are dominant. The subphases AC2a,b differ qualitatively from the phase AC1 of topologically two-dimensional polymer films and it is thus reasonable to expect that the transition between film-like (AC1) and semi-spherical conformations (AC2) is of thermodynamic relevance. Finally, there is the phase of adsorbed globular three-dimensional conformations, AG, where representative conformations are surface-attached globular conformations and look like internally unstructured drops on the surface. The famous wetting transition corresponds to passing the transition lines DG → AG → AC2b → AC1 for fluid droplets, while compact polymers wet the surface in the direction DC → AC2a → AC2b → AC1. The latter process can also be interpreted as a melting transition, which is simply induced by an increase of the surface attraction strength.

In cases where the maximally compact conformation is more stable due to the high symmetry of the intrinsic icosahedral structure (as, e.g., for a chain of length $N = 13$), an additional subphase exists: AC (adsorbed icosahedral compact conformations), which is similar to DC, but polymers are in touch with the surface under the given external parameters in this phase.

13.7 Comparison with lattice results

In order to identify the limitations of both the lattice model for a rather long polymer ($N = 179$ monomers) and the continuum model of a short polymer ($N = 20$), it is instructive to compare the phase diagrams obtained from the sc lattice and the off-lattice models (Figs. 13.2 and 13.12).

The energy of the lattice system is given by Eq. (14.1), which we rewrite here as

$$E_{\mathrm{L}}\left(n_s^{\mathrm{L}}, n_m^{\mathrm{L}}\right) = -\epsilon_s^{\mathrm{L}} n_s^{\mathrm{L}} - \epsilon_m^{\mathrm{L}} n_m^{\mathrm{L}}; \qquad (13.15)$$

n_s^{L} is the number of nearest-neighbor monomer–substrate contacts, n_m^{L} the number of nearest-neighbor but nonadjacent monomer–monomer contacts, and ϵ_s^{L} and ϵ_m^{L} are the respective contact energy scales. The phase diagram shown in Fig. 13.2 is parametrized by temperature T and monomer–monomer interaction strength. The phase diagram corresponds to the specific-heat profile and surface–monomer attraction strength is fixed. But, from the known contact density (see Section 13.2), the specific-heat profile can also be calculated for fixed monomer–monomer interaction strength $\epsilon_m^{\mathrm{L}} = 1$, while varying the surface attraction parameter ϵ_s^{L}, which corresponds to the off-lattice approach in the previous section. Denoting the energy and temperature units in the original lattice model by E' and T', respectively, a simple rescaling yields

$$\frac{E'}{T'} = \frac{E_{\mathrm{L}}}{T} \quad \Leftrightarrow \quad \frac{n_s^{\mathrm{L}} + s n_m^{\mathrm{L}}}{T'} = \frac{\epsilon_s^{\mathrm{L}} n_s^{\mathrm{L}} + n_m^{\mathrm{L}}}{T} \quad \Leftrightarrow \quad T = \frac{T'}{s}, \quad \epsilon_s^{\mathrm{L}} = \frac{1}{s}, \qquad (13.16)$$

where $s = \epsilon_m^{\mathrm{L}}/\epsilon_s^{\mathrm{L}}$ is the ratio of energy scales of intrinsic and surface contacts as introduced in Section 13.2.

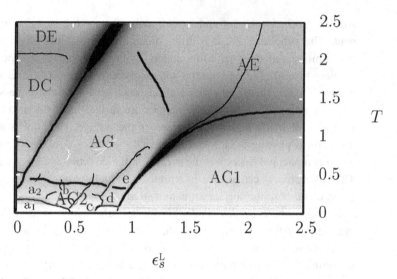

Pseudophase diagram of a lattice polymer with 179 monomers as in Fig. 13.2, but here parametrized by the surface attraction strength ϵ_s^{L} and the temperature T. The color encodes the specific-heat profile; the darker the shade, the larger its value. From [304].

Certain similarities between the phase diagrams obtained from the rescaled lattice approach, shown in Fig. 13.14, and that of the off-lattice model (Fig. 13.12) are obvious. For instance, the adsorption transition line is parametrized in both models by $\epsilon_s \propto T$. However, not only is the slope that depends on the system's geometry and energy scales different. Also, while for the off-lattice model the extrapolation of the transition line seems to go through the origin $\epsilon_s = 0$ and $T = 0$, there is an offset observed in the lattice-system analysis such that the extrapolated transition line roughly crosses $\epsilon_s^{\mathsf{L}} = 0.4$ and $T = 0$. This is due to the intrinsic cuboidal structure of the polymer conformations on the sc lattice that possess planar surfaces at low temperatures even in the bulk. Unlike for off-lattice models, where a compact polymer attains a spherical shape, such a cuboidal conformation is likely to dock at a substrate without substantial conformational rearrangements. Here lies an important difference between lattice and off-lattice models. For sufficiently small surface attraction strengths, the off-lattice model provides a competition between most compact spherical conformations that do not possess planar regions on the polymer surface, and less compact conformations with planar regions that allow for more surface contacts but reduce the number of intrinsic contacts.

This also explains why the wetting transition is more difficult to observe in adsorption studies on regular lattices. On the other hand, AC2 conformations at low T and for ϵ_s between the adsorption and the single-double layering transitions can be observed in both models. Similarly, in both models there exists the AG pseudophase of surface-attached globules.

While for the off-lattice system, apart from the wetting transition, there is only the transition from AC2a (semi-spherical shaped) to AC2b (double-layer structures), on the lattice AC2 comprises a variety of higher-layer subphase transitions (see Fig. 13.6). Decreasing the surface attraction at low temperatures, layer after layer is added until the number of layers is the same as in the most compact conformation. A lattice polymer has no other choice than forming layers in this regime. The layering transition from AC1 to AC2 is very sharp in both models. Also the shape of the transition region from topologically two-dimensional adsorbed to three-dimensional adsorbed conformations looks very similar.

We may conclude that, in particular, the high-temperature pseudophases DE, DC/DG, AG, AE nicely correspond to each other in both models. Noticeable qualitative deviations occur, as expected, in those regions of the pseudophase diagram where compact conformations are dominant and (unphysical) lattice effects are influential. Thus, the choice of the appropriate model depends on the question one wants to answer. Unlike temperatures are not too small and the polymer chain not too short, lattice models are perfectly suitable for the investigation of structural phases. This is particularly true for scaling analyses toward the thermodynamic limit. However, if the focus is more on finite-size effects and the behavior at low temperatures, off-lattice models should generally be preferred.

13.8　Systematic microcanonical analysis of adsorption transitions

We will now take a closer look at the adsorption transition in the phase diagram shown in Fig. 13.12 and we do this by a microcanonical analysis [307,308]. As we have discussed in detail in Section 2.7, the microcanonical approach allows for a unique identification of transition points and a precise description of the energetic and entropic properties of structural transitions in finite systems. The transition bands in canonical pseudophase diagrams are replaced by transition lines. Figure 13.15 shows the microcanonical entropy per monomer $s(e) = N^{-1} \ln g(e)$ as a function of the energy per monomer $e = E/N$ for a polymer with $N = 20$ monomers and a surface attraction strength $\epsilon_s = 5$, as obtained from multicanonical simulations of the model described in Section 13.6.

The entropy curve exhibits the characteristic microcanonical features of a transition with phase coexistence in a small system. For energies right below e_{ads}, the system is in the adsorbed phase AE2 (cf. Fig. 13.12), i.e., the polymer is in contact with the substrate, but monomer–monomer contacts are not particularly favored and thus expanded conformations dominate. For energies between e_{ads} and e_{des}, the system is in the transition region, where $s(e)$ is convex. This is clearly seen by constructing the Gibbs hull $\mathcal{H}_s(e) = s(e_{ads}) + e(\partial s/\partial e)_{e=e_{ads}}$ as the double tangent that touches $s(e_{ads})$ and $s(e_{des})$. Thus, $T_{ads} = (\partial \mathcal{H}_s/\partial e)^{-1} = (\partial s/\partial e)^{-1}_{e=e_{ads}} = (\partial s/\partial e)^{-1}_{e=e_{des}}$ is the microcanonical adsorption temperature defined by Gibbs–Maxwell construction.

The entropic suppression in the adsorption transition region is given by the difference between $\mathcal{H}_s(e)$ and $s(e)$, $\Delta s(e) = \mathcal{H}_s(e) - s(e)$. Then, the surface (or interfacial) entropy, which represents the entropic barrier of the two-state transition, is defined as the maximum deviation $\Delta s_{surf} = \max\{\Delta s(e) \mid e_{ads} \leq e \leq e_{des}\}$. The peak is located at $e = e_{sep}$ and defines the energetic phase-separation point. Finally, the energetic gap between the two macrostates is the latent heat per monomer, $\Delta q = e_{des} - e_{ads} = T_{ads}[s(e_{des}) - s(e_{ads})]$. In the thermodynamic limit, a first-order phase transition will be characterized as usual by

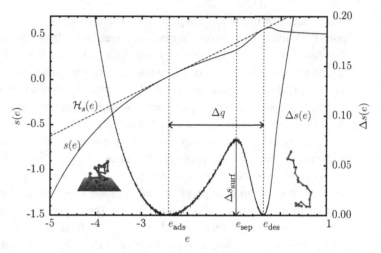

Fig. 13.15 Microcanonical entropy $s(e)$ (up to an unimportant constant) for a 20-mer at $\epsilon_s = 5$, the Gibbs hull $\mathcal{H}_s(e)$, and the difference $\Delta s(e) = \mathcal{H}_s(e) - s(e)$ as functions of the energy per monomer e. The convex adsorption regime is bounded by the energies $e_{ads} = -2.412$ and $e_{des} = -0.369$ of the coexisting phases of adsorbed and desorbed conformations at the adsorption temperature $T_{ads} = 3.885$, as defined via the slope of $\mathcal{H}_s(e)$. The maximum of $\Delta s(e)$, called surface entropy Δs_{surf}, is found at $e_{sep} = -0.962$, which defines the energy of phase separation. The latent heat Δq is defined as the energy being necessary to cross the transition region at the transition temperature T_{ads}. From [307].

$\lim_{N \to \infty} \Delta q = \mathrm{const} > 0$, whereas $\lim_{N \to \infty} \Delta q = 0$ in the case of a second-order transition. However, in both cases we expect the surface entropy to vanish in this limit, $\lim_{N \to \infty} \Delta s_{surf} = 0$, i.e., the microcanonical entropy is always a concave function of energy for infinitely large systems. Before we show for the adsorption transition that the latent heat indeed decreases with system size, we first investigate the origin of the phase separation for chains of finite length and discuss the adhesion strength dependence of surface entropy and microcanonical temperature.

13.8.1 Dependence on the surface attraction strength

In Fig. 13.16(a), the microcanonical entropies of the 20-mer are shown for various surface attraction strengths ϵ_s. Since the high-energy regime is dominated by desorbed conformations, the density of states and hence $s(e)$ are hardly affected by changing the values of ϵ_s. The low-energy tail, on the other hand, increases significantly with ϵ_s. Thus, it is useful to also split the density of states into the contributions of desorbed and adsorbed conformations, $g_{des}(e)$ and $g_{ads}(e)$, respectively, such that $g(e) = g_{des}(e) + g_{ads}(e)$ and $s_{des,ads}(e) = N^{-1} \ln g_{des,ads}(e)$. It is convenient to define the polymer to be adsorbed, if its total surface energy is $E_{surf} < -0.1\, \epsilon_s N$. Since $s_{des}(e)$ corresponds to $s_{\epsilon_s=0}(e)$, for $\epsilon_s > 0$ only the $s_{ads}(e)$ curves were added in Fig. 13.16(a). Both $s_{ads}(e)$ and $s_{des}(e)$ are concave in the whole energy range of the adsorption transition. Thus, convex entropic monotony can only occur in the most sensitive region where adsorbed and desorbed conformations

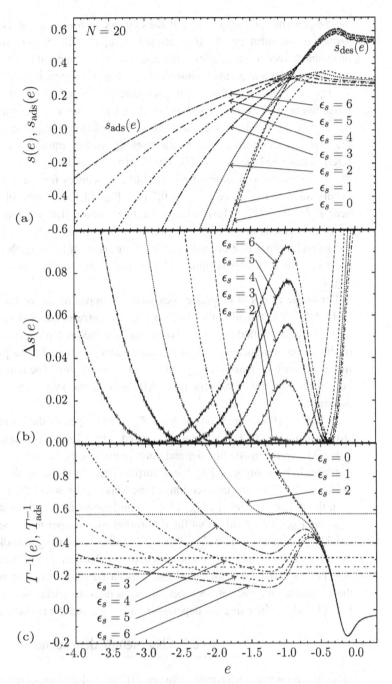

(a) Microcanonical entropies and fractions of adsorbed conformations $s_{ads}(e)$ at various surface attraction strengths $\epsilon_s = 0, 1, \ldots, 6$ for a 20-mer [the fraction of desorbed structures corresponds to $s(e)$ for $\epsilon_s = 0$]; (b) deviations $\Delta s(e)$ from the respective Gibbs hulls (not shown) to illustrate the increase of the surface entropy Δs_{surf} and the latent heat Δq with the attraction strength ϵ_s. Note that $\Delta s_{surf} = \Delta q = 0$ for $\epsilon_s = 0, 1$; (c) caloric inverse temperature curves $T^{-1}(e)$ and Maxwell lines at respective reciprocal transition temperatures T_{ads}^{-1}. From [307].

have equal entropic weight, i.e., at the entropic transition point. Note that for a polymer *grafted* at the substrate, the translational entropy would be very small. The thus far less pronounced increase of $s_{\text{des}}(e)$ at the entropic transition point is not sufficient to induce the convex intruder and no microcanonical peculiarities appear in this case.

Depending on ϵ_s and thus on the energetic location of the crossing point, the adsorption transition appears to be *second-order-like* ($\Delta q = 0$ for $\epsilon_s \lesssim 2$) or *first-order-like* ($\Delta q > 0$ for $\epsilon_s \gtrsim 2$) for a finite, nongrafted chain. Referring to the phase diagram in Fig. 13.12, the first scenario corresponds to the docking/wetting transition from desorbed globules (DG) to adsorbed globules (AG). The $T^{-1}(e)$ curves for $\epsilon_s = 0, 1$ in Fig. 13.16(c) do not at all exhibit microcanonical first-order-like signatures for the adsorption transition that occurs for $\epsilon_s = 1$, e.g., near $T_{\text{ads}} \approx 0.7$ (see Fig. 13.12). Noticeably, the inflection points near $T_{\Theta}^{-1} \approx 0.77$ ($T_{\Theta} \approx 1.3$) indicate the Θ-transition that separates coil-like and globular conformations in the bulk (DE/DG). It is a surprising observation that the adsorption transition becomes first-order-like at the point where it falls together with the Θ-transition ($\epsilon_s \approx 1.8$, $T \approx 1.3$). This is signaled by the saddle point of the corresponding T^{-1} curve in Fig. 13.16(c).

For larger values of ϵ_s, phase coexistence is apparent for the transition from expanded coils (DE) to adsorbed coils (AE2). Here, the corresponding deviations $\Delta s(e)$ from the Gibbs hulls, as plotted in Fig. 13.16(b), become maximal at the crossing point. The curves of the inverse microcanonical temperatures decrease [i.e., T^{-1} as plotted in Fig. 13.16(c) increases] with increasing energy in the transition region. The temperature "bends back," i.e., in the desorption process from AE2 to DE, the system is cooled while energy is increased.

In Fig. 13.16(c), the Maxwell lines T_{ads}^{-1} [the slopes of the corresponding Gibbs constructions in Fig. 13.16(a)] are also inserted. The adsorption temperatures T_{ads} found by Gibbs/Maxwell construction depend roughly linearly on the surface attraction strengths, as has already been suggested by the adsorption transition line in the phase diagram shown in Fig. 13.12. Thus, the intersections of the Maxwell lines and the T^{-1} curves are identical with the extremal points in Fig. 13.16(b) and are located at the respective energies e_{ads}, e_{sep}, and e_{des}. It is obvious that the desorption energies per monomer, e_{des}, converge very quickly to a constant value $e_{\text{des}}^{\epsilon_s \to \infty} \approx -0.35$, when increasing the adhesion strength ϵ_s. On the other hand, the adsorption energies e_{ads} still change rapidly. In consequence, the latent heat per monomer, Δq, increases similarly fast with ϵ_s, i.e., the energetic gap between the coexisting macrostates becomes larger, as does the surface-entropic barrier Δs_{surf} [cf. Fig. 13.16(b)]. Since linearly depending on ϵ_s, e_{ads} and Δq trivially diverge for $\epsilon_s \to \infty$.

13.8.2 Chain-length dependence

Since the adsorption transition between DE and AE2 is expected to be of second order in the thermodynamic limit [315], the microcanonical effects and first-order signatures, as found for the finite system, must disappear for the infinitely large system $N \to \infty$. Therefore, it is also instructive to investigate the chain-length dependence of the microcanonical effects in comparative microcanonical analyses. Figure 13.17(a) shows the microcanonical entropies $s(e)$, the adsorption entropies $s_{\text{ads}}(e)$, and the desorption entropies $s_{\text{des}}(e)$ for

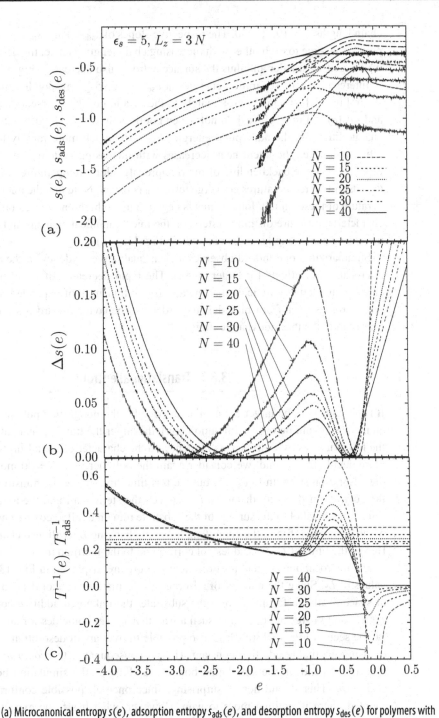

Fig. 13.17 (a) Microcanonical entropy $s(e)$, adsorption entropy $s_{ads}(e)$, and desorption entropy $s_{des}(e)$ for polymers with different chain lengths $N = 10 \ldots 40$ and fixed surface attraction strength $\epsilon_s = 5$ in the adsorption transition regime. The maximum of $s(e)$ and the "convex intruder" begin to disappear with increasing chain length – a first indication of the tendency of the adsorption transition to change its characteristics from first-order to second-order behavior in the thermodynamic limit; (b) deviations $\Delta s(e)$ of $s(e)$ from the Gibbs construction; (c) caloric inverse temperature curves $T^{-1}(e)$ and Maxwell lines at T^{-1}_{ads}, parametrized by chain length N. From [307].

chain lengths $N = 10, \ldots, 40$. The respective slopes of $s_{ads}(e)$ and $s_{des}(e)$ near the crossing points converge to each other with increasing chain length. Hence, the depth of the convex well is getting smaller and thus the surface entropy also decreases [Fig. 13.17(b)].

Interestingly, the separation energies at $e_{sep} \approx -0.95$ [corresponding to the maxima of Δs in Fig. 13.17(b) and approximately to the location of the intersection points of $s_{ads}(e)$ and $s_{des}(e)$ in Fig. 13.17(a)] do not depend noticeably on N. The desorption energies e_{des} change little, but the adsorption energies e_{ads} shift much more rapidly toward the separation point, i.e., the latent heat decreases with increasing system size. Consequently, in Fig. 13.17(c), the backbending of the (reciprocal) caloric temperatures is getting weaker; the adsorption temperatures converge toward a constant. Note that the microcanonical temperature of these finitely long chains is negative in the high-energy region. This is another characteristic feature of finite systems in the microcanonical analysis and disappears with increasing chain lengths, as can be seen in Fig. 13.17(c).

Summarizing, one indeed observes a clear qualitative tendency of the reduction of the microcanonical effects for larger chains. The rapid decreases of latent heat and surface entropy qualitatively indicate that the adsorption transition of expanded polymers (DE to AE2) crosses over from bimodal first-order-like behavior toward a second-order phase transition in the thermodynamic limit.

13.8.3 Translational entropy

It is obvious that the larger the available space for the nongrafted polymer (mimicked by an increased volume of the simulation box), the larger the translational entropy will be. In the following, we will discuss its influence on the character of the adsorption transition in more detail. To this end, we consider again the polymer with $N = 20$ monomers and set the adsorption strength to $\epsilon_s = 5$, but we this time investigate the transition properties of the polymer, if different distances L_z between the steric wall and the attractive substrate (directions parallel to the surface of the substrate remain unaffected) are chosen [307]. It is worth noting that fixing the chain length N, but changing L_z, will also change the density. Hence, the limit of $L_z \to \infty$ does not correspond to the thermodynamic limit.

The microcanonical quantities discussed before are displayed in Fig. 13.18 for various values of L_z. Since the number of adsorbed conformations cannot depend on the amount of space available far away from the substrate, the unknown additive constants in $s(e)$, $s_{ads}(e)$, and $s_{des}(e)$ are chosen in such a way that $s_{ads}(e)$ coincides for all values of L_z, as can be seen in Fig. 13.18(a). It is also possible to overlap all desorption entropies $s_{des}(e)$ by using a suitable additive constant. Hence, consequently, the *conformational* entropy does not depend on the simulation box size as long as the simulation box exceeds the chain size. This should not be surprising, since once all possible conformations can be adopted, there is nothing more to gain. However, what has to be expected is a change of *translational* entropy for desorbed conformations proportional to the logarithm of the box size. As Fig. 13.18(b) shows, both the surface entropy Δs_{surf} and the latent heat Δq increase with L_z. It is a significant qualitative difference compared to the previous analysis of changes by increasing the chain lengths N that the latent heat remains finite for large box

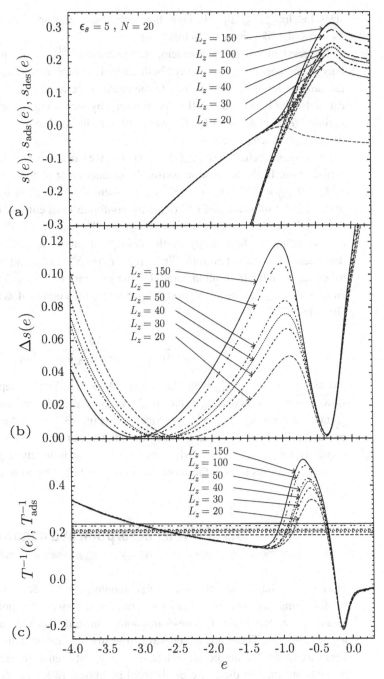

Fig. 13.18 (a) Microcanonical entropies and its fractions for adsorbed and desorbed conformations, $s_{ads}(e)$ and $s_{des}(e)$, for various box sizes $L_z = 20, \ldots, 150$. The shape of both fractions remains unchanged for different box sizes. Only the amount of desorbed conformations increases relative to adsorbed ones for larger boxes; (b) deviations from the respective Gibbs hulls $\Delta s(e)$. An increased $s_{des}(e)$ induces an increase of the surface entropy Δs_{surf} and slightly also of the latent heat Δq; (c) caloric inverse temperature curves $T^{-1}(e)$ and Maxwell lines, parametrized by the distance between attractive and steric wall L_z. From [307].

sizes, i.e., $\lim_{L_z \to \infty} \Delta q \neq 0$. Thus, the adsorption transition of the finite polymer preserves its first-order-like character in this limit.

The resultant caloric inverse temperature curves $T^{-1}(e)$, shown in Fig. 13.18(c), differ only in the energy regime, where both entropic contributions, $s_{\mathrm{ads}}(e)$ and $s_{\mathrm{des}}(e)$, are of the same order of magnitude, i.e., in the coexistence region. As before, the microcanonical first-order backbending effect is amplified by increasing L_z, and only in the transition regime $T(e)$ changes with L_z. The Maxwell lines inserted into Fig. 13.18(c) correspond to the adsorption temperatures.

Knowing the behavior of $s_{\mathrm{ads}}(e)$ and $s_{\mathrm{des}}(e)$, we can estimate T_{ads} by performing a Gibbs construction. In the adsorption phase, the contact point of the Gibbs hull is independent of L_z, $s(e_{\mathrm{ads}}) = s^{\mathrm{trans},\parallel}(e_{\mathrm{ads}}) + s^{\mathrm{conf}}(e_{\mathrm{ads}})$, where $s^{\mathrm{trans},\parallel}(e_{\mathrm{ads}})$ is the translational entropy parallel to the substrate and $s^{\mathrm{conf}}(e_{\mathrm{ads}})$ the conformational entropy of the adsorbed conformations. The other contact point $s(e_{\mathrm{des}}, L_z) = s^{\mathrm{trans},\perp}(e_{\mathrm{des}}, L_z) + s^{\mathrm{trans},\parallel}(e_{\mathrm{des}}) + s^{\mathrm{conf}}(e_{\mathrm{des}})$, corresponding to the entropy in the desorption phase, is a decomposition of the L_z-dependent translational entropy $s^{\mathrm{trans},\perp}(e_{\mathrm{des}}, L_z) = N^{-1} \ln L_z$, and the L_z-independent contributions from the translation parallel to the substrate $s^{\mathrm{trans},\parallel}(e_{\mathrm{des}})$ and the conformational entropy $s^{\mathrm{conf}}(e_{\mathrm{des}})$. The adsorption temperature is then obtained as the inverse slope of the Gibbs hull [307]:

$$T_{\mathrm{ads}} = \frac{\Delta q}{s^{\mathrm{conf}}(e_{\mathrm{des}}) - s^{\mathrm{conf}}(e_{\mathrm{ads}}) + N^{-1} \ln L_z}, \qquad (13.17)$$

where $\Delta q = e_{\mathrm{des}} - e_{\mathrm{ads}}$ is the latent heat. Although Δq also depends on L_z, its growth slows down with increasing values of L_z, because the minimal value of e_{ads} is the finite ground-state energy. Since $e_{\mathrm{des}} \approx \mathrm{const}$ [see Fig. 13.18(b)], Δq converges to a finite constant in the limit $L_z \to \infty$. The conformational entropy $s^{\mathrm{conf}}(e)$ is L-independent. Consequently, $\lim_{L_z \to \infty} T_{\mathrm{ads}} = 0$, which means that for an infinitely large upper half-space and thermal conditions ($T > 0$) the substrate is not recognized by a nongrafted polymer.

13.9 Polymer adsorption at a nanowire

Thus far, we have only considered flat, topologically three-dimensional substrates. In the following, we will investigate conformational phases of a polymer interacting with an attractive, topologically one-dimensional substrate such as a nanostring or nanowire [321–323]. There are many ways to abstract a real system to a nanowire model. There are fabricated atomic nanowires, of course, but even a single stretched polymer with the ends attached to dielectric beads fixed by optical tweezers could be considered as a molecular nanowire. Carbon nanotubes with small radii can also be considered nanowires in an appropriate model [324]. As we will see, one of the most interesting properties of such a polymer–wire complex is that under certain conditions the polymer crystallizes in stable cylindrical shapes with monomer alignments that themselves resemble atomic arrangements known from single-walled carbon or boron nanotubes.

In this conformational phase, the cylindrical polymer hull surrounds the thin wire such that the interior is free of particles. In a hydrodynamic application, for example, molecules could still flow through it. Since the axis of a polymeric tube is always oriented parallel to the direction of the wire, the growth direction of the tube can be controlled. Therefore, one could imagine constructing complex tube systems, which allows for applications beyond those known for conventional nanotubes. Another conceivable application of this structural coincidence is the *systematic* stabilization or functionalization of nanotubes by polymer coating.

13.9.1 Modeling the polymer–nanowire complex

Since we are interested in generic properties of such a system, we model the nanowire as an attractive line-like substrate and employ a coarse-grained semiflexible bead-stick model for the polymer consisting of N monomers. The chain is not grafted to this string and may move freely. The total energy of the system includes three contributions,

$$E = \sum_{i=1}^{N-2} \sum_{j=i+2}^{N} V_{\mathrm{LJ}}(r_{ij}) + \sum_{i=2}^{N-1} V_{\mathrm{bend}}(\cos\theta_i) + \sum_{i=1}^{N} V_{\mathrm{string}}(r_{\perp;i}). \tag{13.18}$$

The interaction between nonadjacent monomers is governed by the standard Lennard-Jones potential,

$$V_{\mathrm{LJ}}(r_{ij}) = 4\epsilon_{\mathrm{m}} \left[\left(\frac{\sigma_{\mathrm{m}}}{r_{ij}} \right)^{12} - \left(\frac{\sigma_{\mathrm{m}}}{r_{ij}} \right)^{6} \right], \tag{13.19}$$

with the distance r_{ij} between nonbonded monomers i and j. The monomer–monomer interaction parameters ϵ_{m} and σ_{m} are set to unity in the following. The bending potential between adjacent bonds reads

$$V_{\mathrm{bend}}(\cos\theta_i) = \kappa(1 - \cos\theta_i), \tag{13.20}$$

with the bending stiffness set to $\kappa = 1/4$. The bending angle θ_i is defined by the covalent bonds connected to the ith monomer. The monomer–string energy is obtained by continuously integrating a standard LJ potential over the string, which may extend infinitely in z direction [321],

$$V_{\mathrm{string}}(r_{\perp;i}) = 4\eta_{\mathrm{f}}\epsilon_{\mathrm{f}} \int_{-\infty}^{\infty} dz \left[\frac{\sigma_{\mathrm{f}}^{12}}{(r_{\perp;i}^2 + z^2)^6} - \frac{\sigma_{\mathrm{f}}^{6}}{(r_{\perp;i}^2 + z^2)^3} \right] = \pi\,\eta_{\mathrm{f}}\epsilon_{\mathrm{f}} \left(\frac{63}{64} \frac{\sigma_{\mathrm{f}}^{12}}{r_{\perp;i}^{11}} - \frac{3}{2} \frac{\sigma_{\mathrm{f}}^{6}}{r_{\perp;i}^{5}} \right), \tag{13.21}$$

where the distance of the ith monomer perpendicular to the string is denoted by $r_{\perp;i}$, and σ_{f} is the van der Waals radius of the string. It can be considered as its effective "thickness" and it is related to the minimum distance r_z^{min} of the monomer–string potential:

$$r_z^{\mathrm{min}} = \left(\frac{693}{480} \right)^{1/6} \sigma_{\mathrm{f}} \approx 1.06\sigma_{\mathrm{f}}. \tag{13.22}$$

The string "charge" density η_f is conveniently chosen in such a way that the minimum monomer–wire energy is $V_{string}(r_{\perp;i}) = -\epsilon_f$, independently of σ_f. This is satisfied by setting $\eta_f = r_z^{min}/2\sigma_f^2 \approx 0.53\sigma_f^{-1}$.

Alternatively, the monomer–string energy can be considered as the limiting case of the interaction of a monomer with a cylinder of radius $R \to 0$, keeping the overall LJ "charge" fixed.

13.9.2 Structural phase diagram

It is reasonable to expect that the structural properties of global energy minima of the polymer–nanowire complex will depend on the effective thickness σ_f of the wire and on the interaction strength ϵ_f between the monomers and the wire. As an example, we will systematically analyze the conformational phases of a polymer with $N = 100$ monomers in the following, but the qualitative features do not strongly depend on the system size. For the identification of lowest-energy conformations, stochastic generalized-ensemble Monte Carlo methods, such as those described in Section 4.6, can be used. The energetic random-walk-like sampling of these methods makes it also possible to access the entropically highly suppressed low-energy conformations.

Figure 13.19 shows the conformational ϵ_f-σ_f phase diagram of lowest-energy conformations. Representative adsorbed polymer structures are also depicted.[8] In the phase diagram, four major structural phases can be identified. "Phase" denotes a domain in the parameter space, where the corresponding conformations share qualitatively the same morphology. For weak attraction ($\epsilon_f \lesssim 3$), two types of crystalline droplets adhered to the string (regions Ge, Gi) can be clearly distinguished. Either the string axis is included inside the droplet (Gi) or passes by externally (Ge). If the adhesion strength ϵ_f of the string increases, compact droplets in Gi melt near $\epsilon_f \approx 3$ and phase B is entered, where polymer conformations extend along the string axis. Near $\epsilon_f \approx 4.5$, a crossover (dashed line in Fig. 13.19) from the multilayer barrel structures to monolayer conformations with strong similarities to single-walled nanotubes occur. This crossover corresponds to a topological transition between three-dimensional compact crystalline and two-dimensional film-like structures. For sufficiently large values of the effective string thickness σ_f, the polymer layers do not completely wrap the tube and stable crescent-shaped "clamshell-like" [325] structures dominate in the region denoted by C.

In region Gi, compact spherical conformations dominate. Increasing the attraction strength ϵ_f in this regime while keeping the effective thickness σ_f fixed, conformations lose their spherical shape at $\epsilon_f \approx 3$ and the cylindrical phase B is entered. This transition can be best characterized by introducing an asymmetry parameter based on the gyration tensor components parallel and perpendicular to the string, $A = r_{\parallel}^{gyr}/r_{\perp}^{gyr} - 1$. In Fig. 13.20, this order parameter is shown as a function of ϵ_f for $\sigma_f = 0.5$. In the spherical regime Gi,

---

[8] This structural phase diagram is a result of extensive analyses of structural properties for more than 150 low-energy conformations with different values of interaction parameters [321].

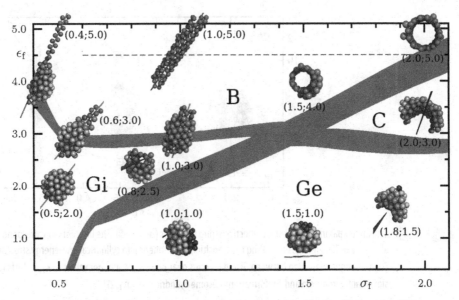

Fig. 13.19 Low-energy structural phases of a polymer–nanowire complex, parametrized by the effective string thickness σ_f and by the string attraction strength ϵ_f. Gray bands (widths correspond to uncertainty) indicate transition lines between compact, crystalline polymer structures with the wire included (Gi) or excluded (Ge), crescent-shaped (C), and barrel-like (B) conformations. The dashed line indicates a topological crossover that separates mono- and multilayer regions. Inset pictures show representative conformations. Monomers with the same coloring (or shadings) belong to the same layer. From [321].

$A \approx 0$. As expected, A increases for $\epsilon_f \gtrsim 3$ and the structures become asymmetric. At this point, it is equally favorable for a monomer to stick to the string, or to form contacts to neighboring monomer layers. The conformations stretch along the string until they form a maximally compact monolayer tube surrounding the string for $\epsilon_f \gtrsim 4.5$.

If the effective thickness σ_f is increased for constant values of the attraction strength $\epsilon_f < 3$, the transition from Gi to Ge is characterized by the different locations of the string relative to the globular droplet. Either it is enclosed in the globule (Gi) or it is excluded and only touches the compact polymer (Ge). For small values of ϵ_f, the transition point can be estimated by assuming a tetrahedral monomer-packing in the crystalline droplet. Then, the radius of a circumsphere of a tetrahedron is $r_o \approx 0.61$, which corresponds to a limiting effective string thickness $\sigma_{f,o} \approx 0.58$. Thus, inserting a string with $\sigma_f < \sigma_{f,o}$ does not break intra-monomer contacts within a compact structure. Above this limiting value, however, the string would cause an energetically disfavored replacement of monomers inside the conformation and is hence "pushed" out of the droplet.

Quantitatively, this transition can be identified by measuring the opening angle α of a given conformation. Projecting the positions of monomers in contact with the string onto a plane perpendicular to the string, α is defined as the angle between the string and two monomers that spans the largest region of the plane with no monomers residing therein. Thus, roughly, conformations with $\alpha < \pi$ correspond to conformations enclosing the string

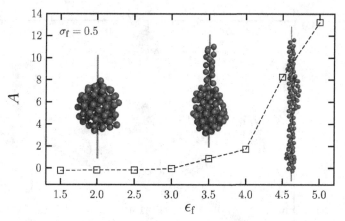

Fig. 13.20 Asymmetry parameter A at small effective string thickness ($\sigma_f = 0.5$). The insets show corresponding conformations at $\epsilon_f = 2.0, 3.5$, and 4.5, illustrating the transition from spherical to cylindrical low-energy structures. The conformational transition between Gi and B occurs near $\epsilon_f = 3.0$. For larger values of ϵ_f, A starts to deviate significantly from zero and conformations become cylindrical. From [321].

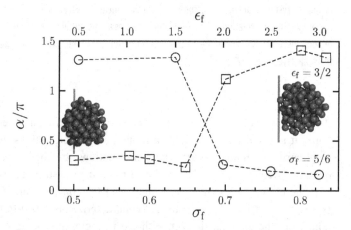

Fig. 13.21 Opening angle α of low-energy conformations for $\epsilon_f = 3/2$ as a function of σ_f (squares, lower scale) and for $\sigma_f = 5/6$ as a function of ϵ_f (circles, upper scale). The inset pictures show corresponding conformations at $\epsilon_f = 3/2$ and $\sigma_f = 1/2, 4/5$, illustrating the separation of the droplet from the string. From [321].

(Gi), whereas $\alpha > \pi$, if the string is located outside the droplet (Ge). Figure 13.21 shows how α changes when crossing the transition line Gi↔Ge horizontally or vertically. Fixing the attraction strength at $\epsilon_f = 3/2$, α increases rapidly from $\sigma_f \approx 0.65$ (squares, lower scale), which is close to the estimate $\sigma_{f,o}$ given above. For larger values of the effective thickness, the string is shifted outward to retain optimal monomer packing. The inset pictures show lowest-energy conformations at $\epsilon_f = 3/2$ and $\sigma_f = 1/2$ (representative for phase Gi) and $\sigma_f = 4/5$ (Ge), respectively. On the other hand, increasing the string attraction strength ϵ_f

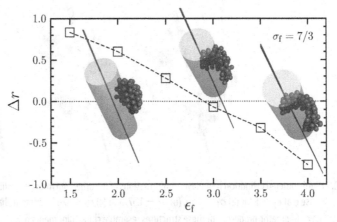

Fig. 13.22 Distance Δr of the center of mass of the polymer from the virtual surface of the cylinder with the radius that corresponds to the minimum position of the string potential ($r_\perp^{min} \approx 1.06\sigma_f$) for $\sigma_f = 7/3$. The intersection of the curve with the dotted line ($\Delta r = 0$) at $\epsilon_f = 2.9$, where the center of mass equals the radius of this cylinder, defines the transition from Ge to C. Pictures show conformations at $\epsilon_f = 2$, 3, and 4. From [321].

while $\sigma_f = 5/6$ is fixed (circles, upper scale), the inclusion of the string, accompanied by a rapid decrease of α, occurs at $\epsilon_f \approx 1.75$.

Starting in Ge and increasing σ_f and ϵ_f above certain threshold values results in the transition toward adsorbed curved conformations (C). The polymer begins to wrap the string and different monomer layers form. It is convenient to quantitatively define this transition to occur at the point where the distance of the center of mass of the polymer from the string, $r_\perp^{com} = N^{-1} | \sum_{i=1}^{N} \vec{r}_{\perp,i} |$, equals the monomer–string potential minimum distance $r_\perp^{min} (\approx 1.06\sigma_f)$, i.e., at $\Delta r = r_\perp^{com} - r_\perp^{min} = 0$. Qualitatively, the center of mass intrudes into the virtual cylinder with radius $r_\perp^{min}$, defined by the inner layer of monomers. In Fig. 13.22, Δr is plotted as a function of ϵ_f at $\sigma_f = 7/3$. The transition point $\Delta r = 0$ is marked by the dotted line, which is intersected at $\epsilon_f = 2.9$, in correspondence to the Ge$\leftrightarrow$C transition line in the phase diagram in Fig. 13.19. The inset pictures represent conformations with $\Delta r = 0.6, -0.1, -0.8$ at $\epsilon_f = 2, 3, 4$.

Finally, increasing ϵ_f further, region B is entered, i.e., ground-state polymer conformations wrap the string completely. If the attraction between a monomer and the string becomes stronger than the interaction between stacked, neighboring monomer layers, regular monolayer films surrounding the string are formed. These resemble single-walled tubes with an ordered arrangement of monomers. Different chiral orientations of the wrapping compete with each other in correspondence with the length scale σ_f associated with the monomer–string interaction σ_f. This structural restriction is very similar to the atomic alignment in boron or carbon nanotubes [324]. Defining the wrapping vector $\mathbf{c} = n\mathbf{a}_1 + m\mathbf{a}_2$, with $\mathbf{a}_1$ and $\mathbf{a}_2$ being the unit vectors of the structure, the limiting "armchair" and "zigzag" structures, corresponding to $m = n$ and $m = 0$, are found for the polymer tubes as well. Examples of nanotube-like polymer conformations with different chiralities are shown for $\sigma_f = 1.50, 1.57$ in Figures 13.23(a) and (b), respectively, and for $\sigma_f = 0.65$ in Fig. 13.23(c) (all at $\epsilon_f = 5$). The alignment of monomers in Fig. 13.23(a) is almost parallel to the string,

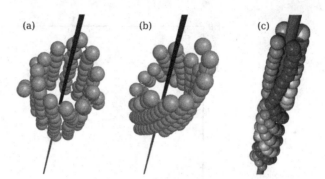

Fig. 13.23 Highly ordered cylindrical monolayer conformations of adsorbed polymers with different wrappings in the barrel phase B at $\epsilon_f = 5$ for (a) $\sigma_f = 1.50$, (b) $\sigma_f = 1.57$, and (c) $\sigma_f = 0.65$ (different shadings facilitate the perception only). Geometric properties of these structures resemble chiral alignments of atomic structures known from single-walled nanotubes. From [321].

whereas the conformation in Fig. 13.23(b) exhibits a noteworthy chiral winding. If the radius of the monolayer polymer tube does not allow for a perfect monomer alignment, defects occur and cause the formation of structural domains with different chiralities within the same conformation, as in the example shown in Fig. 13.23(c).

14 Hybrid protein–substrate interfaces

14.1 Steps toward bionanotechnology

The advancing progress in manipulating soft and solid materials at the nanometer scale opens up new vistas for potential bionanotechnological applications of hybrid organic-inorganic interfaces [278, 326]. This includes, e.g., nanosensors being sensitive to specific biomolecules ("nanoarrays"), as well as organic electronic devices on polymer basis which have, for example, already been realized in organic light-emitting diodes [327]. An important development in this direction is the identification of proteins that can bind to specific compounds. Over the last decade, genetic engineering techniques have been successfully employed to find peptides with affinity for, e.g., metals [273, 328], semi-conductors [274–276], and carbon nanotubes [329]. However, the mechanisms by which peptides bind to these materials are not completely understood; it is, for example, unclear what role conformational changes play in the binding process.

In these mainly experimental studies, it has also been shown that the binding of peptides on metal and semiconductor surfaces depends on the types of amino acids [330] and on the sequences of the residues in the peptide chain [273–276]. These experiments reveal many different interesting and important problems, which are related to general aspects of the question why and how proteins fold. For example, this pertains to the character of the adsorption process, i.e., whether the peptides simply dock to the substrate without notice-able structural changes or whether they perform conformational transitions while binding. Another point is how secondary structures of peptide folds in the bulk influence the binding behavior to substrates. In helical structures, for example, side chains are radially directed and – due to the helical symmetry – residues with a certain distance in the sequence arrange linearly. But, under certain conditions (e.g., in the presence of energetically attractive sub-strates) it could be more favorable, if the protein prefers to take rather flat conformations in order to increase the number of surface contacts.

If native folds resist refolding due to rather little attraction of substrate surfaces, the adsorption propensity is either weak, or a perfect matching of protein structure and the sur-face geometry of the crystal, e.g., its crystal orientation, is required. In this case, the peptide simply docks to the substrate. It is also feasible, however, that the peptide is unstructured under given conditions (temperature, pH value of the surrounding solvent) and the bind-ing process to the substrate is accompanied by refolding processes, i.e., the adsorption process is a coupled binding-folding scenario [287, 331, 332]. Peptide-specific affinity to dock or refold when binding [333, 334] is of great importance in pattern recognition pro-cesses [283, 335–339] such as receptor-ligand binding [285, 286]. From experiments of the adsorption of short peptides at semiconductor substrates, it is known that different surface properties (materials such as Si or GaAs, crystal orientation, etc.) as well as different amino acid sequences strongly influence the binding properties of these peptides at the substrate [274–276, 340]. This specificity is of particular importance for the design of sensory devices and pattern recognition at the nanometer scale. The understanding of the basic principles of substrate–peptide cooperativity and the reasons for binding specificity are major challenges for both experiment and the theory. This problem can be considered as being embedded into a class of similar problems, where the adsorption and docking behavior of polymers is essential, e.g., protein–ligand binding [285], prewetting and layering transitions in polymer solutions as well as dewetting of polymer films [279], molecular pattern formation, electrophoretic polymer deposition and growth [289].

In the following, we first discuss the *substrate specificity* of heteropolymer adhesion by employing a simple hybrid lattice model [284, 303, 306, 309, 318]. After having gained this qualitative insight into the structural binding behavior of heteropolymers, an exemplary realistic hybrid peptide–semiconductor system is investigated to verify the *sequence specificity* of peptide adsorption [274–276, 340].

14.2   Substrate-specific peptide adsorption

14.2.1   Hybrid lattice model

As an example for the importance of substrate-specific properties upon polymer adsorption, we will now investigate conformational transitions of a nongrafted hydrophobic–polar heteropolymer with 103 residues in the vicinity of different substrates [303].

For the study of hybrid peptide-substrate models, we use the HP transcription of the 103-residue protein *cytochrome c*, whose low-energy conformations and thermodynamic properties were extensively studied in the past [38, 40, 43, 44, 341]. The HP sequence contains 37 hydrophobic and 66 polar residues. This lattice peptide resides in a cavity with an attractive substrate (see Fig. 13.1). The distance between the attractive and the steric wall, z_w, is chosen sufficiently large to keep the influence on the unbound heteropolymer small (in the actual example $z_w = 200$). In order to study the specificity of residue binding, we will distinguish three substrates with different affinities to attract the peptide monomers: (a) the type-independent attractive, (b) the hydrophobic, and (c) the polar substrate. The number of corresponding nearest-neighbor contacts between monomers and substrate shall be denoted as n_s^{H+P}, n_s^H, and n_s^P, respectively. In analogy to the polymer-substrate model (13.1), the energy of the hybrid peptide-substrate system can simply be expressed as [303]

$$E_s(n_s, n_{HH}) = -\varepsilon_0(n_s + sn_{HH}), \qquad (14.1)$$

where $n_s = n_s^{H+P}$, n_s^P, or n_s^H, depending on the substrate (we set $\varepsilon_0 = 1$ in the following). n_{HH} denotes the number of intrinsic nearest-neighbor contacts between hydrophobic monomers only. Nearest-neighbor pairs of polar monomers (*PP*) and contacts between polar and hydrophobic residues (*HP*) are considered to be non-energetic in this model (as in the standard HP model). The solubility (or reciprocal solvent parameter) s is, as well as the temperature T, an external parameter. Contact-density chain-growth simulations allow a direct estimation of the degeneracy (or density) $g(n_s, n_{HH})$ of macrostates of the system with given contact numbers n_s and n_{HH} [303].

14.2.2 Influence of temperature and solubility on substrate-specific peptide adsorption

In Figs. 14.1(a)–14.1(c) the gray-scale-encoded profiles of the specific heats for the different substrates are shown (the brighter the larger the value of C_V). The ridges (for accentuation marked by white and gray lines) indicate boundaries of the pseudophases. The gray lines belong to the main transition lines, while the white lines separate pseudophases that strongly depend on specific properties of the heteropolymer, such as its exact number and sequence of hydrophobic and polar monomers. By means of the contact density $g(n_s, n_{HH})$, we can introduce the contact free energy as ($k_B \equiv 1$)

$$F_{T,s}(n_s, n_{HH}) = E_s(n_s, n_{HH}) - T \ln g(n_s, n_{HH}). \tag{14.2}$$

The probability for a macrostate with n_s substrate and n_{HH} hydrophobic contacts is given by $p_{T,s}(n_s, n_{HH}) \sim g(n_s, n_{HH}) \exp(-E_s/T)$. Assuming that the minimum of the free-energy landscape $F_{T,s}(n_s^{(0)}, n_{HH}^{(0)}) \to$ min for given external parameters s and T is related to the class of macrostates with $n_s^{(0)}$ surface and $n_{HH}^{(0)}$ hydrophobic contacts, this class dominates the phase the system resides in. For this reason, it is instructive to calculate all minima of the contact free energy and to determine the associated contact numbers in a wide range of values for the external parameters.

The map of all possible free-energy minima in the range of external parameters $T \in [0, 10]$ and $s \in [-2, 10]$ is shown in Fig. 14.2 for the peptide in the vicinity of a substrate that is equally attractive for both hydrophobic and polar monomers. Solid lines visualize "paths" through the free-energy landscape when changing temperature under constant solvent ($s = $ const) conditions. Let us follow the exemplified trajectory for $s = 2.5$.

Starting at very low temperatures, we know from the pseudophase diagram in Fig. 14.1(a) that the system resides in pseudophase AC1. This means that the macrostate of the peptide is dominated by the class of compact, film-like single-layer conformations. The system obviously prefers surface contacts at the expense of hydrophobic contacts. Nonetheless, the formation of compact hydrophobic domains in the two-dimensional topology is energetically favored, but maximal compactness is hindered by the steric influence of the substrate-binding polar residues.

Increasing the temperature, the system experiences close to $T \approx 0.35$ a sharp first-order-like conformational transition, and a second layer forms (AC2). This is a mainly entropy-driven transition as the extension into the third dimension perpendicular to the substrate surface increases the number of possible peptide conformations. Furthermore, the loss of energetically favored substrate contacts of polar monomers is partly compensated by the energetic gain due to the more compact hydrophobic domains. Increasing the temperature further, the density of the hydrophobic domains reduces and overall compact conformations dominate in the globular pseudophase AG. Reaching AE, the number of

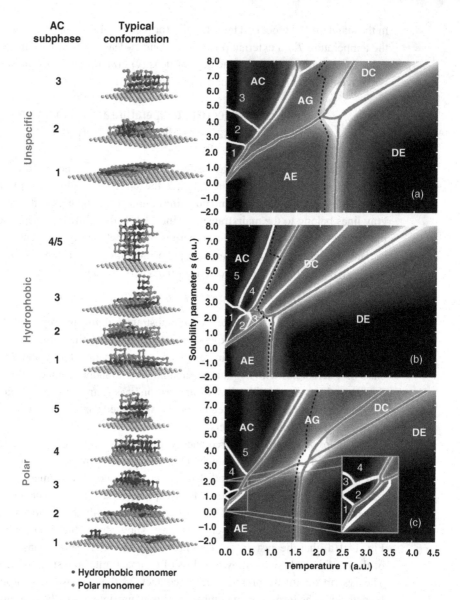

Fig. 14.1 Specific-heat profiles as a function of temperature T and solubility parameter s of the 103-mer near three different substrates that are attractive for (a) all, (b) only hydrophobic, and (c) only polar monomers. White lines indicate the ridges of the profile. Gray lines mark the main "phase boundaries." The dashed black line represents the first-order-like binding/unbinding transition state, where the contact free energy possesses two minima (the adsorbed and the desorbed state). In the left panel typical conformations dominating the associated AC phases of the different systems are shown. After [303].

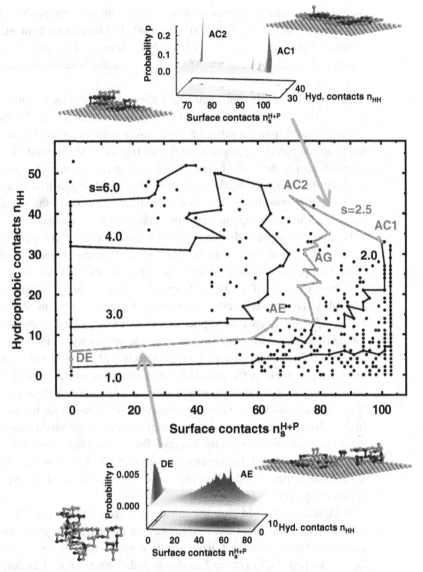

Fig. 14.2 Contact-number map of all free-energy minima for the 103-mer and substrate equally attractive to all monomers. Full circles correspond to minima of the contact free energy $F_{T,s}(n_s^{H+P}, n_{HH})$ in the parameter space $T \in [0, 10]$, $s \in [-2, 10]$. Lines illustrate how the contact free energy changes with the temperature at constant solvent parameter s. For the exemplified solvent with $s = 2.5$, the peptide experiences near $T = 0.35$ a sharp first-order-like layering transition between single- and double-layer conformations (AC1,2). Passing the regimes of adsorbed globules (AG) and expanded conformations (AE), the discontinuous binding/unbinding transition from AE to DE happens near $T = 2.14$. In the DE phase the ensemble is dominated by desorbed, expanded conformations. Representative conformations of the phases are shown next to the respective peaks of the probability distributions.

hydrophobic contacts decreases further, along with the total number of substrate contacts. Extended, dissolved conformations dominate. The transitions from AC2 to AE via AG are comparatively "smooth", i.e., no immediate changes in the contact numbers passing the transition lines are noticed. Therefore, these conformational transitions can be classified as second-order-like.

The situation is different when approaching the unbinding transition line from AE close to $T \approx 2.14$. This transition is accompanied by a dramatic loss of substrate contacts – the peptide desorbs from the substrate and behaves in pseudophase DE like a free peptide, i.e., the substrate and the opposite neutral wall regularize the translational degree of freedom perpendicular to the walls, but rotational symmetries are unbroken (at least for conformations not touching one of the walls). As the probability distribution in Fig. 14.2 shows, the unbinding transition is also first-order-like, i.e., close to the transition line, there is a coexistence of adsorbing and desorbing classes of conformations.

Despite the surprisingly complex phase behavior there are main "phases" that can be distinguished in all three systems. These are separated in Figures 14.1(a)–14.1(c) by gray lines. Comparing the three systems we find that they all possess pseudophases, where adsorbed compact (AC), adsorbed expanded (AE), desorbed compact (DC), and desorbed expanded (DE) conformations dominate. "Compact" here means that the heteropolymer has formed a dense hydrophobic core, while expanded conformations are dissolved, random-coil-like. The sequence and substrate specificity of heteropolymers generates, of course, a rich set of new interesting and selective phenomena not available for homopolymers. One example is the pseudophase of adsorbed globules (AG), which is noticeably present only in those systems where all monomers are equally attractive to the substrate [Fig. 14.1(a)] and where polar monomers favor contact with the surface [Fig. 14.1(b)]. In this phase, the conformations are intermediates in the binding/unbinding region. This means that monomers currently desorbed from the substrate have not yet found their position within a compact conformation. Therefore, the hydrophobic core, which is smaller than in the respective adsorbed phase (i.e., at constant solubility s), appears as a loose cluster of hydrophobic monomers.

In Figures 14.3(a)–14.3(c), the statistical averages of the contact numbers n_s and n_{HH} as well as their variances and covariances for the three systems are plotted for $s = 2$. For comparison the specific heat, whose peaks correspond to the intersected transition lines of Figures 14.1(a)–14.1(c) at $s = 2$, is also included. From Figures 14.3(a) and 14.3(c) we see that the transition from AC to AG near $T \approx 0.4$ is mediated by fluctuations of the intrinsic hydrophobic contacts. The very dense hydrophobic domains in the AC subphases lose their compactness. This transition is absent in the hydrophobic-substrate system [Fig. 14.3(b)]. The signal seen belongs to a hydrophobic layering AC subphase transition, which influences mainly the number of surface contacts n_s^H. The second peak of the specific heats belongs to the transition between adsorbed compact or globular (AC/AG) and expanded (AE) conformations. This behavior is similar in all three systems. Remarkably, it is accompanied by a strong anti-correlation between surface and intrinsic contact numbers, n_s and n_{HH}. Not surprisingly, the hydrophobic contact number n_{HH} fluctuates more strongly than the number of surface contacts, but apparently in a different way. Dense conformations with hydrophobic core (and therefore many hydrophobic contacts) possess a relatively

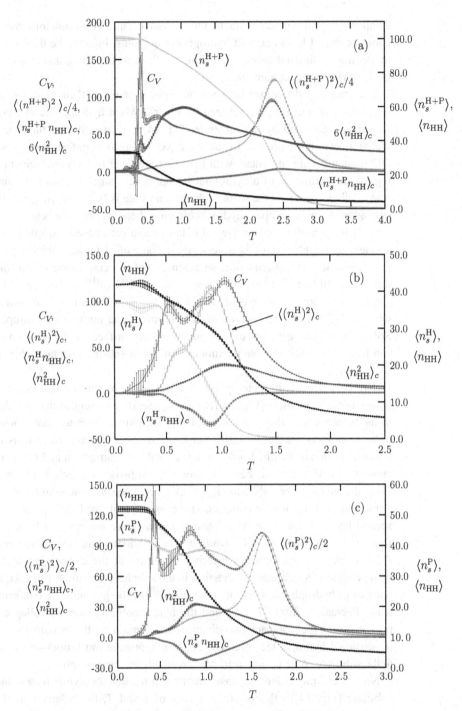

Fig. 14.3 Temperature dependence of specific heat, correlation matrix components, and contact number expectation values of the 103-mer for surfaces attractive for (a) all, (b) only hydrophobic, and (c) only polar monomers at $s = 2$. From [303].

small number of surface contacts. On the other hand, conformations with many surface contacts cannot form compact hydrophobic domains. Finally, the third specific heat peak marks the binding/unbinding transition, which is, as expected, due to a strong fluctuation of the surface contact number.

The strongest difference between the three systems is their behavior in pseudophase AC, which is roughly parametrized by $s > 5T$. When hydrophobic and polar monomers are equally attracted by the substrate [Fig. 14.1(a)], we observe three AC subphases in the parameter space plotted. In subphase AC1, film-like conformations dominate, i.e., all 103 monomers are in contact with the substrate. Due to the good solvent quality in this region, the formation of a hydrophobic core is less attractive than the maximal deposition of all monomers at the surface, the ground state is $(n_s^{H+P}, n_{HH})_{min} = (103, 32)$. In fact, instead of a single compact hydrophobic core there are nonconnected hydrophobic clusters. At least on the used simple cubic lattice and the chosen sequence, the formation of a single hydrophobic core is necessarily accompanied by an unbinding of certain polar monomers and, in consequence, an extension of the conformation into the third spatial dimension. In fact, this happens when entering AC2 [$(n_s^{H+P}, n_{HH})_{min} = (64, 47)$], where a single hydrophobic two-layer domain has formed at the expense of losing surface contacts. In AC3, the heteropolymer has maximized the number of hydrophobic contacts. Solely, local arrangements of monomers on the surface of the very compact structure can lead to the still possible maximum number of substrate contacts. $F_{T,s}$ is minimal for $(n_s^{H+P}, n_{HH})_{min} = (40, 52)$.

The behavior of the heteropolymer adsorbed at a surface that is only attractive to hydrophobic monomers [Fig. 14.1(b)] is apparently different in the AC phase, compared to the behavior near the type-independently attractive substrate. Since surface contacts of polar monomers are energetically not favored, the subphase structure is determined by the competition of two hydrophobic forces: substrate attraction and formation of intrinsic contacts. In AC1, the number of hydrophobic substrate contacts is maximal for the single hydrophobic layer, $(n_s^{HH}, n_{HH})_{min} = (37, 42)$. The *single* two-dimensional hydrophobic domain is also maximally compact, at the expense of displacing polar monomers into a second layer. In subphase AC2 intrinsic contacts are entropically broken with minimal free energy for $35 \leq n_{HH} \leq 40$, while $n_s^{HH} = 37$ remains maximal. Another AC subphase, AC3, exhibits a hydrophobic layering transition at the expense of hydrophobic substrate contacts. Much more interesting is the subphase transition from AC1 to AC5. The number of hydrophobic substrate contacts n_s^{HH} of the ground-state conformation dramatically decreases (from 37 to 4) and the hydrophobic monomers collapse in a one-step process from the compact two-dimensional domain to the maximally compact three-dimensional hydrophobic core. The conformations are mushroom-like structures grafted at the substrate. AC4 is similar to AC5, with advancing desorption.

No less exciting is the subphase structure of the heteropolymer interacting with a polar substrate [Fig. 14.1(c)]. For small values of s and T, the behavior of the heteropolymer is dominated by the competition between polar monomers contacting the substrate and hydrophobic monomers favoring the formation of a hydrophobic core, which, however, also requires cooperativity of the polar monomers. In AC1, film-like conformations ($n_s^P = 66$, $n_{HH} = 31$) with disconnected hydrophobic clusters dominate. Entering AC2, hydrophobic contacts are energetically favored and a second hydrophobic layer forms at the expense of a reduction of polar substrate contacts [$(n_s^P, n_{HH})_{min} = (61, 37)$]. In AC3, the upper layer is mainly hydrophobic [$(n_s^P, n_{HH})_{min} = (53, 45)$], while the poor quality of the solvent (s large) and the comparatively strong hydrophobic force let the conformation further collapse [AC4: $(n_s^P, n_{HH})_{min} = (42, 52)$] and the steric cooperativity forces more polar monomers to break the contact to the surface and to form a shell surrounding the hydrophobic core [$(n_s^P, n_{HH})_{min} = (33, 54)$ in AC5].

After these general considerations on the adsorption behavior of heteropolymers on simple-cubic lattices, we will investigate sequence-specific adhesion properties of semiconductor-binding synthetic peptides in the following.

14.3 Semiconductor-binding synthetic peptides

Let us now discuss a real-world example of some synthetic peptides that exhibit distinct binding properties to semiconductor substrates. For this purpose, we are going to study the solution and adsorption behavior of synthetic 12-residue peptides, whose adhesion properties to surfaces of GaAs and Si crystals were the subject of extensive experiments [274–276, 340].

One of the major results was that the adsorption of these peptides to semiconductor substrates strongly depends on intrinsic properties of the hybrid peptide-semiconductor system. It was found, for example, that by testing a library of about a billion random peptides containing 12 amino acids, only a few sequences could be extracted that exhibit a particularly good binding behavior at gallium arsenide (GaAs) surfaces [274]. One of these sequences, AQNPSDNNTHTH,[1] was investigated in more detail in further atomic-force microscopy (AFM) experiments with respect to different semiconductor substrates, crystal orientations, and sequence mutation and permutation [275, 276, 340]. It is this sequence and its variations that we want to look at more carefully. The sequences of these peptides are listed in Table 14.1.

A suitable quantity for the determination of the binding affinity that can be obtained by cluster analysis of AFM images is the peptide adhesion coefficient (PAC), defined as the percentage of surface coverage, after drying and washing of the samples that were originally in contact with the peptide solution [275, 276]. To reduce the dependence on the peptide concentration, it is useful to introduce the calibrated PAC (cPAC). It is defined as the ratio of PACs measured for the binding of peptides to silicon (Si) and GaAs substrates with (100) surface crystal orientation[2] under identical conditions:

$$cPAC_{GaAs(100)}^{Si(100)} = PAC_{Si(100)}/PAC_{GaAs(100)}. \qquad (14.3)$$

---

[1] For the amino acid code, see Fig. 1.2.

[2] (100) is the so-called Miller index associated with the crystal orientation of a substrate. The crystal orientation at the surface of the substrate is also an important factor that influences the binding affinity of molecules [275, 276].

Table 14.1 Sequences and their cPAC values for adsorption to Si(100) surfaces [276].

Label	Sequence	$cPAC^{Si(100)}_{GaAs(100)}$
S1	AQNPSDNNTHTH	0.04
S2	AQNPSDNNTATA	0.21
S3	TNHDHSNAPTNQ	0.94
S4	AQAPSDAATHTH	0.29

GaAs(100) is chosen as a reference substrate, because the peptides listed in Table 14.1 bind comparatively well to this substrate. The cPAC values for the peptides discussed here in more detail are also given in Table 14.1.

The peptide S1, which was found to possess strong adhesion properties at GaAs(100) surfaces [274, 275], binds poorly to Si(100). This is reflected by the small cPAC value 0.04. Exchanging the basic histidines (H) against the nonpolar alanines (A), resulting in sequence S2, leads to a slightly reduced binding affinity to GaAs(100) (which can be considered as a polar substrate), but the adsorption strength to bare, non-oxidized Si(100) (which typically behaves "hydrophobic") increases noticeably (cPAC ≈ 0.21). The most dramatic effect observed in the experiments was that of a certain "random" permutation (S3) of the original sequence S1 that made the peptide almost equally attractive to the Si(100) and GaAs(100) substrates (cPAC ≈ 0.94) [275]. Noteworthy but much less pronounced is the improved binding affinity, if the asparagines (N) of S1 are replaced by alanines (A), as in sequence S4.

The main question is how these differences in binding behavior occur. From the experiments, it is known that the bound peptides form clusters on the substrate [276], but it is unlikely that the peptides aggregate before binding to the surface, because the hydrophobicity of the peptides studied is low and the peptide concentration was extremely low, in the nanomolar range. Measurements of circular dichroism (CD) spectra suggest that all four experimentally studied small peptides are, as expected, largely unstructured in solution [276].

It has been argued that the adhesion propensity of peptides to various surfaces can be partially explained in terms of individual adhesion properties of their constituent amino acids [277, 342]. However, the amino acid composition alone cannot explain the PAC values obtained experimentally for the peptides studied here. In fact, some of these peptides share exactly the same amino acid composition, but still have quite different adhesion properties (S1 and S3). In order to explain the adhesion properties, it might thus be necessary to take structural characteristics into account, although, as already indicated, CD measurements at room temperature do not reveal any clear structural differences between these peptides.

To this end, it is useful to discuss the aqueous solution behavior of these peptides first, and look for possible structural differences by means of a computational thermodynamic analysis. Thereafter, we will introduce a simple model for the interaction of peptides with Si(100) surfaces [340] and discuss the thermodynamics of adsorption on the basis of the folding characteristics found for the free peptides in solution.

14.4 Thermodynamics of semiconductor-binding peptides in solution

Since we are interested in rather specific problems, a mesoscopic coarse-grained model of the type we employed successfully in the past chapters is not sufficient in the present context, because secondary structures are of relevance here. For this purpose, a simplified efficient all-atom model[3] [223, 271, 272] was used to obtain the results to be discussed in the following [343].

As before, thermodynamic quantities are obtained by canonical statistics of the simulation data. Each conformation $\mathbf{X}$ is uniquely defined by the set of degrees of freedom, i.e., the set of dihedral backbone and side-chain angles of each amino acid $j = 1, \ldots, M$: $\xi_j = \phi_j, \psi_j, \chi_j^{(1)}, \ldots$ ($\omega_j = 180°$ is fixed in the model used). The Boltzmann probability for a conformation $\mathbf{X}$ is $p(\mathbf{X}) = \exp[-E(\mathbf{X})/RT]/Z$ and the partition function is given by

$$Z = \int \mathcal{D}\mathbf{X}\, e^{-E(\mathbf{X})/RT}, \tag{14.4}$$

where $\mathcal{D}\mathbf{X}$ is the formal integral measure for all possible conformations in the space of the degrees of freedom. The gas constant takes the value $R \approx 1.99 \times 10^{-3}$ kcal/K mole. The statistical average of any quantity O is obtained by

$$\langle O \rangle = \frac{1}{Z} \int \mathcal{D}\mathbf{X}\, O(\mathbf{X}) e^{-E(\mathbf{X})/RT}. \tag{14.5}$$

Cooperative structural activity is typically signaled by a peak in the statistical fluctuations of system-relevant quantities, such as the energy. Figure 14.4 shows how the heat capacity $C_V(T) = d\langle E \rangle/dT$ varies with temperature for the sequences S1–S4. The qualitative behavior of the three sequences S1, S2, and S4 is virtually identical. For these sequences, the specific heat exhibits a broad peak with maximum around 280 K. However, the heat capacity curve of S3 is qualitatively different and suggests a different folding behavior that becomes even more exciting, if one recalls the particularly poor binding of this sequence to Si(100) substrates. In general, given the shortness of the peptides, it is reasonable to conclude that the peaks indicate the transition from random coils at high temperature into an ordered secondary structure. The insets in Fig. 14.4 depict low-energy structures of these peptides. From these it is apparent that the folding behavior of S3 is different as it has the tendency to form a β-hairpin, whereas the other sequences prefer to fold into α-helical

---

[3] The model contains all atoms of the peptide chain, including H atoms, but no explicit water molecules. Bond angles, bond lengths, and peptide torsion angles (180°) are fixed. The only degrees of freedom of each amino acid are the Ramachandran angles ϕ, ψ and a number of side-chain torsion angles [223, 271, 272]. Results were obtained by simulated tempering simulations [343].

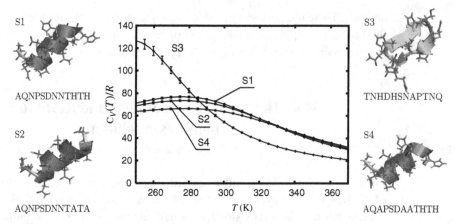

S1

AQNPSDNNTHTH

S2

AQNPSDNNTATA

S3

TNHDHSNAPTNQ

S4

AQAPSDAATHTH

Fig. 14.4 Temperature dependence of the heat capacity C_V in units of the gas constant R for the sequences S1–S4. Also shown are typical low-energy conformations of the peptides. Note the changed location of proline (P) in sequence S3, if compared to the other sequences, and the different type of its secondary structure. After [343].

conformations. Thus, we realize that the nature of the secondary structures is of particular interest for our forthcoming discussion.

In order to quantify the tendencies of these peptides to form secondary structures, it is useful to investigate the average α-helical and β-strand content as a function of temperature. We define a residue as α-helical if its Ramachandran angles ϕ and ψ satisfy $\phi \in (-90°, -30°)$ and $\psi \in (-77°, -17°)$, and $\langle n_\alpha \rangle$ denotes the fraction of the 10 inner residues that are α-helical. Similarly, $\langle n_\beta \rangle$ is the average β-strand content and it is defined as the fraction of the 10 inner residues with Ramachandran angles satisfying $\phi \in (-150°, -90°)$ and $\psi \in (90°, 150°)$.

In the temperature regime, where the heat capacity curves exhibit a peak, the secondary-structure content of S1, S2, and S4 changes more rapidly than that of S3. The α-helix content, $\langle n_\alpha \rangle$ [Fig. 14.5(a)], and the β-strand content, $\langle n_\beta \rangle$ [Fig. 14.5(b)], clearly confirm the earlier assumption that the sequences S1, S2, and S4 have the tendency to fold into α-helical structures. The α-helix content increases and the β-strand content decreases toward lower temperatures. For S3, the situation is reverse; $\langle n_\alpha \rangle$ saturates at low level at lower temperatures, but $\langle n_\beta \rangle$ is much larger in the folding region. This suggests folding into a β-like shape. However, it is worth noting that both $\langle n_\alpha \rangle$ and $\langle n_\beta \rangle$ have rather small values throughout the entire temperature space of interest. This means, for example, that of all amino acids in the ensemble of dominant peptide conformations at room temperature, about 80% are neither in an α nor in a β state according to our definition. In other words, the peptides are widely unfolded, which is fully consistent with the CD analysis of the solution behavior of these peptides [276]. This suggests that S1–S4 are largely unstructured at room temperature.

A closer inspection of the low-energy structures shown in Fig. 14.4 reveals that in the helical structures the α-helix does not span the entire chain, but rather the region between residues 3 to 12. The N-terminal section of the sequence is not α-helical, as there is a

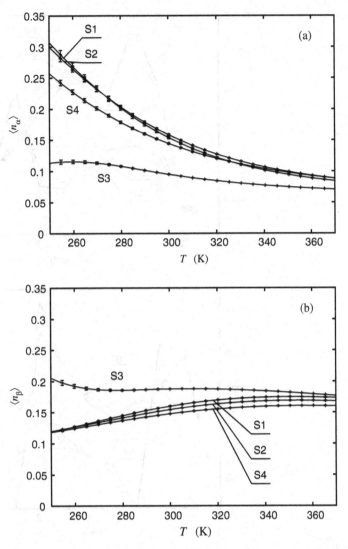

Fig. 14.5 Temperature dependence of (a) the α-helix content $\langle n_\alpha \rangle$ and (b) the β-strand content $\langle n_\beta \rangle$, for the sequences S1–S4. From [343].

proline (P) at position 4. Because of its two covalent bonds with the backbone, the side-chain of proline is most rigid and helical states of its backbone are highly suppressed. The β-hairpin structure identified for S3 has its turn at residues 6 and 7. The second strand, spanning residues 8–12, is not perfect but broken in the vicinity of the proline at position 9.

Let us now turn to a more detailed structural description at $T = 299$ K, the temperature at which the CD measurements were taken [276]. This discussion will focus mainly on S1 and S3, as the double mutant S2 and the triple mutant S4 show a behavior very similar to that of S1. In order to obtain average structural information about the behavior of each amino acid in the 299 K ensemble, it is certainly reasonable to examine how the

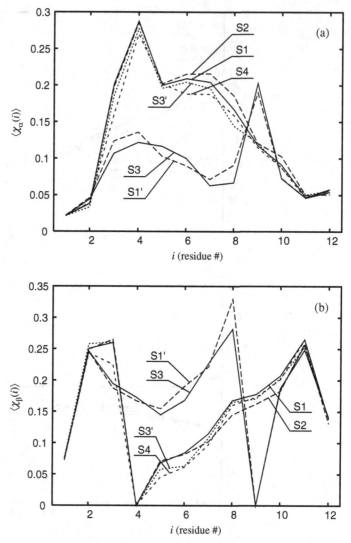

Fig. 14.6 Secondary-structure profiles for S1–S4, and S1′ and S3′ at $T = 299$ K. (a) The probability that residue i is in the α-helix state, $\langle \chi_\alpha(i) \rangle$, against i. (b) The probability that residue i is in the β-strand state, $\langle \chi_\beta(i) \rangle$, against i. The lines are only guides to the eye. From [343].

α-helix and β-strand contents vary along the chains. Let $\chi_\alpha(i) = 1$ if residue i is in the α-helix state and $\chi_\alpha(i) = 0$ otherwise, so that $\langle \chi_\alpha(i) \rangle$ is the probability of finding residue i in α-helix state, and let $\chi_\beta(i)$ denote the corresponding function for a β-strand state. Figure 14.6 shows $\langle \chi_\alpha(i) \rangle$ and $\langle \chi_\beta(i) \rangle$ against i for S1–S4 at $T = 299$ K. The low-energy conformations of S1, S2, and S4 contain an α-helix starting near position 3 and ending at the C terminus. The α-helix probability profile in Fig. 14.6(a) reveals that the stability of this α-helix is not uniform along the chain; its N-terminal part is most stable, whereas the stability decreases significantly toward the C terminus. For S3, it can be seen from

Fig. 14.6(b) that the values of $\langle\chi_\beta(i)\rangle$ are similar in the two regions that make the strands of the β-hairpin. An exception is Pro9, for which $\langle\chi_\beta(i)\rangle$ is strictly zero (proline has a fixed $\phi = -65°$ in the model, which falls outside the ϕ interval in our β-strand definition). We also note that the two end residues tend to be unstructured for all four sequences, with relatively small values of both $\langle\chi_\alpha(i)\rangle$ and $\langle\chi_\beta(i)\rangle$.

To gauge the importance of the proline location, it may be a good idea to study a variant of S3 with Asp4 (D) and Pro9 (P) interchanged. We will call this sequence S3′: TNHPHSNADTNQ. The $\langle\chi_{\alpha,\beta}(i)\rangle$ profiles are also included in Fig. 14.6. It is remarkable that the behavior of S3′ so closely resembles that of S1, S2, and S4. The S3′ profiles are nearly identical to those for S1, S2, and S4. In the reshuffling of S1 to get S3, it is the change of proline position that appears to be extraordinarily important.

Seeking further confirmation of this amazing observation, we also introduce a new sequence obtained by interchanging Pro4 (P) and Thr9 (T) in S1. This sequence shall be denoted by S1′: AQNTSDNNPHTH. As expected, it turns out that this transposition of S1 leads to a behavior similar to that of S3, as is illustrated by Fig. 14.6, which confirms the importance of the position of proline in all these sequences.

But how important is all this for the understanding of the adsorption behavior at a silicon substrate? After these intriguing preliminary considerations of the peptide behavior in solvent, we are now in a position to extend the peptide model by introducing the interaction with the substrate and discuss the apparently different binding behavior of these short peptides.

14.5 Modeling a hybrid peptide–silicon interface

14.5.1 Introduction

There is an enormous technological interest in understanding adhesion and self-assembly processes of polymers, proteins, or protein-like synthetic peptides at crystalline and amorphous solid materials such as metals [273, 317, 328, 344–348], and semiconductors [274–276, 349], but also carbon [350, 351], carbon nanotubes [329, 352], mica [353, 354], and silica [355–358]. However, the precise modeling of such a hybrid system is intricate. Given the large number of possible types of atom–atom interaction between organic and inorganic matter, it is impossible to introduce a generic microscopic model for interfaces of macromolecules and solid substrates. We have already discussed polymer and protein adsorption on larger scales and found that mesoscopic models can be useful for the description of conformational phases. However, specific systems such as the one we discuss here require specific input and thus microscopic details of the interaction cannot completely be spared. On the other hand, entirely microscopic models are computationally demanding.

The trade-off may be a model that consists of a combination of coarse-grained and atomic interactions. Such a model will not only be computationally feasible, it will also allow to treat essential specific properties of hybrid interfaces of proteins and solid materials. In the following, we will focus on silicon substrates with (100) orientation. One reason is that the binding behavior of the peptides S1, S2, and S4 on one hand and S3 on the other is so different [275,276] (the corresponding peptide adhesion coefficients are listed in Table 14.1) that the study of (100) silicon is intriguing. In contrast to the reference substrate mentioned in the last section, gallium-arsenide, a bare, not oxidized (100) silicon surface repels water molecules, i.e., it is effectively hydrophobic. This is fortunate, because it will be sufficient to introduce an implicit-solvent model. Polar solvent molecules will not essentially influence the binding of a peptide to an effectively hydrophobic substrate. A polar substrate such as GaAs is attractive to water molecules and, in such cases, the influence of single water molecules and even water layers upon the peptide binding propensity is expected to be strong and the additional simulation of explicit water molecules can hardly be avoided [347,348].

The goal is to investigate and compare silicon-binding properties of four of the synthetic peptides with 12 residues introduced in the last section: S1, S3, S1′, and S3′ [340]. Silicon (Si) is one of the technically most important semiconductors, as it serves, for example, as carrier substrate for almost all electronic circuits. For this reason, electronic properties of Si are well investigated. Also surface properties of Si have been the subject of numerous studies. This includes, for example, oxidation processes in air [359, 360] and water [361, 362], as well as the formation of hydride surface structures, frequently in connection with etching processes [349, 363–366]. The binding characteristics to Si substrates of small organic compounds and, for example, their influence on surface restructuring of the particularly reactive Si(100)-2×1 surface have also found broad attraction [349, 367–373]. Before the peptide–Si(100) model can be introduced, some remarks about the influence of the environment on the expected peptide adsorption process are necessary.

14.5.2   Si(100), oxidation, and the role of water

It is important to review properties of silicon surfaces and the sample preparation in experiments, because this information is necessary input for modeling the peptide–Si(100) interface. In these experiments [275,276], the Si(100) surfaces were first cleaned in a solution of ammonium fluoride (NH_4F) and hydrofluoric acid (HF) and then the adsorption experiments were performed in distilled water. This standard procedure ensures that the Si surface is widely free of oxide, which causes strong hydrophobic properties [359, 361]. The initial Si–F bonds after etching are replaced by Si–H bonds in the rinsing process in de-ionized water. Although the oxidation proceeds also in water [361, 362], it was found that the hydrophobicity of the Si samples is widely sustained during the peptide adsorption process.

It is also known that Si surfaces are comparatively rough after HF treatment [363]. This renders a detailed modeling intricate, even more as the reactivity of the surface is strongly influenced by roughness effects (such as, e.g., steps). Si(100)-2×1 surfaces are known to form Si–Si dimers on top of the surface [349] with highly reactive dangling bonds. From the considerations and the experimental preparations described above, it seems plausible that these bonds are mainly passivated by hydrogen, forming hydride layers [349,361,363].

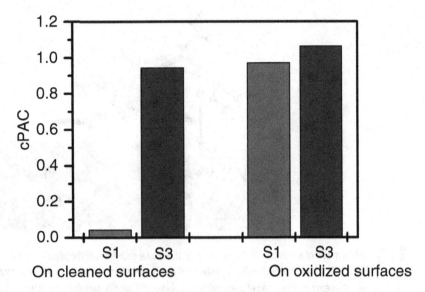

Fig. 14.7 Experimental cPAC values for S1 and S3 at bare and oxidized Si(100) substrates. From [340].

It should be emphasized that under these conditions the surface structure of Si(100) is substantially different from oxidized Si(100), which is polar and in effect hydrophilic [359]. This is nicely demonstrated in Fig. 14.7, where for S1 and S3 the cPAC values are shown for bare and oxidized Si(100) samples.[4] These data were obtained from AFM experiments of the peptides S1 and S3 interacting with Si(100) and GaAs(100) substrates in solution [340]. The main result is that the binding of S1 to oxidized GaAs(100) and Si(100) surfaces is virtually independent of the substrate type, which is widely screened by the top oxygen layer. The different adhesion propensities to the "bare" (hydrated) substrates lead to the conclusion that oxidation does not strongly progress on timescales of the peptide adsorption process.

From these considerations, one may finally conclude that the key role of water is to slow down the oxidation process of the Si(100) surface, but up to screening effects its influence upon the peptide binding process is rather limited. In particular, we do not expect that stable water layers form between adsorbate and substrate. These characteristic properties of etched Si(100) surfaces in de-ionized water effectively enter into the definition of the hybrid model which will then serve as the basis for the analysis and interpretation of the specificity of peptide adhesion on these substrates.

14.5.3 The hybrid model

Following the idea to treat the interaction between the peptide and the substrate on a coarse-grained, effective level, but to maintain atomic resolution for the peptide, the representation of the substrate is simplified and it is considered to consist only of atomic layers with

---

[4] As mentioned earlier, etched and oxidized GaAs(100) reference substrates, respectively, were chosen for calibration purposes.

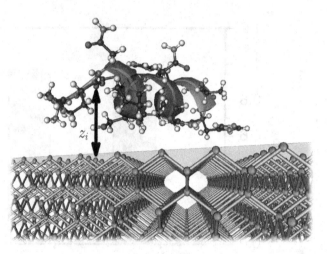

Fig. 14.8 Hybrid model of a peptide–Si(100) interface. In this model each atom of the peptide interacts with the surface layer of the Si(100) crystal. The perpendicular distance of the ith atom from the surface is denoted by z_i. The image shows the lowest-energy conformation of the peptide S1. Because this peptide sustains its α-helical structure that it also adopts in solution, the binding to the substrate is supposed to be weak. It should also be noted that the population of this conformation is very low at room temperature.

surface-specific atomic density. The substrate and its surface structure itself is fixed and thus its energy does not enter into the model at all.[5] Therefore, the energy of a single peptide with conformation $\mathbf{X}$ (where dihedral backbone and side-chain angles are the degrees of freedom) and interacting with the substrate, whose surface structure is characterized by the Miller index (hkl), is generally written as

$$E(\mathbf{X}) = E_{\mathrm{pep}}(\mathbf{X}) + E_{\mathrm{pep\text{-}sub}}^{\mathrm{Si}(hkl)}(\mathbf{z}). \qquad (14.6)$$

Here, $\mathbf{z} = (z_1, z_2, \ldots, z_N)$ is the perpendicular distance vector of all N peptide atoms from the surface layer of the substrate (see Fig. 14.8). The effect of the surrounding solvent is implicitly contained in the force field parameters. The all-atom peptide model [223, 271, 272] is the same that has been used in the study of the peptides in solution in Section 14.4.

The interaction of the peptide with the substrate is modeled in a simplified way, i.e., each peptide atom feels the mean field of the atomic substrate layers. The atomic density of these layers is dependent on the surface characteristics, i.e., it depends on the crystal orientation (the Miller index hkl) of the substrate at the surface. We make the following assumptions for setting up the model. According to our considerations about Si(100) surface properties in de-ionized water, the Si(100) surface can be considered to be hydrophobic. This has the effect that it is not favorable for water molecules to reside between the adsorbed peptide and the substrate. Furthermore, polarization effects between side chains and substrate are not expected.

---

[5] These simplifications cannot be justified, if local, microscopic effects of binding are of interest. In reality, the surface atoms of the substrate will respond to the presence of solvent molecules and the peptide. Depending on the system, even covalent bonds might form (chemisorption).

Table 14.2 Atomic van der Waals radii σ and energy depths ε used in the simulations of peptide adsorption at a (100) silicon substrate.

Atom	σ [Å]	ε [kcal/mole]
Si	2.0 [375]	0.05 [355]
H	1.2 [376]	0.04 [376]
C	1.8 [375, 376]	0.05 [376]
O	1.5 [375, 376]	0.08 [376]
N	1.6 [375, 376]	0.09 [376]

Since dangling bonds on the Si(100) surface are probably saturated by covalent bonds to hydrogen atoms (due to the HF etching process), it is also assumed that covalent bonds between peptide and surface atoms do not form. Thus, the surface is also considered to be uncharged [374]. Si dimers sticking off the substrate and the hydration effect are expected to weakly screen the peptide from the substrate, but are not explicitly taken into account.

Based on these assumptions, it is sufficient to use a generic noncovalent Lennard-Jones approach [340, 348, 350, 351] for modeling the interaction between peptide atoms and surface layer,

$$E_{\text{pep-sub}}^{\text{Si}(hkl)}(\mathbf{z}) = 2\pi \rho^{\text{Si}(hkl)} \sum_{i=1}^{I} \varepsilon_{i,\text{Si}} \sigma_{i,\text{Si}}^2 \left[\frac{2}{5} \left(\frac{\sigma_{i,\text{Si}}}{z_i} \right)^{10} - \left(\frac{\sigma_{i,\text{Si}}}{z_i} \right)^4 \right], \qquad (14.7)$$

where $\rho^{\text{Si}(hkl)}$ is the atomic density of the Si(hkl) surface layer. Similar to the derivations presented in Sections 13.6 and 13.9, this form of the potential is obtained by performing a two-dimensional integration for the Lennard-Jones interactions of a peptide atom with any point of an atomic layer of the substrate, and then summing the atomic layers that shall be considered in the simulation.[6]

Si has diamond structure and since we assume the surface to be ideally flat and the surface layer to be identical to the parallel atomic layers in the interior of the crystal (i.e., the surface is neither relaxed, reconstructed, nor rough), $\rho^{\text{Si}(100)} \approx 0.068$ Å^{-2}. The non-covalent interaction between the peptide atoms and the Si substrate is parametrized by force-field parameters $\varepsilon_{i,\text{Si}} = \sqrt{\varepsilon_i \varepsilon_{\text{Si}}}$ and $\sigma_{i,\text{Si}} = \sigma_i + \sigma_{\text{Si}}$ and thus depends on the energy depths ε_i and van der Waals radii σ_i of the individual atoms. Suitable parameter values are listed in Table 14.2.

---

[6] The simulation results for this model that will be discussed in the following section were obtained by multi-threaded multicanonical and parallel tempering Monte Carlo simulations under the assumption that the peptide only interacts with the surface layer ($I = 1$). A simulation box of dimension [50 Å]^3 with periodic boundary conditions parallel to the substrate was used. In perpendicular direction, peptide mobility is restricted by the Si substrate residing by definition at $z = 0$. The influence of the wall parallel to the substrate is simply steric, i.e., the atoms experience hard-wall repulsion at $z = z_{\max} = 50$ Å [340].

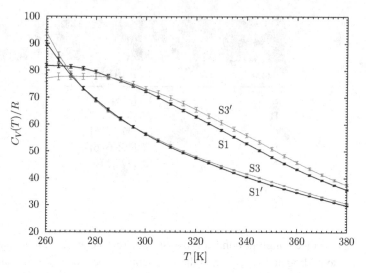

Fig. 14.9 Heat capacities of the peptides S1, S3, S1′, and S3′ interacting with a de-oxidized (100) silicon substrate.

14.6 Sequence-specific peptide adsorption at silicon (100) surface

It is instructive and sufficient to discuss the adsorption behavior of the sequences S1, S3, S1′, and S3′ now. If our assumption is correct that the structural behavior of these peptides depends sensitively on the proline position in the amino acid sequence, then not only the folding behavior in solvent is affected, but also the adsorption process might substantially be influenced by this.

14.6.1 Thermal fluctuations and deformations upon binding

The adsorption of the peptides at the semiconductor surface is a conformational pseudophase transition and accompanied by structural changes of the peptides during the adsorption process. The energetic response of the peptides upon binding is shown in Fig. 14.9, where the heat capacity curves are plotted for each of the peptides. The peaks for S1 and S3′ and the increase toward lower temperatures for S3 and S1′ indicate energetic activity that signals the onset of a crossover between random-coil structures in solvent and adsorbed conformations at the substrate.

Most remarkably, the bundle of the four heat capacity curves splits into the two groups of pairs S1, S3′ and S3, S1′. To repeat this, sequences of these pairs differ mainly by the location of the amino acid proline. Thus, it seems that changes of the thermodynamic behavior of these synthetic peptides are virtually controlled by proline only. The knowledge that individual amino acids or segments of the sequence are particularly relevant for a certain structural behavior is of obvious value for the design of sequences with specific functions. The example discussed here is certainly simple, but any systematic attempt to design drugs or technological applications on protein basis must take these specific properties into account.

Considering the previous argument about structural properties of these peptides in solvent only, we also expect that the adsorbed conformations are not supposed to exhibit clear symmetries. However, good binding properties are only possible if the peptide forms comparatively flat structures, enabling it to maximize the number of substrate contacts. Thus, at room temperature, surface-attached peptides are expected to be compact without noticeable internal structure. The deformation of the peptides due to adsorption is apparent if we analyze the gyration tensor components perpendicular and parallel to the substrate separately. Let the square gyration radius of the heavy atoms be

$$R_{\text{gyr}}^2 = \frac{1}{N_h} \sum_{i=1}^{N_h} (\mathbf{x}_i - \mathbf{x}_0)^2, \tag{14.8}$$

where the sum runs over the N_h heavy (non-hydrogen) atoms of the peptide, $\mathbf{x}_i$ is the position of the ith heavy atom, and $\mathbf{x}_0 = \sum_{i=1}^{N_h} \mathbf{x}_i / N_h$. For identifying asymmetries in the peptide conformations and orientations, we use the simple decomposition $R_{\text{gyr}}^2 = R_\perp^2 + R_\parallel^2$, where $R_\perp$ and $R_\parallel$ are the components perpendicular and parallel to the substrate, respectively. The temperature dependence of the thermal averages of the components is shown in Fig. 14.10 for the peptides in the presence of the Si(100) substrate and, for comparison, in solvent without substrate. The "thickness" $\langle R_\perp \rangle$ of the structures in the perpendicular direction is, in the temperature interval shown, much smaller than the average extension without substrate [Fig. 14.10(a)]. This is different for the component $\langle R_\parallel \rangle$ parallel to the substrate, i.e., the planar extension is larger than in the bulk case [Fig. 14.10(b)]. An interesting feature is that S1 and S3′ behave similarly, as well as S3 and S1′. This is the same grouping that has already been observed in the discussion of the heat capacity of the peptides (see Fig. 14.9) or, already, in the substrate-free case (cf. Fig. 14.6). Although folding in solvent and adsorption are different, if not unrelated, structural transitions, the same pairs of peptides, which behave qualitatively similarly, form.

14.6.2 Secondary-structure contents of the peptides

As expected, adsorbed peptides do not exhibit clear structures at room temperature. In Fig. 14.11, the respective α-helix and β-strand contents of the adsorbed conformations, $\langle n_\alpha \rangle_b$ and $\langle n_\beta \rangle_b$, respectively, are shown.[7] Although noticeable differences for the mentioned peptide groups are found, there is no significant population of α-helical or β-sheet structures, at least at room temperature. Nonetheless, there is a tendency that residues of S1 and S3′ are rather in α and residues of S3 and S1′ in β state. The small secondary-structure contents are quite similar to what was observed for the peptides in solution (without substrate; see Fig. 14.5). It is obvious that the presence of the Si(100) substrate does not lead to a stabilization of secondary structures here. Such stabilization can occur, however, and has been reported, for example, for a synthetic peptide binding at silica nanoparticles [358].

---

[7] Here a peptide conformation is considered to be adsorbed to the substrate, if at least 2% of the heavy atoms are in "contact" with the substrate, i.e., reside within a 5Å distance from the substrate.

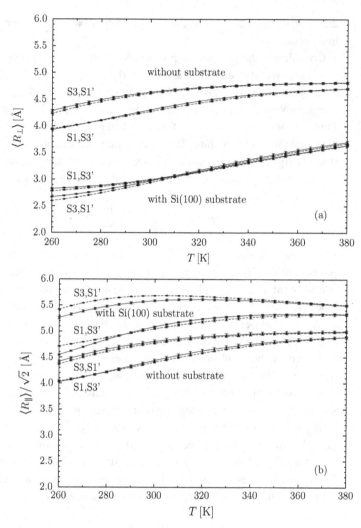

Fig. 14.10 Gyration tensor components of the peptides (a) perpendicular and (b) parallel to the substrate. For comparison, results for the peptides in solvent only (i.e., without substrate) are also shown.

In Fig. 14.11, low-energy conformations and orientations of the peptides, an α-helix for S1 [Fig. 14.11(a)] and a β-strand for S3 [Fig. 14.11(b)], are also inserted. The respective peptide structures have strong similarities with the low-energy conformations of the peptides in solution (see Fig. 14.4). Proline lies in both lowest-energy conformations very close to the substrate and influences the orientation of the peptide on the substrate. Together with the mutual reversal of the structural properties after changing the proline positions this confirms the prediction that not only the folding in solution, but also the adsorption properties of the investigated sequences strongly depend on the respective proline positions in the sequences [340].

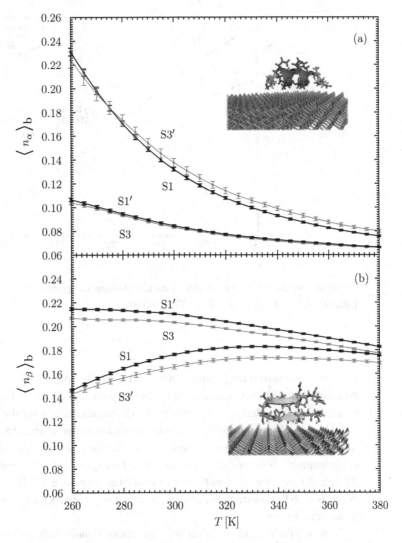

Fig. 14.11 (a) α-helix and (b) β-strand content of the *adsorbed* peptides as functions of temperature.

14.6.3 Order parameter of adsorption and nature of adsorption transition

To quantify the degree of adsorption, it is convenient to introduce a parameter that counts the number of contacts of the protein with the surface of the substrate, as we have already done in the discussion of adsorption transition of polymers. An atom of the peptide is defined to be in contact with the substrate if it is located within a distance ≤ 5 Å above the substrate. Hydrogen atoms can be ignored so that if we denote the number of bound heavy atoms by n_{h} and their total number by N_{h}, the order parameter of adsorption is simply defined by

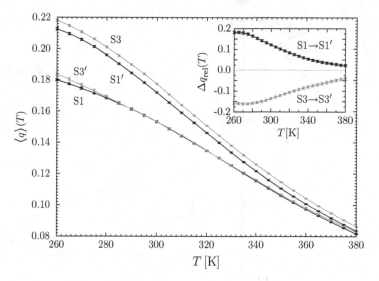

Fig. 14.12 Temperature dependence of the mean "order" parameter of adsorption $\langle q \rangle$ and of its relative change Δq_{rel} upon pairwise mutation of S1 (to S1') and S3 (to S3'), respectively.

$$q = \frac{n_{\mathrm{h}}}{N_{\mathrm{h}}}. \tag{14.9}$$

It is obviously zero if the peptide is desorbed and its hypothetical maximum of unity would be reached if all heavy atoms were in contact with the substrate. Figure 14.12 shows the temperature dependence of the canonical expectation value q for the four sequences studied. The first observation is that the pairwise mutation that changed S1 to S1' and S3 to S3' has precisely the effect upon adsorption that we have already guessed from our previous considerations. With proline at position 4 in the sequence, the adsorption characteristics of S3' are virtually identical to those of the original sequence S1. Both peptides form fewer contacts with the substrate than the other pair of sequences S3, S1', where proline is located at the 9th position.

These results not only confirm the experimental findings that S3 binds better to Si(100) than S1, they also predict that respective pairwise mutations with respect to the proline position will reverse the binding affinities. This is depicted in the inset of Fig. 14.12, where the relative change of the order parameter,

$$\Delta q_{\mathrm{rel}}^{\mathrm{S}x \to \mathrm{S}x'} = \frac{\langle q \rangle_{\mathrm{S}x'} - \langle q \rangle_{\mathrm{S}x}}{\langle q \rangle_{\mathrm{S}x}}, \quad x = 1, 3 \tag{14.10}$$

is plotted. At room temperature (300 K), $\Delta q_{\mathrm{rel}}^{\mathrm{S}1 \to \mathrm{S}1'} \approx +0.11$ and $\Delta q_{\mathrm{rel}}^{\mathrm{S}3 \to \mathrm{S}3'} \approx -0.15$, i.e., the adsorption propensity of S1 is improved by proline mutation by about the same value as the binding capability of S3 is decreased upon proline mutation. The relative parameter Δq_{rel} can be interpreted in the same way as the cPAC value 14.3 that is measured in the experiments. Indeed, the experimentally obtained cPAC exhibit the same feature of quantitatively similar changes in both directions of mutation. For S1→S1', $\Delta \mathrm{cPAC}(\mathrm{S}1 \to \mathrm{S}1') = \mathrm{cPAC}(\mathrm{S}1') - \mathrm{cPAC}(\mathrm{S}1) \approx +0.27$, whereas for S3→S3', $\Delta \mathrm{cPAC}(\mathrm{S}3 \to \mathrm{S}3') = \mathrm{cPAC}(\mathrm{S}3') -$

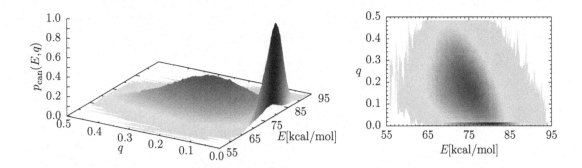

Fig. 14.13 Perspective (left) and top (right) view of the unnormalized canonical probability distribution $p_{can}(E, q)$ for S3 at $T = 300$ K.

cPAC(S3) ≈ -0.25 [340]. This is a nice example of consistent information about the system behavior obtained from both simulation and experiment.

Finally, let us compare the adsorption behavior with what we had found in Chapter 13 for simplified hybrid lattice models of polymers and peptides near attractive substrates. The adhesion of the peptides at the Si(100) substrate exhibits very similar features. Exemplified for peptide S3, Fig. 14.13 shows the plot of the canonical probability distribution $p_{can}(E, q) \sim \langle \delta(E - E(\mathbf{X}))\delta(q - q(\mathbf{X})) \rangle$ at room temperature ($T = 300$ K). The peak at $(E, q) \approx (80.5 \text{ kcal/mole}, 0.0)$ corresponds to conformations that are not in contact with the substrate. It is separated from another peak near $(E, q) \approx (74.5 \text{ kcal/mole}, 0.2)$ and belongs to conformations with about 17% of the heavy atoms with distances ≤ 5 Å from the substrate surface (compare with Fig. 14.12). That means adsorbed and desorbed conformations coexist and the gap in between the peaks separates the two pseudophases in q-space, which causes a kinetic free-energy barrier.[8] Thus, the adsorption transition is a first-order-like pseudophase transition in q, but since both structural phases (adsorbed and desorbed) coexist almost at the same energy, the transition in E space is weakly of first order.[9]

---

[8] Here, the free energy is considered as a function of system energy E and order parameter q and is defined as $F(E, q) = -RT \ln p_{can}(E, q) + \text{const}$.

[9] As a microcanonical analysis shows, the adsorption transition of S1, for example, is rather second-order-like. For such small systems, energetic gaps are typically too tiny for first-order signals. In this context, see the general discussion in Section 6.4.

Concluding remarks and outlook

Biomolecular research has so many facets that it is impossible to cover all important aspects of this highly interdisciplinary field in a single book. However, by definition, generic physics-based approaches have the potential to introduce concepts and tools that enable systematic and consistent investigations of complex systems, even in cases where these systems (cell systems, individual cells, molecular composites, biomolecules, solvent molecules, etc.) do not seem to possess any similarities. The physical concepts are based on quantum and classical theories, intertwined by the basic theory of complexity under the influence of thermodynamic effects: statistical mechanics. All biological, biochemical, and biophysical processes are caused by the interaction of basic units such as atoms, chemical groups, molecules. None of those processes can be thought of as being disconnected from cooperative ordering (or disordering) effects, and for our understanding of these effects on nanoscopic to mesoscopic scales only the basic theory of statistical physics is available to unravel the macroscopic consequences of these processes. The macroscopic description is what we call thermodynamics.

The currently most successful theoretical tool to investigate and to analyze thermal fluctuations statistically is the computer simulation. The computer has not replaced the human brain, but it has changed the way we deal with complex problems. Mathematical methods that formerly have been used with great success to reveal what happens to infinitely large systems, if thermal fluctuations are integrated out, turn out to be almost useless when trying to deal with finitely large, or effectively small, systems. Mathematical methods have increasingly made a place for algorithmic solutions, and the most efficient tool to deal with algorithms is the computer. This change of paradigms in statistical physics has consequences that go much deeper. These algorithmic methods enable us to re-think the basic approaches to statistical physics, because we can now study quantities such as the density of states or the configurational entropy that have not been easily accessible at all before.

This book cannot offer a complete compilation of statistical mechanics concepts and computational tools, and it only covers a selection of problems in biomolecular research. However, the methods for the analysis of structural transitions of finite systems that are introduced in the book are general and systematic, many of the modeling approaches are generic, and the examples presented provide an insight into the fascinating world of cooperative behavior in miniature systems as seen through the theoretical physicist's eyes. The most appropriate computer simulation methodologies for thermodynamic investigations of molecular systems are efficient stochastic importance-sampling methods such as Monte Carlo methods. The most popular ones are described in the book, but with the exception of multicanonical sampling no attempt is made to discuss features of sophisticated variants of these methods. Certainly the most striking example is the class of existing replica-exchange methods, whose particular popularity is due to the rather simple and straightforward implementation on parallel computers. The reason why in this book more emphasis has been dedicated to multicanonical sampling is that the sampling strategy is effectively based on the gradient of the microcanonical entropy, i.e., on the shape of the inverse microcanonical temperature curve as a function of energy. This is particularly interesting for the microcanonical inflection-point analysis of transitions that is also described in detail in this book.

A major topic of the entire book is dedicated to the investigation of the usefulness of coarse-grained modeling for the identification and classification of structural phases of polymer and protein systems. In the future, there will even be an increased necessity for the employment of sufficiently simplified models, because the understanding of structure and function of rather small systems such as individual proteins will give way to studies of more complex systems ranging from larger composites of molecular units such as aggregates to entire cells and cell systems. It is certainly an illusion to assume that entire cells and cell clusters up to tissues can be simulated on atomic scales by conventional computer simulations in the near future (consider that there are about 10^{13}–10^{14} atoms in each cell). So the major modeling effort will continue to be to find reasonable approximations on coarse scales. One of the key messages of this book is that this is no disaster at all, because cooperative effects that guide structure formation processes on scales that are larger than the scale of individual units (such as atoms, molecules, cells, etc.) do not necessarily require a precise description on scales as small as the size of the individual constituents. This holds true for the "dynamics" (better: kinetics) of such processes as well.

In this context, it is also important to remember that not a single process is decisive for the stability and function of a biological system, but the multiplicity of such processes. Only the statistical information about many such events is relevant for the understanding of molecular and cellular structures. No individual process or event is like any other. To mention only one example, even in a healthy system, folding, unfolding, and misfolding events of proteins occur at any time. This makes it so difficult to understand these processes, because we are lacking parameters that clearly separate these "pathways" in a general way. It must also be kept in mind that thermodynamics is essential, i.e., fluctuations occur and actually have to occur to allow for both the change between structural phases and the stabilization of structural phases. Simply stated, a single computer simulation of a million-atom system, even though it might require months of computation on a supercomputer, is likely to be insufficient to supply the information that helps understand how a system behaves under thermal fluctuations. Fortunately, the answer to thermodynamic questions can often be found by simply simulating the core effects that govern structure formation processes. In this regard, a Monte Carlo simulation of an adequately coarse-grained model, running on a decent computer cluster, can be much more efficient and more accurate than an atomic molecular dynamics simulation with complicated heat-bath coupling that wastes millions of CPU hours of supercomputer time. For this reason, it is fair to believe that in the future large efforts will be undertaken to improve the quality of coarse-grained models and the efficiency of generalized-ensemble methods for their simulation. Successful computational research done in condensed matter physics in the last decades, specifically for magnetic systems, guides the way. This book is an attempt to transfer and adapt those concepts to macromolecular systems.

Three major structure formation processes, where use of these concepts has been made, have been discussed in this book. First, properties of structural phases of individual polymers and proteins have been investigated on various levels of model abstraction, ranging from hydrophobic-core-based lattice models to atomic-scale models. The second application of modern statistical concepts has been dedicated to the class of multiple-chain polymer and protein systems. The understanding of features of aggregation transitions is particularly relevant in connection with notorious neurodegenerative diseases such as Alzheimer's. The third structural transition that we have investigated in more detail is the adsorption of individual macromolecules to solid substrates. Specificity in adsorption properties on the molecular or on the substrate side (or by the combination of both) is the key to various technological applications. It is likely that a large fraction of future nanotechnology will be based on specifically designed hybrid organic/inorganic interfaces, which puts this field into the focus of macromolecular research as well.

This book is not a standalone reference. It is a guide to the fascinating world of computational macromolecular research, but it must be considered in connection with a widespread complementary literature that covers other aspects of the same field. The reference list in this book is too short to seriously reflect the recent progress in the field, but the given references are good starting points for individual research. Several sections of this book offer suggestions for the statistical and thermodynamic interpretation of conformational transitions. Understanding these processes is a core task of active, ongoing science and a large interdisciplinary effort. Progress of systematic research in virtually all areas of the natural sciences and new discoveries will influence the way we think about ideas, concepts, and interpretations. This is also true on the methodological side. The book gives a short introduction to the computational methods that the author considers particularly useful, but given the many advances on small and large scales in the recent past, the selection must remain incomplete and many improvements of the basic algorithms are not covered at all. Readers are encouraged to fill the gap on their own, by consulting the more technical literature.

Methods, models, statistical analyses, and suggestions for interpretations of results have been presented in a form that is supposed to enable a quick start for the next generation of scientists who are interested in this field, but the author also hopes that certain aspects might be interesting for experienced scientists as well. These two goals have provided the impetus for writing this book, and the reader may decide whether it is a useful contribution to biologically motivated research.

References

[1] T. E. Creighton, *Proteins: Structure and Molecular Properties* (New York: Freeman, 1993).

[2] A. V. Finkelstein and O. B. Ptitsyn, *Protein Physics* (London: Academic Press, 2002).

[3] C. Branden and J. Tooze, *Introduction to Protein Structure* (New York: Garland, 1999).

[4] B. L. de Groot, T. Frigato, V. Helms, and H. Grubmüller, J. Mol. Biol., **333** (2003), 279.

[5] M. L. Zeidel, S. V. Ambudkar, B. L. Smith, and P. Agre, Biochemistry, **31** (1992), 7436.

[6] B. L. de Groot and H. Grubmüller, Science, **294** (2001), 2353.

[7] H. Lodish, D. Baltimore, A. Berk, S. L. Zipursky, P. Matsudaira, and J. Darnell, *Molecular Cell Biology* (New York: Freeman, 1995).

[8] K. P. C. Vollhardt and N. E. Schore, *Organic Chemistry* (New York: Freeman, 2003).

[9] C. B. Anfinsen, Science, **181** (1973), 223.

[10] www.rcsb.org

[11] T. Ooi, M. Obatake, G. Nemethy, and H. A. Scheraga, Proc. Natl. Acad. Sci. (USA), **84** (1987), 3086.

[12] K. A. Dill, Biochemistry, **24** (1985), 1501; K. F. Lau and K. A. Dill, Macromolecules, **22** (1989), 3986.

[13] C. Tang, Physica A, **288** (2000), 31.

[14] F. H. Stillinger, T. Head-Gordon, and C. L. Hirshfeld, Phys. Rev. E, **48** (1993), 1469; F. H. Stillinger and T. Head-Gordon, Phys. Rev. E, **52** (1995), 2872.

[15] B. Berger and T. Leighton, J. Comp. Biol., **5** (1998), 27; P. Crescenzi, D. Goldman, C. Papadimitriou, A. Piccolboni, and M. Yannakakis, J. Comp. Biol., **5** (1998), 423.

[16] A. Irbäck and E. Sandelin, J. Chem. Phys., **108** (1998), 2245.

[17] A. Irbäck and C. Troein, J. Biol. Phys., **28** (2002), 1.

[18] H. Cejtin, J. Edler, A. Gottlieb, R. Helling, H. Li, J. Philbin, C. Tang, and N. Wingreen, J. Chem. Phys., **116** (2002), 352.

[19] R. Schiemann, M. Bachmann, and W. Janke, J. Chem. Phys., **122** (2005), 114705.

[20] R. Schiemann, M. Bachmann, and W. Janke, Comp. Phys. Comm., **166** (2005), 8.

[21] K. Yue and K. A. Dill, Phys. Rev. E, **48** (1993), 2267; Proc. Natl. Acad. Sci. (USA), **92** (1995), 146.

[22] T. C. Beutler and K. A. Dill, Prot. Sci., **5** (1996), 2037.

[23] R. Unger and J. Moult, J. Mol. Biol., **231** (1993), 75.

[24] N. Krasnogor, W. E. Hart, J. Smith, and D. A. Pelta, in *Proceedings of the Genetic and Evolutionary Computation Conference (GECCO99)*, ed. W. Banzhaf, J. Daida, A. E. Eiben, M. H. Garzon, V. Honavar, M. Jakiela, and R. E. Smith (San Francisco: Morgan Kaufmann, 1999), p. 1596.

[25] Y. Cui, W. H. Wong, E. Bornberg-Bauer, and H. S. Chan, Proc. Natl. Acad. Sci. (USA), **99** (2002), 809.

[26] N. Lesh, M. Mitzenmacher, and S. Whitesides, in *Proceedings of the Seventh Annual International Conference on Computational Molecular Biology (RECOMB03)*, ed. M. Vingron, S. Istrail, P. Pevzner, and M. Waterman (New York: ACM, 2003), p. 188.

[27] T. Jiang, Q. Cui, G. Shi, and S. Ma, J. Chem. Phys., **119** (2003), 4592.

[28] F. Seno, M. Vendruscolo, A. Maritan, and J. R. Banavar, Phys. Rev. Lett., **77** (1996), 1901.

[29] R. Ramakrishnan, B. Ramachandran, and J. F. Pekny, J. Chem. Phys., **106** (1997), 2418.

[30] A. Irbäck, C. Peterson, F. Potthast, and E. Sandelin, Phys. Rev. E, **58** (1998), R5249.

[31] L. W. Lee and J.-S. Wang, Phys. Rev. E, **64** (2001), 056112.

[32] G. Chikenji, M. Kikuchi, and Y. Iba, Phys. Rev. Lett., **83** (1999), 1886, and references therein.

[33] M. N. Rosenbluth and A. W. Rosenbluth, J. Chem. Phys., **23** (1955), 356.

[34] D. Aldous and U. Vazirani, in *Proceedings of the 35th Annual Symposium on Foundations of Computer Science*, Santa Fe (Los Alamitos: IEEE, 1994), p. 492.

[35] P. Grassberger, Phys. Rev. E, **56** (1997), 3682.

[36] H. Frauenkron, U. Bastolla, E. Gerstner, P. Grassberger, and W. Nadler, Phys. Rev. Lett., **80** (1998), 3149; U. Bastolla, H. Frauenkron, E. Gerstner, P. Grassberger, and W. Nadler, Proteins, **32** (1998), 52.

[37] P. Grassberger and W. Nadler, in *Computational Statistical Physics – From Billiards to Monte Carlo*, ed. K. H. Hoffmann and M. Schreiber (Berlin: Springer, 2002), p. 169, and references therein.

[38] H.-P. Hsu, V. Mehra, W. Nadler, and P. Grassberger, J. Chem. Phys., **118** (2003), 444; Phys. Rev. E, **68** (2003), 21113.

[39] M. Bachmann and W. Janke, Phys. Rev. Lett., **91** (2003), 208105.

[40] M. Bachmann and W. Janke, J. Chem. Phys., **120** (2004), 6779.

[41] R. J. Najmanovich, J. L. de Lyra, and V. B. Henriques, Physica A, **249** (1998), 374.

[42] K. Yue, K. M. Fiebig, P. D. Thomas, H. S. Chan, E. I. Shakhnovich, and K. A. Dill, Proc. Natl. Acad. Sci. (USA), **92** (1995), 325.

[43] E. E. Lattman, K. M. Fiebig, and K. A. Dill, Biochemistry, **33** (1994), 6158.

[44] L. Toma and S. Toma, Prot. Sci., **5** (1996), 147.

[45] S. Miyazawa and R. L. Jernigan, J. Mol. Biol., **256** (1996), 623.

[46] S. Schnabel, M. Bachmann, and W. Janke, Phys. Rev. Lett., **98** (2007), 048103.

[47] S. Schnabel, M. Bachmann, and W. Janke, J. Chem. Phys., **126** (2007), 105102.

[48] O. Kratky and G. Porod, J. Colloid. Sci., **4** (1949), 35.

[49] J. E. Lennard-Jones, Proc. Phys. Soc., **43** (1931), 461.

[50] R. B. Bird, C. F. Curtiss, R. C. Armstrong, and O. Hassager, *Dynamics of Polymeric Liquids*, 2nd ed. (New York: Wiley, 1987).

[51] A. Milchev, A. Bhattacharaya, and K. Binder, Macromolecules, **34** (2001), 1881.

[52] S. Schnabel, T. Vogel, M. Bachmann, and W. Janke, Chem. Phys. Lett., **476** (2009), 201.

[53] S. Schnabel, M. Bachmann, and W. Janke, J. Chem. Phys., **131** (2009), 124904.

[54] K. Kremer and G. S. Grest, J. Chem. Phys., **92** (1990), 5057.

[55] I. Carmesin and K. Kremer, Macromolecules, **21** (1988), 2819.

[56] H. P. Deutsch and K. Binder, J. Chem. Phys., **94** (1991), 2294.

[57] R. B. Laughlin and D. Pines, Proc. Natl. Acad. Sci. (USA), **97** (2000), 28.

[58] R. B. Laughlin, D. Pines, J. Schmalian, B. P. Stojković, and P. Wolynes, Proc. Natl. Acad. Sci. (USA), **97** (2000), 32.

[59] R. P. Feynman and A. R. Hibbs, *Quantum Mechanics and Path Integrals* (New York: McGraw-Hill, 1965).

[60] H. Kleinert, *Path Integrals in Quantum Mechanics, Statistics, Polymer Physics, and Financial Markets*, 5th ed. (Singapore: World Scientific, 2009).

[61] S. Schnabel, D. T. Seaton, D. P. Landau, and M. Bachmann, Phys. Rev. E, **84** (2011), 011127.

[62] D. S. Gaunt and A. J. Guttmann, *Asymptotic Analysis of Coefficients*, in *Phase Transitions and Critical Phenomena*, ed. C. Domb and M. S. Green (London: Academic Press, 1974), p. 181.

[63] J. L. Cardy and A. J. Guttmann, J. Phys. A: Math. Gen., **26** (1993), 2485.

[64] See, e.g., H. E. Stanley, *Introduction to Phase Transitions and Critical Phenomena* (New York: Oxford University Press, 1987); A. J. Guttmann, *Asymptotic Analysis of Power-Series Expansions*, in *Phase Transitions and Critical Phenomena*, ed. C. Domb and J. L. Lebowitz (London: Academic Press, 1989), p. 3.

[65] D. MacDonald, D. L. Hunter, K. Kelly, and N. Jan, J. Phys. A: Math. Gen., **25** (1992), 1429.

[66] D. MacDonald, S. Joseph, D. L. Hunter, L. L. Moseley, N. Jan, and A. J. Guttmann, J. Phys. A: Math. Gen., **33** (2000), 5973.

[67] M. Chen and K. Y. Lin, J. Phys. A: Math. Gen., **35** (2002), 1501.

[68] S. Caracciolo, A. S. Causo, and A. Pelissetto, Phys. Rev. E, **57** (1998), R1215.

[69] R. Guida and J. Zinn-Justin, J. Phys. A: Math. Gen., **31** (1998), 8103.

[70] M. Vendruscolo and E. Domany, Folding & Design, **2** (1997), 295; Folding & Design, **3** (1998), 329.

[71] E. G. Emberly, J. Miller, C. Zeng, N. S. Wingreen, and C. Tang, Proteins, **47** (2002), 295.

[72] H. Li, R. Helling, C. Tang, and N. Wingreen, Science, **273** (1996), 666.

[73] M. Bachmann and W. Janke, Acta Phys. Pol. B, **34** (2003), 4689.

[74] See, e.g., R. Kubo, Rep. Prog. Phys., **29** (1966), 255.

[75] D. Frenkel and B. Smit, *Understanding Molecular Simulation*, 2nd ed. (San Diego: Academic Press, 2002).

[76] J. Schluttig, M. Bachmann, and W. Janke, J. Comput. Chem., **29** (2008), 2603.

[77] See, e.g., L. P. Kadanoff, Physica A, **163** (1990), 1.

[78] D. P. Landau and K. Binder, *A Guide to Monte Carlo Simulations in Statistical Physics*, 3rd ed. (New York: Cambridge University Press, 2009).

[79] K. Qi and M. Bachmann, preprint (2013).

[80] R. G. Miller, Biometrika, **61** (1974), 1.

[81] B. Efron, *The Jackknife, the Bootstrap, and Other Resampling Plans* (Philadelphia: SIAM, 1982).

[82] W. Janke, *Statistical Analysis of Simulations: Data Correlations and Error Estimation*, in *Proceedings of the Winter School "Quantum Simulations of Complex Many-Body Systems: From Theory to Algorithms,"* John von Neumann Institute for Computing, Jülich, NIC Series vol. **10**, ed. J. Grotendorst, D. Marx, and A. Muramatsu (Jülich: NIC, 2002), p. 423.

[83] P. Bézier, Automatisme, **13** (1968), 391.

[84] W. J. Gordon and R. F. Riesenfeld, J. Assoc. Comput. Machin., **21** (1974), 293.

[85] N. Metropolis, A. W. Rosenbluth, M. N. Rosenbluth, A. H. Teller, and E. Teller, J. Chem. Phys., **21** (1953), 1087.

[86] A. M. Ferrenberg and R. H. Swendsen, Phys. Rev. Lett., **63** (1989), 1195.

[87] S. Kumar, D. Bouzida, R. H. Swendsen, P. A. Kollman, and J. M. Rosenberg, J. Comput. Chem., **13** (1992), 1011.

[88] R. H. Swendsen and J.-S. Wang, Phys. Rev. Lett., **57** (1986), 2607.

[89] K. Hukushima and K. Nemoto, J. Phys. Soc. Jpn., **65** (1996), 1604.

[90] K. Hukushima, H. Takayama, and K. Nemoto, Int. J. Mod. Phys., C **7** (1996), 337.

[91] C. J. Geyer, in *Computing Science and Statistics*, Proceedings of the 23rd Symposium on the Interface, ed. E. M. Keramidas (Fairfax Station: Interface Foundation, 1991), p. 156.

[92] E. Marinari and G. Parisi, Europhys. Lett., **19** (1992), 451.

[93] A. P. Lyubartsev, A. A. Martsinovski, S. V. Shevkunov, and P. N. Vorontsov-Velyaminov, J. Chem. Phys., **96** (1992), 1776.

[94] B. A. Berg and T. Neuhaus, Phys. Lett. B, **267** (1991), 249; Phys. Rev. Lett., **68** (1992), 9.

[95] W. Janke, Physica A, **254** (1998), 164; B. A. Berg, Fields Inst. Comm., **26** (2000), 1; Comp. Phys. Commun., **153** (2003), 397.

[96] B. A. Berg, *Markov Chain Monte Carlo Simulations* (Singapore: World Scientific, 2004).

[97] G. M. Torrie and J. P. Valleau, J. Comp. Phys., **23** (1976), 187.

[98] T. Çelik and B. A. Berg, Phys. Rev. Lett., **69** (1992), 2292.

[99] F. Wang and D. P. Landau, Phys. Rev. Lett., **86** (2001), 2050; Phys. Rev. E, **64** (2001), 056101.

[100] G. Favrin, A. Irbäck, and F. Sjunnesson, J. Chem. Phys., **114** (2001), 8154.

[101] N. Madras and A. D. Sokal, J. Stat. Phys., **50** (1988), 109.

[102] J. C. Guillou and J. Zinn-Justin, Phys. Rev. Lett., **39** (1977), 95; Phys. Rev. B, **21** (1980), 3976; A. Pelissetto and E. Vicari, Phys. Rep., **368** (2002), 549.

[103] T. Prellberg and J. Krawczyk, Phys. Rev. Lett., **92** (2004), 120602.

[104] F. James, Comp. Phys. Commun., **60** (1990), 329.

[105] W. Janke, *Pseudo Random Numbers: Generation and Quality Checks*, in *Proceedings of the Winter School "Quantum Simulations of Complex Many-Body Systems: From Theory to Algorithms,"* John von Neumann Institute for Computing, Jülich, NIC Series vol. **10**, ed. J. Grotendorst, D. Marx, and A. Muramatsu (Jülich: NIC, 2002), p. 447.

[106] M. Matsumoto and T. Nishimura, ACM Trans. Model. Comp. Sim., **8** (1998), 3.

[107] G. Marsaglia, A. Zaman, and W. W. Tsang, Statist. Probab. Lett., **9** (1990), 35.

[108] L. Verlet, Phys. Rev., **159** (1967), 98.

[109] M. P. Allen and D. J. Tildesley, *Computer Simulation of Liquids* (New York: Oxford University Press, 1987).

[110] D. C. Rapaport, *The Art of Molecular Dynamics Simulation*, 2nd ed. (Cambridge: Cambridge University Press, 2004).

[111] T. Schlick, *Molecular Modeling and Simulation: An Interdisciplinary Guide*, 2nd ed. (Heidelberg: Springer, 2010).

[112] M. E. Tuckerman, *Statistical Mechanics and Molecular Simulation* (New York: Oxford University Press, 2010).

[113] S. Nosé, J. Chem. Phys., **81** (1984), 511; Mol. Phys., **52** (1984), 255; Mol. Phys., **57** (1986), 187; Mol. Phys., **100** (2002), 191.

[114] W. G. Hoover, Phys. Rev. A, **31** (1985), 1695; Phys. Rev. A, **34** (1986), 2499.

[115] G. J. Martyna and M. L. Klein, J. Chem. Phys., **97** (1992), 2635.

[116] H. C. Andersen, J. Comput. Phys., **52** (1983), 24.

[117] For the difficulty of characterizing the type of the coil-globule transition, see, e.g., the review by I. M. Lifshitz, A. Yu. Grosberg, and A. R. Khokhlov, Rev. Mod. Phys., **50** (1978), 683.

[118] A. R. Khokhlov, Physica A, **105** (1981), 357.

[119] P. D. de Gennes, *Scaling Concepts in Polymer Physics* (Ithaca: Cornell University Press, 1979).

[120] B. Duplantier, J. Phys. (France), **43** (1982), 991; J. Chem. Phys., **86** (1987), 4233.

[121] B. Duplantier, Europhys. Lett., **1** (1986), 491.

[122] M. J. Stephen, Phys. Lett. A, **53** (1975), 363.

[123] J. Hager and L. Schäfer, Phys. Rev. E, **60** (1999), 2071.

[124] P. Grassberger and R. Hegger, J. Chem. Phys., **102** (1995), 6881.

[125] M. C. Tesi, E. J. Janse van Rensburg, E. Orlandini, and S. G. Whittington, J. Phys. A: Math. Gen., **29** (1996), 2451.

[126] M. C. Tesi, E. J. Janse van Rensburg, E. Orlandini, and S. G. Whittington, J. Stat. Phys., **82** (1996), 155.

[127] F. Rampf, W. Paul, and K. Binder, Europhys. Lett., **70** (2005), 628.

[128] F. Rampf, W. Paul, and K. Binder, J. Polym. Sci.: Part B: Polym. Phys., **44** (2006), 2542.

[129] D. F. Parsons and D. R. M. Williams, J. Chem. Phys., **124** (2006), 221103; Phys. Rev. E, **74** (2006), 041804.

[130] N. B. Wilding, M. Müller, and K. Binder, J. Chem. Phys., **105** (1996), 802.

[131] H. Frauenkron and P. Grassberger, J. Chem. Phys., **107** (1997), 9599.

[132] A. Z. Panagiotopoulos, V. Wong, and M. A. Floriano, Macromolecules, **31** (1998), 912.

[133] Q. Yan and J. J. de Pablo, J. Chem. Phys., **113** (2000), 5954.

[134] M. A. Anisimov and J. V. Sengers, Mol. Phys., **103** (2005), 3061.

[135] M. P. Taylor, W. Paul, and K. Binder, J. Chem. Phys., **131** (2009), 114907; Phys. Rev. E, **79** (2009), 050801(R).

[136] J. Gross, T. Neuhaus, T. Vogel, and M. Bachmann, J. Chem. Phys., **138** (2013), 074905.

[137] T. Vogel, M. Bachmann, and W. Janke, Phys. Rev. E, **76** (2007), 061803.

[138] F. L. McCrackin, J. Mazur, and C. M. Guttman, Macromolecules, **6** (1973), 859.

[139] W. Bruns, Macromolecules, **17** (1984), 2826.

[140] J. Batoulis and K. Kremer, Europhys. Lett., **7** (1988), 683.

[141] H. Meirovitch and H. A. Lim, J. Chem. Phys., **92** (1990), 5144.

[142] K. Kremer, *Computer Simulation Methods for Polymer Physics*, in *Monte Carlo and Molecular Dynamics of Condensed Matter Systems*, ed. K. Binder and G. Cicotti (Bologna: Editrice Compositori, 1996), p. 669.

[143] M. P. Taylor and J. E. G. Lipson, J. Chem. Phys., **109** (1998), 7583.

[144] A. L. Owczarek and T. Prellberg, Europhys. Lett., **51** (2000), 602.

[145] P. J. Flory, *Principles of Polymer Chemistry* (Ithaca: Cornell University Press, 1953).

[146] W. Janke, Phys. Rev. B, **55** (1997), 3580.

[147] A. Daanoun, C. F. Tejero, and M. Baus, Phys. Rev. E, **50** (1994), 2913.

[148] P. Bolhuis and D. Frenkel, Phys. Rev. Lett., **72** (1994), 2211.

[149] C. F. Tejero, A. Daanoun, H. N. W. Lekkerkerker, and M. Baus, Phys. Rev. Lett., **73** (1994), 752.

[150] M. G. Noro and D. Frenkel, J. Chem. Phys., **113** (2000), 2941.

[151] C. Rascón, G. Navascués, and L. Mederos, Phys. Rev. B, **51** (1995), 14899.

[152] S. M. Ilett, A. Orrock, W. C. K. Poon, and P. N. Pusey, Phys. Rev. E, **51** (1995), 1344.

[153] N. Asherie, A. Lomakin, and G. B. Benedek, Phys. Rev. Lett., **77** (1996), 4832.

[154] F. Y. Naumkin and D. J. Wales, Mol. Phys., **96** (1999), 1295.

[155] W. Jiang, J. Chuang, J. Jakana, P. Weigele, J. King, and W. Chiu, Nature, **439** (2006), 612.

[156] T. Hugel, J. Michaelis, C. L. Hetherington, P. J. Jardine, S. Grimes, J. M. Walter, W. Falk, D. L. Anderson, and C. Bustamante, PLoS Biol., **5** (2007), 558.

[157] J. P. K. Doye and F. Calvo, J. Chem. Phys., **116** (2002), 8307.

[158] J. A. Northby, J. Chem. Phys., **87** (1987), 6166.

[159] E. G. Noya and J. P. K. Doye, J. Chem. Phys., **124** (2006), 104503.

[160] P. A. Frantsuzov and V. A. Mandelshtam, Phys. Rev. E, **72** (2005), 037102.

[161] Y. Zhou, M. Karplus, J. M. Wichert, and C. K. Hall, J. Chem. Phys., **107** (1997), 10691.

[162] F. Calvo, J. P. K. Doye, and D. J. Wales, J. Chem. Phys., **116** (2002), 2642.

[163] D. T. Seaton, S. J. Mitchell, and D. P. Landau, Braz. J. Phys., **38** (2008), 48.

[164] D. T. Seaton, T. Wüst, and D. P. Landau, Comp. Phys. Comm., **180** (2009), 587.

[165] D. T. Seaton, T. Wüst, and D. P. Landau, Phys. Rev. E, **81** (2010), 011802.

[166] W. Paul, T. Strauch, F. Rampf, and K. Binder, Phys. Rev. E, **75** (2007), 060801(R).

[167] P. J. Steinhardt, D. R. Nelson, and M. Ronchetti, Phys. Rev. B, **28** (1983), 784.

[168] N. W. Johnson, Canad. J. Math., **18** (1966), 169.

[169] J. P. K. Doye, D. J. Wales, and R. S. Berry, J. Chem. Phys., **103** (1995), 4234; J. P. K. Doye and D. J. Wales, J. Phys. B.: At. Mol. Opt. Phys., **29** (1996), 4589.

[170] L. Cheng and J. Yang, J. Phys. Chem. A, **111** (2007), 5287.

[171] G. Caliskan, C. Hyeon, U. Perez-Salas, R. M. Briber, S. A. Woodson, and D. Thirumalai, Phys. Rev. Lett., **95** (2005), 268303; C. Hyeon, R. I. Dima, and D. Thirumalai, J. Chem. Phys., **125** (2006), 194905.

[172] J. A. Abels, F. Moreno-Herrero, T. van der Heijden, C. Dekker, and N. H. Dekker, Biophys. J., **88** (2005), 2737; P. A. Wiggins, T. van der Heijden, F. Moreno-Herrero, A. Spakowitz, R. Phillips, J. Widom, C. Dekker, and P. C. Nelson, Nature Nanotech., **1** (2006), 137.

[173] C. Yuan, H. Chen, X. Wen Lou, and L. A. Archer, Phys. Rev. Lett., **100** (2008), 018102.

[174] T. E. Cloutier and J. Widom, Mol. Cell, **14** (2004), 355.

[175] J. Yan and J. F. Marko, Phys. Rev. Lett., **93** (2004), 108108.

[176] P. A. Wiggins, R. Phillips, and P. C. Nelson, Phys. Rev. E, **71** (2005), 021909; P. A. Wiggins and P. C. Nelson, Phys. Rev. E, **73** (2006), 031906.

[177] D. A. Sivak and P. L. Geissler, J. Chem. Phys., **136** (2012), 045102.

[178] H.-P. Hsu, W. Paul, and K. Binder, Europhys. Lett., **92** (2010), 28003; H.-P. Hsu and K. Binder, J. Chem. Phys., **136** (2012), 024901.

[179] F. Affouard, M. Kröger, and S. Hess, Phys. Rev. E, **54** (1996), 5178.

[180] M. Kröger, Phys. Rep., **390** (2004), 453.

[181] R. G. Winkler, J. Chem. Phys., **133** (2010), 164905.

[182] R. Everaers, S. K. Sukumaran, G. S. Grest, C. Svaneborg, A. Sivasubramanian, and K. Kremer, Science, **303** (2004), 823.

[183] J. A. Martemyanova, M. R. Stukan, V. A. Ivanov, M. Müller, W. Paul, and K. Binder, J. Chem. Phys., **122** (2005), 174907.

[184] D. T. Seaton, S. Schnabel, D. P. Landau, and M. Bachmann, Phys. Rev. Lett., **110** (2013), 028103.

[185] M. D. Yoder, N. T. Keen, and F. Jurnak, Science, **260** (1993), 1503.

[186] H.-P. Hsu, V. Mehra, and P. Grassberger, Phys. Rev. E, **68** (2003), 037703.

[187] A. Irbäck, C. Peterson, F. Potthast, and O. Sommelius, J. Chem. Phys., **107** (1997), 273.

[188] A. Irbäck, C. Peterson, and F. Potthast, Phys. Rev. E, **55** (1997), 860.

[189] M. Bachmann, H. Arkın, and W. Janke, Phys. Rev. E, **71** (2005), 031906.

[190] J. N. Onuchic, Z. Luthey-Schulten, and P. G. Wolynes, Annu. Rev. Phys. Chem., **48** (1997), 545.

[191] C. Clementi, A. Maritan, and J. R. Banavar, Phys. Rev. Lett., **81** (1998), 3287.

[192] J. N. Onuchic and P. G. Wolynes, Curr. Opin. Struct. Biol., **14** (2004), 70.

[193] R. Du, V. S. Pande, A. Yu. Grosberg, T. Tanaka, and E. S. Shakhnovich, J. Chem. Phys., **108** (1998), 334.

[194] V. S. Pande and D. S. Rokhsar, Proc. Natl. Acad. Sci. (USA), **96** (1999), 1273.

[195] U. H. E. Hansmann, M. Masuya, and Y. Okamoto, Proc. Natl. Acad. Sci. (USA), **94** (1997), 10652.

[196] B. A. Berg, H. Noguchi, and Y. Okamoto, Phys. Rev. E, **68** (2003), 036126.

[197] P. G. Wolynes, *Spin Glass Ideas and the Protein Folding Problems*, in *Directions in Condensed Matter Physics*, ed. D. L. Stein, Vol. **6**: Spin Glasses and Biology (Singapore: World Scientific, 1992), p. 225.

[198] V. S. Pande, A. Yu. Grosberg, C. Joerg, and T. Tanaka, Phys. Rev. Lett., **76** (1996), 3987.

[199] E. Pitard and E. I. Shakhnovich, Phys. Rev. E, **63** (2001), 041501.

[200] A. Kallias, M. Bachmann, and W. Janke, J. Chem. Phys., **128** (2008), 055102.

[201] D. Sherrington and S. Kirkpatrick, Phys. Rev. Lett., **35** (1975), 1792; S. F. Edwards and P. W. Anderson, J. Phys. F: Metal Phys., **5** (1975), 965; G. Parisi, Phys. Rev. Lett., **43** (1979), 1754.

[202] S. K. Kearsley, Acta Cryst. A, **45** (1989), 208.

[203] K. A. Dill and H. S. Chan, Nature Struct. Biol., **4** (1997), 10.

[204] H. Zhou and Y. Zhou, Biophys. J., **82** (2002), 458.

[205] S. E. Jackson and A. R. Fersht, Biochemistry, **30** (1991), 10428.

[206] A. R. Fersht, *Structure and Mechanisms in Protein Science: A Guide to Enzyme Catalysis and Protein Folding* (New York: Freeman, 1999).

[207] Y. Ueda, H. Taketomi, and N. Gō, Int. J. Pept. Res., **7** (1975), 445.

[208] N. Gō, Annu. Rev. Biophys. Bioeng., **12** (1983), 183.

[209] S. Takada, Proc. Natl. Acad. Sci. (USA), **96** (1999), 11698.

[210] T. Head-Gordon and S. Brown, Curr. Opin. Struct. Biol., **13** (2003), 160.

[211] J.-E. Shea, J. N. Onuchic, and C. L. Brooks III, Proc. Natl. Acad. Sci. (USA), **96** (1999), 12512.

[212] C. Clementi, H. Nymeyer, and J. Onuchic, J. Mol. Biol., **298** (2000), 937.

[213] S. B. Ozkan, I. Bahar, and K. A. Dill, Nature Struct. Biol., **8** (2001), 765.

[214] M. Cieplak and T. X. Hoang, Proteins: Struct., Funct., and Genet., **44** (2001), 20.

[215] N. Koga and S. Takada, J. Mol. Biol., **313** (2001), 171.

[216] L. Li and E. I. Shakhnovich, Proc. Natl. Acad. Sci. (USA), **98** (2001), 13014.

[217] H. Kaya and H. S. Chan, J. Mol. Biol., **326** (2003), 911.

[218] H. Kaya and H. S. Chan, Phys. Rev. Lett., **90** (2003), 258104.

[219] J. Schonbrun and K. A. Dill, Proc. Natl. Acad. Sci. (USA), **100** (2003), 12678.

[220] H. S. Chan and K. A. Dill, Proteins: Struct., Funct., and Genet., **30** (1998), 2.

[221] E. Shakhnovich, Chem. Rev., **106** (2006), 1559.

[222] M. Bachmann, Phys. Proc., **3** (2010), 1387.

[223] A. Irbäck and S. Mohanty, Biophys. J., **88** (2005), 1560.

[224] J. M. Scholtz, E. J. York, J. M. Stewart, and R. L. Baldwin, J. Am. Chem. Soc., **113** (1991), 5102.

[225] T. Bereau, M. Bachmann, and M. Deserno, J. Am. Chem. Soc., **132** (2010), 13129.

[226] T. Bereau, M. Deserno, and M. Bachmann, Biophys. J., **100** (2011), 2764.

[227] T. Bereau and M. Deserno, J. Chem. Phys., **130** (2009), 235106.

[228] S. T. R. Walsh, H. Cheng, J. W. Bryson, H. Roder, and W. F. DeGrado, Proc. Natl. Acad. Sci. (USA), **96** (1999), 5486.

[229] S. A. Sabeur, F. Hamdache, and F. Schmid, Phys. Rev. E, **77** (2008), 020802(R).

[230] H. Noguchi and K. Yoshikawa, J. Chem. Phys., **109** (1998), 5070.

[231] T. X. Hoang, A. Trovato, F. Seno, J. R. Banavar, and A. Maritan, Proc. Natl. Acad. Sci. (USA), **101** (2004), 7960.

[232] K. Wolff, M. Vendruscolo, and M. Porto, Gene, **422** (2008), 47.

[233] Y. Snir and R. D. Kamien, Science, **307** (2005), 1067; Phys. Rev. E, **75** (2007), 051114.

[234] H. Hansen-Goos, R. Roth, K. Mecke, and S. Dietrich, Phys. Rev. Lett., **99** (2007), 128101.

[235] J. P. Kemp and Z. Y. Chen, Biomacromolecules, **2** (2001), 389.

[236] D. C. Rapaport, Phys. Rev. E, **66** (2002), 011906.

[237] J. R. Banavar, A. Flammini, D. Marenduzzo, A. Maritan, and A. Trovato, J. Phys.: Condens. Matter, **15** (2003), S1787.

[238] J. R. Banavar and A. Maritan, Rev. Mod. Phys., **75** (2003), 23.

[239] A. Maritan, C. Micheletti, A. Trovato, and J. R. Banavar, Nature, **406** (2000), 287.

[240] S. Auer, M. A. Miller, S. V. Krivov, C. M. Dobson, M. Karplus, and M. Vendruscolo, Phys. Rev. Lett., **99** (2007), 178104.

[241] T. Vogel, T. Neuhaus, M. Bachmann, and W. Janke, Europhys. Lett., **85** (2009), 10003.

[242] T. Vogel, T. Neuhaus, M. Bachmann, and W. Janke, Eur. Phys. J. E, **30** (2009), 7.

[243] T. Vogel, T. Neuhaus, M. Bachmann, and W. Janke, Phys. Rev. E, **80** (2009), 011802.

[244] O. Gonzalez and J. H. Maddocks, Proc. Natl. Acad. Sci. (USA), **96** (1999), 4769.

[245] T. Neuhaus, O. Zimmermann, and U. H. E. Hansmann, Phys. Rev. E, **75** (2007), 051803.

[246] J. Gsponer and M. Vendruscolo, Prot. & Pept. Lett., **13** (2006), 287.

[247] A. Irbäck and S. Mitternacht, Proteins: Struct., Funct., and Bioinform., **71** (2008), 207.

[248] H. Lin, R. Bhatia, and R. Lal, FASEB J., **15** (2001), 2433.

[249] A. Quist, I. Doudevski, H. Lin, R. Azimova, D. Ng, B. Frangione, B. Kagan, J. Ghiso, and R. Lal, Proc. Natl. Acad. Sci. (USA), **102** (2005), 10427.

[250] H. A. Lashuel and P. T. Lansbury Jr., Quart. Rev. Biophys., **39** (2006), 167.

[251] D. H. E. Gross, *Microcanonical Thermodynamics* (Singapore: World Scientific, 2001).

[252] D. H. E. Gross and J. F. Kenney, J. Chem. Phys., **122** (2005), 224111.

[253] D. H. E. Gross, Physica E, **29** (2005), 251.

[254] C. Junghans, M. Bachmann, and W. Janke, Phys. Rev. Lett., **97** (2006), 218103.

[255] C. Junghans, M. Bachmann, and W. Janke, J. Chem. Phys., **128** (2008), 085103.

[256] W. Thirring, Z. Physik, **235** (1970), 339.

[257] M. Schmidt, R. Kusche, T. Hippler, J. Donges, W. Kronmüller, B. von Issendorff, and H. Haberland, Phys. Rev. Lett., **86** (2001), 1191.

[258] M. Pichon, B. Tamain, R. Bougault, and O. Lopez, Nucl. Phys. A, **749** (2005), 93c.

[259] O. Lopez, D. Lacroix, and E. Vient, Phys. Rev. Lett., **95** (2005), 242701.

[260] W. Janke, Nucl. Phys. B (Proc. Suppl.), **63A–C** (1998), 631.

[261] H. Behringer and M. Pleimling, Phys. Rev. E, **74** (2006), 011108.

[262] D. J. Wales and R. S. Berry, Phys. Rev. Lett., **73** (1994), 2875; D. J. Wales and J. P. K. Doye, J. Chem. Phys., **103** (1995), 3061.

[263] S. Hilbert and J. Dunkel, Phys. Rev. E, **74** (2006), 011120; J. Dunkel and S. Hilbert, Physica A, **370** (2006), 390.

[264] W. Janke, *Histograms and All That*, in *Computer Simulations of Surfaces and Interfaces*, NATO Science Series, II. Mathematics, Physics, and Chemistry, vol. **114**, ed. D. P. Landau, A. Milchev, and B. Dünweg (Dordrecht: Kluwer, 2003), p. 137.

[265] C. Junghans, M. Bachmann, and W. Janke, Europhys. Lett., **87** (2009), 40002.

[266] C. Junghans, W. Janke, and M. Bachmann, Comp. Phys. Commun., **182** (2011), 1937.

[267] M. Balbirnie, R. Grothe, and D. S. Eisenberg, Proc. Natl. Acad. Sci. (USA), **98** (2001), 2375.

[268] J. Gsponer, U. Haberthür, and A. Caflisch, Proc. Natl. Acad. Sci. (USA), **100** (2003), 5154.

[269] B. Strodel, C. S. Whittleston, and D. J. Wales, J. Am. Chem. Soc., **129** (2007), 16005.

[270] K. L. Osborne, M. Bachmann, and B. Strodel, Proteins: Struct. Func. Bioinf., **81** (2013), 1141.

[271] A. Irbäck, B. Samuelsson, F. Sjunnesson, and S. Wallin, Biophys. J., **85** (2003), 1466.

[272] A. Irbäck and S. Mohanty, J. Comput. Chem., **27** (2006), 1548.

[273] S. Brown, Nature Biotechnol., **15** (1997), 269.

[274] S. R. Whaley, D. S. English, E. L. Hu, P. F. Barbara, A. M. Belcher, Nature, **405** (2000), 665.

[275] K. Goede, P. Busch, and M. Grundmann, Nano Lett., **4** (2004), 2115.

[276] K. Goede, M. Grundmann, K. Holland-Nell, and A. G. Beck-Sickinger, Langmuir, **22** (2006), 8104.

[277] R. L. Willett, K. W. Baldwin, K. W. West, and L. N. Pfeiffer, Proc. Natl. Acad. Sci. (USA), **102** (2005), 7817.

[278] J. J. Gray, Curr. Opin. Struct. Biol., **14** (2004), 110.

[279] G. Reiter, Phys. Rev. Lett., **87** (2001), 186101.

[280] J. Forsman and C. E. Woodward, Phys. Rev. Lett., **94** (2005), 118301.

[281] S. Metzger, M. Müller, K. Binder, and J. Baschnagel, J. Chem. Phys., **118** (2003), 8489.

[282] E. Nakata, T. Nagase, S. Shinkai, and I. Hamachi, J. Am. Chem. Soc., **126** (2004), 490.

[283] T. Bogner, A. Degenhard, and F. Schmid, Phys. Rev. Lett., **93** (2004), 268108.

[284] A. Swetnam and M. P. Allen, Phys. Rev. E, **85** (2012), 062901.

[285] E. Balog, T. Becker, M. Oettl, R. Lechner, R. Daniel, J. Finney, and J. C. Smith, Phys. Rev. Lett., **93** (2004), 028103.

[286] M. Ikeguchi, J. Ueno, M. Sato, and A. Kidera, Phys. Rev. Lett., **94** (2005), 078102.

[287] N. Gupta and A. Irbäck, J. Chem. Phys., **120** (2004), 3983.

[288] C.-H. Cheng and P.-Y. Lai, Phys. Rev. E, **71** (2005), 060802(R).

[289] G. M. Foo and R. B. Pandey, Phys. Rev. Lett., **80** (1998), 3767; Phys. Rev. E, **61** (2000), 1793.

[290] G. Binnig, C. F. Quate, and Ch. Gerber, Phys. Rev. Lett., **56** (1986), 930.

[291] M. Rief, H. Clausen-Schaumann, and H. Gaub, Nature Struct. Biol., **6** (1999), 346.

[292] D. E. Smith, S. Tans, S. Smith, S. Grimes, D. L. Anderson, and C. Bustamante, Nature, **413** (2001), 748.

[293] M. Salomo, K. Kegler, C. Gutsche, M. Struhalla, J. Reinmuth, W. Skokow, U. Hahn, and F. Kremer, Colloid. Polym. Sci., **284** (2006), 1325.

[294] R. Hegger and P. Grassberger, J. Phys. A, **27** (1994), 4069.

[295] T. Vrbová and S. G. Whittington, J. Phys. A, **29** (1996), 6253; J. Phys. A, **31** (1998), 3989; T. Vrbová and K. Procházka, J. Phys. A, **32** (1999), 5469.

[296] Y. Singh, D. Giri, and S. Kumar, J. Phys. A, **34** (2001), L67; R. Rajesh, D. Dhar, D. Giri, S. Kumar, and Y. Singh, Phys. Rev. E, **65** (2002), 056124.

[297] M. S. Causo, J. Chem. Phys., **117** (2002), 6789.

[298] J. Krawczyk, T. Prellberg, A. L. Owczarek, and A. Rechnitzer, Europhys. Lett., **70** (2005), 726.

[299] J.-H. Huang and S.-J. Han, J. Zhejiang Univ. Sci., **5** (2004), 699.

[300] J. Luettmer-Strathmann, F. Rampf, W. Paul, and K. Binder, J. Chem. Phys., **128** (2008), 064903.

[301] M. Bachmann and W. Janke, Phys. Rev. Lett., **95** (2005), 058102.

[302] M. Bachmann and W. Janke, Phys. Rev. E, **73** (2006), 041802.

[303] M. Bachmann and W. Janke, Phys. Rev. E, **73** (2006), 020901(R).

[304] M. Möddel, M. Bachmann, and W. Janke, J. Phys. Chem. B, **113** (2009), 3314.

[305] L. Wang, T. Chen, X. Lin, Y. Liu, and H. Liang, J. Chem. Phys., **131** (2009), 244902.

[306] A. D. Swetnam and M. P. Allen, Phys. Chem. Chem. Phys., **11** (2009), 2046.

[307] M. Möddel, W. Janke, and M. Bachmann, Phys. Chem. Chem. Phys., **12** (2010), 11548.

[308] M. Möddel, W. Janke, and M. Bachmann, Macromolecules, **44** (2011), 9013.

[309] Y. W. Li, T. Wüst, and D. P. Landau, Phys. Rev. E, **87** (2013), 012706.

[310] N. Källrot and P. Linse, Macromolecules, **40** (2007), 4669.

[311] F. Celestini, T. Frisch, and X. Oyharcabal, Phys. Rev. E, **70** (2004), 012801.

[312] J. Krawczyk, T. Prellberg, A. L. Owczarek, and A. Rechnitzer, J. Stat. Mech. (2004), P10004.

[313] P. Benetatos and E. Frey, Phys. Rev. E, **70** (2004), 051806.

[314] M. Breidenreich, R. R. Netz, and R. Lipowsky, Europhys. Lett., **49** (2000), 431; Eur. Phys. J. E, **5** (2001), 403.

[315] E. Eisenriegler, K. Kremer, and K. Binder, J. Chem. Phys., **77** (1982), 6296.

[316] E. Eisenriegler, *Polymers Near Surfaces: Conformation Properties and Relation to Critical Phenomena* (Singapore: World Scientific, 1993).

[317] R. Braun, M. Sarikaya, and K. Schulten, J. Biomater. Sci. Polym. Ed., **13** (2002), 747.

[318] M. Bachmann and W. Janke, Lect. Notes Phys., **736** (2008), 203.

[319] C. Borgs and W. Janke, Phys. Rev. Lett., **68** (1992), 1738; W. Janke, Phys. Rev. B, **47** (1993), 14757.

[320] W. Janke, *First-Order Phase Transitions*, in *Computer Simulations of Surfaces and Interfaces*, NATO Science Series, II. Mathematics, Physics and Chemistry, vol. **114**, ed. D. P. Landau, A. Milchev, and B. Dünweg (Dordrecht: Kluwer, 2003), p. 111.

[321] T. Vogel and M. Bachmann, Phys. Rev. Lett., **104** (2010), 198302.

[322] T. Vogel and M. Bachmann, Phys. Proc., **4** (2010), 161.

[323] T. Vogel and M. Bachmann, Comp. Phys. Commun., **182** (2011), 1928.

[324] T. Vogel, T. Mutat, J. Adler, and M. Bachmann, Commun. Comp. Phys., **13** (2013), 1245.

[325] A. Milchev and K. Binder, J. Chem. Phys., **114** (2001), 8610.

[326] M. Sarikaya, C. Tamerler, A. K.-Y. Jen, K. Schulten, and F. Baneyx, Nature Mat., **2** (2003), 577.

[327] B. Geffroy, P. le Roy, and C. Prat, Polym. Int., **55** (2006), 572.

[328] K.-I. Sano and K. Shiba, J. Am. Chem. Soc., **125** (2003), 14234.

[329] S. Wang, E. S. Humphreys, S.-Y. Chung, D. F. Delduco, S. R. Lustig, H. Wang, K. N. Parker, N. W. Rizzo, S. Subramoney, Y.-M. Chiang, and A. Jagota, Nature Mat., **2** (2003), 196.

[330] R. L. Willett, K. W. Baldwin, K. W. West, and L. N. Pfeiffer, Proc. Natl. Acad. Sci. (USA), **102** (2005), 7817.

[331] H. J. Dyson and P. E. Wright, Curr. Opin. Struct. Biol., **12** (2002), 54.

[332] B. A. Shoemaker, J. J. Portman, and P. G. Wolynes, Proc. Natl. Acad. Sci. (USA), **97** (2000), 8868.

[333] V. P. Zhdanov and B. Kasemo, Proteins, **30** (1998), 168.

[334] V. Castells, S. Yang, and P. R. Van Tassel, Phys. Rev. E, **65** (2002), 031912.

[335] M. Muthukumar, J. Chem. Phys., **103** (1995), 4723.

[336] D. Bratko, A. K. Chakraborty, and E. I. Shakhnovich, Chem. Phys. Lett., **280** (1997), 46.

[337] A. J. Golumbfskie, V. S. Pande, and A. K. Chakraborty, Proc. Natl. Acad. Sci. (USA), **96** (1999), 11707.

[338] Y. A. Kriksin, P. G. Khalatur, and A. R. Khokhlov, J. Chem. Phys., **122** (2005), 114703.

[339] M. S. Moghaddam and H. S. Chan, J. Chem. Phys., **125** (2006), 164909.

[340] M. Bachmann, K. Goede, A. G. Beck-Sickinger, M. Grundmann, A. Irbäck, and W. Janke, Angew. Chem. Int. Ed., **49** (2010), 9530.

[341] R. Backofen and S. Will, Constraints, **11** (2006), 5.

[342] B. R. Peelle, E. M. Krauland, K. D. Wittrup, and A. M. Belcher, Langmuir, **21** (2005), 6929.

[343] S. Mitternacht, S. Schnabel, M. Bachmann, W. Janke, and A. Irbäck, J. Phys. Chem. B, **111** (2007), 4355.

[344] J. Qian, R. Hentschke, and W. Knoll, Langmuir, **13** (1997), 7092.

[345] L. Delle Site, C. F. Abrams, A. Alavi, and K. Kremer, Phys. Rev. Lett., **89** (2002), 156103.

[346] N. Kantarci, C. Tamerler, M. Sarikaya, T. Haliloglu, and P. Doruker, Polymer, **46** (2005), 4307.

[347] P. Schravendijk, N. van der Vegt, L. Delle Site, and K. Kremer, ChemPhysChem, **6** (2005), 1866.

[348] P. Schravendijk, L. M. Ghiringhelli, L. Delle Site, and N. F. A. van der Vegt, J. Phys. Chem. C, **111** (2007), 2631.

[349] J. M. Buriak, Chem. Rev., **102** (2002), 1271.

[350] W. A. Steele, Surf. Sci., **36** (1973), 317.

[351] R. Hentschke, Macromol. Theory Simul., **6** (1997), 287.

[352] S. De Miranda Tomásio and T. R. Walsh, Mol. Phys., **105** (2007), 221.

[353] H. Heinz, H. J. Castelijns, and U. W. Suter, J. Am. Chem. Soc., **125** (2003), 9500.

[354] M. L. Sushko, A. L. Shluger, and C. Rivetti, Langmuir, **22** (2006), 7678.

[355] H. Heinz, H. Koerner, K. L. Anderson, R. A. Vaia, and B. L. Farmer, Chem. Mater., **17** (2005), 5658.

[356] A. Rimola, M. Sodupe, S. Tosoni, B. Civalleri, and P. Ugliengo, Langmuir, **22** (2006), 6593.

[357] A. Rimola, S. Tosoni, M. Sodupe, and P. Ugliengo, ChemPhysChem, **7** (2006), 157.

[358] M. Lundqvist, P. Nygren, B.-H. Jonsson, and K. Broo, Angew. Chem. Int. Ed., **45** (2006), 8169.

[359] R. G. Frieser, J. Electrochem. Soc., **121** (1974), 669.

[360] A. Hemeryck, N. Richard, A. Estève, and M. Djafari Rouhani, Surf. Sci., **601** (2007), 2339.

[361] L. Ling, S. Kuwabara, T. Abe, and F. Shimura, J. Appl. Phys., **73** (1993), 3018.

[362] M. K. Weldon, B. B. Stefanov, K. Raghavachari, and Y. J. Chabal, Phys. Rev. Lett., **79** (1997), 2851.

[363] Y. J. Chabal and K. Raghavachari, Phys. Rev. Lett., **53** (1984), 282.

[364] E. Yablonovitch, D. L. Allara, C. C. Chang, T. Gmitter, and T. B. Bright, Phys. Rev. Lett., **57** (1986), 249.

[365] Y. J. Chabal, G. S. Higashi, K. Raghavachari, and V. A. Burrows, J. Vac. Sci. Technol. A, **7** (1989), 2104.

[366] R. L. Cicero, M. R. Linford, and C. E. D. Chidsey, Langmuir, **16** (2000), 5688.

[367] R. Konecny and D. J. Doren, J. Am. Chem. Soc., **119** (1997), 11098.

[368] R. A. Wolkow, G. P. Lopinski, and D. J. Moffatt, Surf. Sci., **416** (1998), L1107.

[369] G. P. Lopinski, D. J. Moffatt, D. D. M. Wayner, M. Z. Zgierski, and R. A. Wolkow, J. Am. Chem. Soc., **121** (1999), 4532.

[370] K. Seino and W. G. Schmidt, Surf. Sci., **585** (2005), 191.

[371] J.-H. Wang, F. Bacalzo-Gladden, and M. C. Lin, Surf. Sci., **600** (2006), 1113.

[372] S. H. Jang, S. Jeong, and J. R. Hahn, J. Phys. Chem. C, **111** (2007), 340.

[373] F. Gou, M. A. Gleeson, and A. W. Kleyn, Surf. Sci., **601** (2007), 76.

[374] J. Tien, A. Terfort, and G. M. Whitesides, Langmuir, **13** (1997), 5349.

[375] S. S. Batsanov, Inorg. Mater., **37** (2001), 871.

[376] T. A. Halgren, J. Am. Chem. Soc., **114** (1992), 7827.

Index